Polar Ice and Global Warming in Cryosphere Regions

Polar Ice and Global Warming in Cryosphere Regions is based on recent and past climate variabilities data gathered through satellites and spatial-temporal analysis to explain the role of global warming on cryosphere regions such as high-latitude Himalaya, Arctic and Antarctic regions, and the surrounding Southern Ocean and Arctic Ocean. Through several case studies the book describes the atmospheric processes and their interactions with high-latitude regions toward a better understanding of climate variability. Understanding cryosphere regions helps readers develop plausible models for disaster risk management and policy on different polar events.

Features

- Presents a thorough review on climate variability over the Southern Ocean and Antarctica, and the impact of climate variability and global warming on cryosphere regions
- Explains how the inferred climatological environmental conditions using natural archives may shed light on climate scenarios in cryosphere regions
- Includes case studies on globally connected geoscientific phenomena in the Himalayan, Arctic, and Antarctic regions
- Discusses the use of natural archives to explain the current climate scenario in the cryosphere regions

Intended for researchers, academics, and graduate students following oceanography, meteorology, or environmental studies, and those working on projects related to climate change in governmental organizations, institutions, and global NGOs, this book outlines ways in which readers can initiate plans and policies to help mitigate the effects of global warming in these regions.

Maritime Climate Change: Physical Drivers and Impact
Series Editor: Neloy Khare

As global climate change continues to unfold, the two-way links between the tropical oceans and the poles will play key determining factors in these sensitive regions' climatic evolution. Now is the time to take a detailed look at how the tropical oceans and the poles are coupled climatically. The signatures of environmental and climatic conditions are well preserved in many natural archives available over land and ocean. Many efforts have been made to unravel such mysteries of climate through many natural geological archives from tropics to the polar region. This series makes an effort to cover in pertinent time various depositional regimes, different proxies: planktic, benthic, pollens and spores, invertebrates, geochemistry, sedimentology, etc., and emerged teleconnections between the poles and tropics at regional and global scales, besides sea-level changes and neo-tectonism. This book series will review theories and methods, analyse case studies, and identify and describe the evolving spatial-temporal variations in climate, providing a better process-level understanding of these patterns. It will discuss significantly, generalizable insights that improve our understanding of climatic evolution across time—including for the future. It aims to serve all professionals, students and researchers, and scientists alike in academia, industry, government, and beyond.

Climate Change in the Arctic: An Indian Perspective
Neloy Khare

Climate Change and Geodynamics in Polar Regions
Edited by Neloy Khare

Polar Ice and Global Warming in Cryosphere Regions
Edited by Neloy Khare

Polar Ice and Global Warming in Cryosphere Regions

Edited by
Neloy Khare

CRC Press
Taylor & Francis Group
Boca Raton London New York

CRC Press is an imprint of the
Taylor & Francis Group, an **informa** business

Designed cover image: © iStock Photo

First edition published 2024
by CRC Press
2385 NW Executive Center Drive, Suite 320, Boca Raton FL 33431

and by CRC Press
4 Park Square, Milton Park, Abingdon, Oxon, OX14 4RN

CRC Press is an imprint of Taylor & Francis Group, LLC

Library of Congress Cataloging-in-Publication Data
Names: Khare, Neloy, editor.
Title: Polar ice and global warming in cryosphere regions / edited by Neloy Khare.
Description: First edition. | Boca Raton, FL : CRC Press, 2024. |
Series: Maritime climate change | Includes bibliographical references and index. |
Summary: ""Polar Ice and Global Warming in Cryosphere Regions" is based on recent and past climate variabilities data gathered through satellites and spatial-temporal analysis to explain the role of global warming on cryosphere regions such as high-latitude Himalaya, Arctic and Antarctic regions, and the surrounding Southern Ocean and Arctic Ocean. The featured case studies describe the atmospheric processes and their interactions with high-latitude regions helping readers better understand climate variability, and develop plans and policies for disaster risk management"– Provided by publisher.
Identifiers: LCCN 2023029327 (print) | LCCN 2023029328 (ebook) |
ISBN 9781032426433 (hbk) | ISBN 9781032427478 (pbk) | ISBN 9781003364115 (ebk)
Subjects: LCSH: Cryosphere. | Climatic changes–Himalaya Mountains. |
Climatic changes–Polar regions. | Cryosphere–Remote sensing. | Climatic changes–Remote sensing.
Classification: LCC QC880.4.C79 P65 2024 (print) | LCC QC880.4.C79
(ebook) | DDC 551.34/3–dc23/eng/20231012
LC record available at https://lccn.loc.gov/2023029327
LC ebook record available at https://lccn.loc.gov/2023029328

ISBN: 978-1-032-42643-3 (hbk)
ISBN: 978-1-032-42747-8 (pbk)
ISBN: 978-1-003-36411-5 (ebk)

DOI: 10.1201/9781003364115

Typeset in Times Roman
by Newgen Publishing UK

In sweet memories

This book is dedicated to
my beloved mother Late Smt. Kusum Khare
who left for her heavenly abode on December 10, 2023

Contents

Foreword

The cryosphere is composed of the frozen parts of the Earth and encompasses snow and ice on land, ice caps, glaciers, permafrost, and sea ice. It acts like a heat-reflecting sheet and thus influences the heating and cooling system of Earth. The system eventually controls air temperatures, ocean currents, sea levels, and storm patterns. Undoubtedly this sphere helps maintain Earth's climate by reflecting incoming solar radiation back into space. Due to the unprecedented rise in global temperature, the whole cryosphere is experiencing substantial warming causing many adverse effects on the cryosphere ecosystem in a variety of ways. Besides the adverse impact over the cryosphere, it also in turn creates further influence over the global climate system. It is therefore essential to address the vital aspects of global warming over the cryosphere with special emphasis on the Himalayas, the Antarctic, and the Arctic ecosystem.

Polar Ice and Global Warming in Cryosphere Regions aims to focus on all such assessments over the cryosphere regions of the Himalayas, the Antarctic, and the Antarctic. The book is divided into three themes dedicated to each above-mentioned cryosphere. The introductory chapter by Khare highlights the importance of the cryosphere in the present context of global warming followed by four chapters dedicated to the Himalayan region which include the assessment of the recession rate and surface velocity of Gangotri glacier Garhwal Himalaya India by Sharma et al., followed by the modelling of a glacial lake outburst flood of Sona-Sar Lake in the Jhelum basin of Kashmir Himalayas, India, using HEC-RAS by Ahmed et al., a thorough in-depth review of the Hindukush Karakoram Himalayan Glaciers by Kulkarni et al., and an assessment of the response of global warming on glaciers of Garhwali Himalaya, India, using remote sensing and GIS by Kumar et al.

Subsequently, Chowdhury et al. provide an exhaustive account of the understanding of global warming through polar ice. This chapter paves the way for another four chapters devoted to the Antarctic cryosphere including an evaluation of the Antarctic sea ice through a climate change perspective by Das coupled with the effect of climate change on the multifractal properties of the sea-ice concentration around the Indian Antarctic stations Maitri and Bharti by Dwivedi. While Mishra and Srivastava explain the spatial-temporal variations in Antarctic snow-ice energy balance with special reference to the glacial-metrological parameters and reflectance, Pednekar provides the spatial and temporal variability of physical parameters in Prydz Bay (East Antarctica) for climate change.

The book ends with attention focused on the Arctic cryosphere in a bid to understand the impact of climate change over the Arctic region through three special chapters dedicated to understanding the predictability of East Arctic sea ice in a warming environment by Dwivedi, the spatio-temporal variation in surface mass balance patterns of Vestre Brøggerbreen glacier, Ny-Ålesund, Svalbard, and its relationship with climate by Jat et al., and an insight into Arctic amplification and monsoon variability by Chattopadhyay et al.

This book aptly consolidates recent scientific findings and insights related to the ongoing climate change and its consequential impacts over the cryosphere region (the Himalaya, the Antarctic, and the Arctic) through an integrated approach. This book will act as a ready reference to all avid researchers, professionals, and students and emerge as a good source of information about the importance of the cryosphere.

Dr. M. Ravichandran
New Delhi
Secretary, Ministry of Earth Sciences
May 12, 2023

Preface

Glaciers are known to exist on almost every continent. These are dynamic natural resources that are influenced by environmental changes and can change global and local ecosystems. As per an estimate of the geographical distribution of these glaciers, approximately 91% are found in Antarctica and 8% in Greenland, whereas less than 0.5% is in North America (about 0.1% in Alaska), and 0.2% in Asia. In contrast, South America, Europe, Africa, New Zealand, and Indonesia witness less than 0.1% of all glaciers.

The cryosphere is an all-encompassing term for those portions of Earth's surface where water is in solid form, including sea ice, lake ice, river ice, snow cover, glaciers, ice caps, ice sheets, and frozen ground. Thus, there is an apparent overlap with the hydrosphere. The cryosphere is the part of the Earth's climate system that includes solid precipitation, icebergs, glaciers and ice caps, ice sheets, ice shelves, permafrost, and seasonally frozen ground. The main features of the cryosphere include snow, ice, frozen ground and permafrost, glaciers that include alpine glaciers, ice caps, ice sheets that cover Greenland and Antarctica, ice shelves, icebergs, and sea ice. Thus, the cryosphere includes the Himalayas and both polar regions. The cryosphere is the part of the planet which is made up of solid water or ice. It has been shrunk due to human activities that release greenhouse gases such as carbon dioxide which lead to global warming. Global warming causes the ice to melt, which reduces the extent of the cryosphere.

Extending over 2,000 km across South Asia, the Hindu Kush Himalayan region includes all or parts of Afghanistan, Bangladesh, Bhutan, China, India, Nepal, and Pakistan. Rising temperatures due to climate change are causing glaciers worldwide to shrink in volume and mass, a phenomenon known as glacial retreat. There are approximately 15,000 glaciers in the Himalayas. Each summer, these glaciers release meltwater into the Indus, Ganges, and Brahmaputra Rivers. Approximately 500 million people depend upon water from these three rivers.

Glaciers in the eastern and central regions of the Himalayas appear to be retreating at rates that have accelerated over the past century and are comparable to those in other parts of the world. In the western Himalayas, glaciers are more stable and may even be increasing in size. Siachen Glacier is the second-longest glacier outside the polar regions and the largest glacier in the Himalayas-Karakoram region.

As far as polar regions are concerned, the polar ice caps are dome-shaped sheets of ice found near the North and South Poles. They form because high-latitude polar regions receive less heat from the sun than other areas on Earth. As a result, average temperatures at the poles can be very cold. As stated earlier, the polar ice caps are found, clearly, in the polar regions – they start at the extreme poles of our planet and stretch outwards, covering almost the entirety of Antarctica, the Arctic Ocean, the majority of Greenland, a good portion of northern Canada and some of Siberia and Scandinavia.

About three-quarters of Earth's freshwater is stored in glaciers. Therefore, glacier ice is the second-largest reservoir of water on Earth and the largest reservoir of freshwater on Earth.

Global warming is the long-term warming of the planet's overall temperature. Though this warming trend has been going on for a long time, its pace has significantly increased in the last 100 years due to the burning of fossil fuels. As the human population has increased, so has the volume of fossil fuels burned. Thus, global warming is due to the long-term heating of the Earth's surface observed since the pre-industrial period (between 1850 and 1900) due to human activities, primarily fossil fuel burning, which increases heat-trapping greenhouse gas levels in the Earth's atmosphere. In other words, global warming is an aspect of climate change, referring to the long-term rise of the planet's temperatures. It is caused by increased concentrations of greenhouse gases in the atmosphere, mainly from human activities such as burning fossil fuels and farming. These greenhouse gases such as carbon dioxide, methane, and ozone trap the incoming radiation from the sun. This effect creates a natural "blanket", which prevents the heat from escaping back into the atmosphere. This effect is called the greenhouse effect.

Contrary to popular belief, greenhouse gases are not inherently bad. The greenhouse effect is vital for life on Earth. Without this effect, the sun's radiation would be reflected into the atmosphere, freezing the surface and making life impossible. However, when greenhouse gases in excess amounts get trapped, serious repercussions begin to appear. The polar ice caps begin to melt, leading to a rise in sea levels. Furthermore, the greenhouse effect is accelerated when polar ice caps and sea ice melt. This is due to the fact that the ice reflects 50–70% of the sun's rays back into space, but without ice, the solar radiation gets absorbed. Seawater reflects only 6% of the sun's radiation back into space. What's more frightening is the fact that the poles contain large amounts of carbon dioxide trapped within the ice. If this ice melts, it will significantly contribute to global warming.

Climate change is the single biggest health threat facing humanity. Climate impacts are already harming health, through air pollution, disease, extreme weather events, forced displacement, pressures on mental health, and increased hunger and poor nutrition in places where people cannot grow or find enough food.

Global warming can result in many serious alterations to the environment, eventually impacting human health. It can also cause a rise in sea level, leading to the loss of coastal land, a change in precipitation patterns, increased risks of droughts and floods, and threats to biodiversity. It is an established fact that a warmer world, even by a half-degree Celsius, has more evaporation, leading to more water in the atmosphere. Such changing conditions put our agriculture, health, water supply, and more at risk.

A related scenario when this phenomenon goes out of control is the runaway greenhouse effect. This scenario is essentially like an apocalypse, and all too real. Though this has never happened in the Earth's entire history, it is speculated to have occurred on Venus. Millions of years ago, Venus was thought to have an atmosphere like that of the Earth. But due to the runaway greenhouse effect, surface temperatures around the planet began rising. If this occurs on Earth, the runaway greenhouse effect will lead to many unpleasant scenarios, and the temperatures will rise hot enough for oceans to evaporate. Once the oceans evaporate, the rocks will start to sublimate under heat. To prevent such a scenario, proper measures must be taken to stop climate change.

The polar ice caps are melting as global warming causes climate change. We lose Arctic sea ice at a rate of almost 13% per decade, and over the past 30 years, the oldest and thickest ice in the Arctic has declined by a stunning 95%. If emissions continue to rise unchecked, the Arctic could be ice-free in the summer by 2040. It is important to understand that unprecedented changes occurring in the Arctic do not remain confined to the Arctic; rather, sea-ice loss has far-reaching effects around the world.

In view of the above, it is indubitable that the cryosphere exerts an important influence on Earth's climate, owing to its high surface reflectivity (albedo). This property gives it the ability to reflect a large fraction of solar radiation back into space and influences how much solar energy is absorbed by land and oceans.

Therefore, *Polar Ice and Global Warming in Cryosphere Regions* aptly covers all the important aspects of the Himalayan cryosphere as well as the cryosphere of both the North (Arctic) and South Poles (Antarctic) with the aim of understanding how the unprecedented global warming is impacting these cryosphere regions and what the repercussions of the warmed up cryosphere will be on geographically distinct regions.

The book begins with an introductory chapter on the cryosphere and changing climates by Khare, which is followed by chapters dedicated to the Himalayan cryosphere.

In their chapter, Sharma et al. assess the recession rate and surface velocity of Gangotri glacier Garhwal Himalaya India. The Gangotri is one of the largest glaciers in the Indian Himalayan Region (IHR), which has always been an issue of discussion due to its dynamic nature over the years. In the current study, we have estimated the recession rate and surface velocity of the Gangotri glacier by using satellite data and a kinematic GPS survey, respectively. The retreating rate of the Gangotri glacier has been estimated using the satellite data of the Landsat Series from 1980 to 2018. Based on satellite data interpretation, the total retreat and area loss of Gangotri glacier from 1980 to 2018 is 920.66 ± 37.04 m (24.23 ± 0.97 m/year) and 53.27 ± 2.1 km^2, respectively. The results also indicate that from 1980 to 1990 the recession rate was higher, 32.01 ± 4.43 m/year, while during the more recent years from 2010 to 2018, the lowest recession rate of 13.59 ± 3.65 m/year was observed. The surface velocity results of Gangotri glacier derived from the kinematic GPS survey for September to October 2016 were 2.28 ± 0.03 m/month, with an average annual velocity of around 27.36 ± 0.03 m/year. The results also indicate that the northern portion of the glacier exhibits a higher velocity (3.64 ± 0.02 m/month) than compared to the southern portion (1.11 ± 0.01 m/month). Their results conclude that the Gangotri glacier is dynamic throughout its length as well as its surface velocity, and the retreating rate has slowed down during recent years, probably due to prevailing local climatic conditions and an increase of debris thickness on the surface of the glacier. The decrease in the retreating rate supports the view that, apart from global warming, the retreat rate of the glacier is also governed by local aspects and glacier characteristics.

Ahmed et al. modelled a glacial lake outburst flood of Sona-Sar Lake in the Jhelum basin of Kashmir Himalayas, India, using HEC-RAS. The glacial lake outburst mountain communities live in different parts of the Himalayan region. The frequency of GLOF events in the Himalayan region is increasing under climate change-driven glacier recession and glacial lake expansion. Thus, it has become imperative to access the hazards and risks associated with these glacial lakes under such circumstances. Their study has focused on the pro-glacial lake, namely "Son-Sar" located in the Lidder catchment of the Jhelum basin. This lake has been identified as potentially dangerous and therefore it has been selected for GLOF modelling. They analysed the changes in the glacial lake during the years 1972 to 2022 using multi-temporal satellite data (Landsat and Sentinel). The depth, volume, and peak discharge were quantified using various empirical equations available in the scholarly literature. Hydrodynamic model HEC-RAS version 5.0.7 was used to route the breach hydrograph through the channel up to a distance of 23.4 km to evaluate the probable GLOF impacts on the downstream area. The areas of the glaciers were 0.59 km^2 and 0.26 km^2 in 1972, which reduced to 0.45 km^2 and 0.18 km^2 in 2022. The snouts of the glaciers retreated by 236 m and 242 m during the study period, with annual retreat rates of 4.72 m/year and 4.84 m/year, respectively. In 1972, the lake area was estimated as 0.08 km^2, which expanded to 0.19 km^2 in 2022. The analysis of the GLOF hydrograph revealed that the flood wave will reach the nearest settlement (Chandanwari) located at a distance of 7.2 km within 1 hour and 50 minutes after the breach initiation with a peak discharge of 1366 m^3 s^{-1} with a maximum flood depth of 6.4 m and a flood velocity of 6 min/s. In the worst-case scenario, the total inundated area resulting from the potential flood was estimated as 4.10 km^2. Among the six evaluated sites, the maximum area was inundated at site 2 (Betab Valley) with a total inundated area of 0.57 km^2, whereas the minimum inundated area (0.056 km^2) was estimated at site 1 (Chandanwari). The results of the present study have a bearing on the understanding of the risks associated with future GLOF hazards in the Jhelum basin of north-western Himalaya.

Kulkarni et al. throw light on the Hindukush Karakoram Himalayan glaciers. It is an established fact that the Hindukush-Karakoram-Himalaya is one of the largest reservoirs of ice and snow, supporting the lives of billions of people. Climate change is significantly impacting the glaciers in the region, leading to widespread glacier recession. Over the years, scientific studies on glacier dynamics have gained attention worldwide. In their chapter, a comprehensive review of the glaciers in the Hindukush Karakoram Himalaya (HKH) has been provided. The study outlines the state of knowledge on glaciers in the HKH including the glacier inventories, glacier volume, and glacier mass balance. It further discusses the newly developed techniques, remotely sensed data, and different field-based and modelling datasets. The study is focused on comprehensively presenting the major previous findings for the three major basins in the HKH including the Indus, Ganga, and Brahmaputra. The Indus basin marks the highest hydrological significance compared to Ganga and Brahmaputra in terms of glacier-stored water and melts contribution. The present analysis shows that the total glacier volume in HKH varies from 2377 Gt to 6715.96 Gt, with the highest reserves in the Indus basin and the lowest in the Ganga basin. The average mass balance of the region based on field measurements is –0.61 m.w.e. a^{-1} for the period 1975 to 2020, with an accelerated mass loss after 2000. Glacier dynamics involve a series of processes that are integral parts of the hydrological cycle. Changes in the climate are greatly impacting these processes, leading to their alteration and thereby affecting the

overall hydrological cycle. The present study not only presents the existing knowledge on the HKH glaciers but also highlights the need for further studies to improve
the overall understanding of the glaciers in the region.

Kumar et al. have studied the response of global warming on glaciers of Garhwal
Himalaya through a remote sensing and GIS approach. To take up such studies
the temporal response of global warming and the rise in global temperature on the
Gangotri, Meru, Thelu, and Swetvarn glaciers of the Garhwal Himalaya region during
the last 50 years were analysed. The monitoring of these glaciers was carried out
using high- and medium-resolution satellite images of CORONA (KHA), LANDSAT
(MSS, TM, ETM+, OLI), and LISS IV for different years. The study explains the
continuous retreat in these glaciers with uneven rates since 1968. According to them,
Gangotri glacier has lost 1460 m of its total length with a 17.80 m a^{-1} retreat rate from
1935 to 2017. Meru glacier has receded 416 m of its total length from 1976 to 2019
with a 9.67 m a^{-1} retreat rate. Between 1968 and 2019 Swetvarn has lost ~ 623 m total
length with a 12.21 m a^{-1} retreat rate and Thelu has lost ~ 590 m total length with an
11.56 m a^{-1} rate of retreat. The ablation zone shows an upward shift of 160 m for the
Swetvarn glacier and 145 m for the Thelu glacier. The Chorabari glacier also lost ~
306.43 ± 61 m total length and it retreated 8.62 m a^{-1} during 1976–1990, 7.00 m a^{-1}
during 1990–2001, 5.02 m a^{-1} during 2001–2012, 10.66 m a^{-1} during 2012–2017, and
considerable growth has been monitored since 2012. Their analysis explains that the
melting of the glaciers is mainly controlled by: (1) catastrophic variation in rainfall
and temperature, (2) gradual increase in temperature and decrease in precipitation,
and (3) increase in rainfall and decrease in snowfall.

While focusing on the polar regions to assess the impact of global warming,
Chowdhary et al. attempt to understand global warming through polar ice. To further elaborate on these aspects, especially over the Antarctic region, it is believed
that the pulsating sea ice in the Antarctic is the largest cryospheric seasonal variation on Earth. It plays a key role in inhibiting and modifying the exchange of mass,
momentum, and energy between the atmosphere and ocean in the south polar region,
affecting the local weather and climate in the long term. Sea ice also plays a decisive
role in maintaining the habitat of the diverse and unique Southern Ocean ecosystem.
Therefore, it is very important to understand, explore, exploit, and model ice cover in
the Southern Ocean in a climate change scenario.

Dash, in his chapter, discusses the methods of monitoring sea ice in the Antarctic,
its seasonal cycle, trend, and the impact of different (atmospheric and oceanic) drivers
on its growth and decay. Finally, projections, findings, and limitations of Coupled
Model Intercomparison Project Phase 5 (CMIP5) are discussed.

Meanwhile, Dwivedi analysed the multifractal properties of the long-term sea ice
concentration (SIC) time series over the Indian Antarctic stations Maitri and Bharati.
The SIC time series for the period 1900–2014 is generated using the historical data of
15 CMIP6 models for this purpose. It is shown that the q-order generalized Hurst exponent h(q) is a nonlinear decreasing function of q, which confirms the multifractality
of the SIC time series over the Maitri and Bharati stations. The shape and width
of the multifractal spectrum of the SIC time series vary significantly for different
CMIP6 models, with the spectrum being either left- or right-tailed. The strength
of the multifractality of the SIC time series is quantified in terms of multifractal

spectrum width $\Delta\alpha$ and change in singularity dimension $\Delta f(\alpha)$. The analysis of the multimodel mean SIC time series of the historical period reveals that the strength of multifractality of the SIC around the Bharati region is stronger as compared to that of the Maitri region. However, a complete reversal of multifractal properties is noticed in analysing the SIC time series of high greenhouse gas concentration SSP5–8.5 future scenario data for the period 2015–2100. It is found that, as a result of climate change, not only the strength of multifractality in the Maitri region will be higher as compared to the Bharati region, but an overall decrease in the strength of multifractality of the SIC is also observed.

On the other hand, Mishra and Srivastava analysed the spatial and temporal variations in the Antarctic snow-ice energy balance and also reviewed the glacial-metrological parameters and reflectance. Based on the studies conducted by the Snow and Avalanche Study Establishment (SASE) on the study of snow-ice albedo over the Antarctic ice shelf during the fifteenth expedition in January–February 1996, glacio-meteorological data were collected manually using snow-met instruments. Further, two automatic weather stations (AWSs) were installed over the continental shelf and continental ice surface in Dronning Maud Land in East Antarctica during the seventeenth Indian Scientific Expedition to Antarctica in 2000. High-quality continuous recording of glacial-met parameters such as surface temperature, ambient temperature, relative humidity, wind direction, wind speed, atmospheric pressure, snow depth, albedo, cloud amount and type, moisture content, etc. were undertaken throughout the year. The studies were aimed at (i) qualitative and quantitative analysis of the dependence of albedo on various snow-met parameters from the year 1996 to 2001 and (ii) investigation of spatial and temporal variations in various energy fluxes, i.e. short wave, long wave, latent heat, sensible heat flux, and net energy balance on different snow-ice surfaces in Antarctica from 1996 to 2001. Measurements of the dependence of albedo in different snow and ice media on solar elevation angle, cloud cover, liquid water, grain size, etc. can be interpreted in terms of single scattering and multiple scattering radiative transfer theory. A reduction in albedo values has been observed at a high solar zenith angle. The presence of clouds has been observed to increase albedo values linearly. A decrease in albedo values with an increase in melt-water amount has been observed for the natural melt process on the Antarctic glacier. A simple energy balance model based on snow-met parameters is formulated to calculate the net energy exchange over the snow-ice surfaces. Surface roughness and albedo of the different snow-ice media have been accounted for in the model. Summer and winter energy budgets are also compared and discussed for the continental ice surface. The difference in the energy balance over the continental shelf, continental ice, and snow are mainly due to differences in their reflectance, surface roughness, and extinction coefficient. The average energy budget over continental ice was found to be negative during both summer and winter periods. The results obtained from the analysis of AWS data indicate that the primary difference in summer and winter energy budgets over continental ice is in the net radiation and sensible heat fluxes during the corresponding periods.

Interestingly, Pednekar in his chapter presents the variability of the physical parameter in Prydz Bay which is bounded by the open sea to the north surrounded by a clockwise Prydz gyre and a vast polar ice sheet to the south. Instrumental Seals data supported scientific communities to describe the variability in the Prydz Bay south of the polar frontal zone. An attempt has been made to highlight the changes that occurred in space and time in Prydz Bay based on previous studies. The major water masses that exist in Prydz Bay are briefly explained and demonstrated using a potential temperature and salinity diagram. The vertical sections of potential temperature showed the occurrence of CDW below 200 m between 65°E to 75°E near to the slope of Prydz Bay along the 66.3°S transect. The presence of mCDW flows onto Prydz Bay occurs further south below 100 m in pockets in the transect of 67°S between 72°E to 78.5°E and 74°E. Time and space analyses have shown the entrance of CDW into the Prydz Bay near the shelf break during summer having warmer and saltier water characteristics as identified. There exists variability in the extension of mCDW each year in both isopycnal surfaces on a spatial and temporal scale. The distribution of −1.7°C isotherm on potential temperature and salinity highlights the annual variability in space and time and the extent of the mCDW signal to the interior of the bay as isotherm −1.7°C extended more southward to 68.5°S with small pockets up to 69°S. The continental shelf region of the bay is influenced by the signature of warm mCDW responsible for the climate change in Prydz Bay due to the strong winds blowing from the north-east to the south-west in the Larsemann Hills region.

The last couple of chapters focus on the Arctic region to understand teleconnections between the Arctic region and other parts of the world, especially the tropical regions. Such an assessment becomes imperative because it is believed that the warming in the Arctic is taking place three times faster than the global average pace owing to the melting of snow and ice, which exposes a darker surface and increases the amount of solar energy absorbed in these areas, which is known as the albedo effect.

In this direction, the fidelity of the Coupled Model Intercomparison Project Phase 6 (CMIP6) models in correctly simulating the sea ice area (SIA) of the eastern Arctic region around 0°E–35°E, 65°N–85°N is evaluated against the satellite observations by Dwivedi. The region includes the Indian Arctic station, Himadri. It is found that the multi-model mean seasonal variability of SIA of the region matches well with the observed satellite data, but there exists a large inter-model spread. The SIA time series shows a decreasing trend both in the historical period (years 1900–2014) and high greenhouse gas (GHG) concentration projection scenario SSP5-8.5 data (years 2015–2100). However, the rate of decay of SIA is very rapid (approximately five times faster) in the SSP5-8.5 scenario. It is shown that there will be no sea ice around the Himadri Indian Arctic station by the end of the 21st century. The generalized Hurst exponent and Predictability Index are used as quantifiers of predictability of the SIA of the east Arctic region in a changing climate scenario. The SIA predictability of the east Arctic region will decrease with an increase in the GHG concentration during the years 2015–2100. On the other hand, the SIA predictability around the Himadri Indian Arctic station will increase in a warming environment.

To further augment the understanding of the Arctic response to ongoing warming, Jat et al. attempt to understand the spatio-temporal variation in surface mass balance patterns of Vestre Brøggerbreen glacier, Ny-Ålesund, Svalbard, and its relationship with climate since 2011. The main objective of their study is to discover the pattern of the temporal and spatial variability in ablation, snow cover, and specific mass balance of Vestre Brøggerbreen glacier in Ny-Ålesund, Svalbard. The study was based on in situ observations using stakes, the global navigation satellite system (GNSS) for precise positioning of stakes, and mapping of the glacierized area since the inception of glaciological studies in the Ny-Ålesund area of Svalbard archipelago. The observation of the 4.7 km^2 glacierized area was carried out during the spring and autumn seasons, which included ablation at stakes, the precise position of the stakes, snow density profiling, snow cover thickness over the glacier, and meteorological parameters from the nearest weather station at Ny-Ålesund. After data processing and analysis at the GIS platform (ArcGIS 10.2) surface mass balance, variability in the ablation pattern, and its correlation with the meteorological parameters have been estimated. One decade of observations shows annual mass balances in Vestre Brøggerbreen are negative except for 2014, whereas a positive trend was observed in Vestre Brøggerbreen-I. The meteorological parameters such as average air temperature and precipitation corroborate the patterns of mass balance.

On the contrary, to assess Arctic amplification and monsoon variability, Chattopadhyay et al. explored the role of a reduction in sea ice temperature due to Arctic amplification in the modulation of summer monsoon circulation over the Indian region. The study is based on model simulations in which the latitudinal extent of Arctic sea ice is reduced in one experiment and increased in another. The reduction or increase in the sea-ice extent is based on observation of sea-ice concentration data during 1981–2015. The month with the minimum (maximum) sea-ice extent is used to force the GFS model to represent the warming (cooling) scenario due to Arctic amplification. The twin experiment is conducted to see if the monsoon response is the opposite in warming and cooling scenarios. Surprisingly, the monsoon weakens in both scenarios, indicating that monsoon circulation is related to the equator-to-pole heat gradients rather than an equatorial phenomenon. The analysis also suggests a substantial change in the mid-latitude circulation, affecting the tropical–extratropical interaction, heat and momentum fluxes transport, and teleconnections. The most significant result is that the analysis represents an increase in rainfall over the Himalayan region and the foothills of Himalaya and the north-eastern parts of India. The recent rise in extreme events over these regions could indicate such changes in sea-ice concentration.

Dr. Neloy Khare
New Delhi
Advisor, Ministry of Earth Sciences
May 12, 2023

About the Editor

Dr Neloy Khare is currently an adviser and Scientist 'G' to the Government of India at MoES. He has a very distinctive acumen not only in administration but also in quality science and research in his areas of expertise covering a large spectrum of geographically distinct locations including the Antarctic, Arctic, Southern Ocean, Bay of Bengal, Arabian Sea, Indian Ocean, etc. Dr Khare has 30 years of experience in the field of palaeoclimate research using palaeobiology, palaeontology, working in roles such as teaching, science management, administration, and coordination for scientific programmes (including the Indian Polar Programme). He has also been conferred as an honorary professor and adjunct professor by many Indian universities. Having completed his PhD in the tropical marine region and Doctor of Science (DSc) in southern high-latitude marine regions towards environmental/climatic implications, he has made significant contributions in the field of palaeoclimatology of southern high-latitude regions (the Antarctic and the Southern Ocean) using micropalaeontology as a tool and various proxies including foraminifera (microfossils). These studies, coupled with his palaeoclimatic reconstructions from tropical regions, helped uncover causal linkages and teleconnections between the processes taking place in southern high latitudes and climate variability occurring in tropical regions.

He has a very impressive list of publications to his credit (125 research articles in national and international scientific journals; three special issues of national scientific journals as guest editor; authored/edited many books; 130 abstracts have been contributed to various seminars; 23 popular science articles; and five technical reports). The government of India and many professional bodies have bestowed him with prestigious awards for his scientific contributions to climate change research, oceanography, polar science, and southern oceanography. The most coveted award is the Rajiv Gandhi National Award – 2013 conferred by the Honourable President of India. Others include the ISCA Young Scientist Award, BOYSCAST Fellowship, CIES French Fellowship, Krishnan Gold Medal, Best Scientist Award, Eminent Scientist Award, ISCA Platinum Jubilee Lecture, IGU Fellowship, etc. Dr Khare has made tremendous efforts to popularize ocean science and polar science across the country by way of delivering lectures, radio talks, and publishing popular science articles.

Dr Khare sailed in the Arctic Ocean as a part of "Science PUB" in 2008 during the International Polar Year campaign for scientific exploration and became the first Indian to sail in the Arctic Ocean.

Contributors

Syed Towseef Ahmad
Department of Geography and Disaster
  Management
School of Earth and Environmental
  Sciences
University of Kashmir
Srinagar, India

Rayees Ahmed
Department of Geography and Disaster
  Management
School of Earth and Environmental
  Sciences
University of Kashmir
Srinagar, India

Roja Asharaf
DST-Centre for Excellence in
  Climate Change
Divecha Centre for Climate Change, IISc
Bengaluru, India

V. Bharti
Defence Geoinformatics Research
  Establishment (DGRE), Him Parisar,
  Sector 37 A, Chandigarh, India

Harish Bisht
Centre of Advanced Study in Geology
Kumaun University, Nainital
Uttarakhand, India

Vikash Chandra
Polar Studies Division
Geological Survey of India
Faridabad, India

Rajib Chattopadhyay
Indian Institute of Tropical
  Meteorology (IITM)
Ministry of Earth Sciences
Pune, India

Shabnam Choudhary
National Centre for Polar and Ocean
  Research
Vasco da Gama, Goa, India

Mihir Kumar Dash
Centre for Oceans, Rivers,
  Atmosphere and Land
  Sciences (CORAL)
Indian Institute of Technology
Kharagpur, India

P. Datt
Defence Geoinformatics Research
  Establishment (DGRE), Him Parisar,
  Sector 37 A, Chandigarh, India

Amit Dharwadkar
Geological Survey of India
Northern Region, Lucknow, India

Chetan Anand Dubey
Department of Geology
University of Lucknow
Lucknow, India

Sharat Dutta
Polar Studies Division
Geological Survey of India
Faridabad, India

Suneet Dwivedi
K Banerjee Centre of Atmospheric and
  Ocean Studies and M N Saha Centre
  of Space Studies
University of Allahabad
Allahabad, UP, India

Deepak Y. Gajbhiye
Polar Studies Division
Geological Survey of India
Faridabad, India

Pawan Kumar Gautam
Department of Geology
University of Lucknow
Lucknow UP, India

Pulak Guhathakurta
Office of Climate Research and Services
India Meteorological Department
Ministry of Earth Sciences
Pune, India

H.S. Gusain
Institute of Technology Management
    (ITM), Landour Cantt, Mussoorie,
    Uttarakhand, India

Surendra Jat
Polar Studies Division
Geological Survey of India
Faridabad, India

Sumaira Javaid
Department of Geography and Regional
    Development
School of Earth and Environmental
    Sciences
University of Kashmir
Srinagar, India

Mahesh Kalshetti
Indian Institute of Tropical
    Meteorology (IITM)
Ministry of Earth Sciences
Pune, India

Rajkumar Kashyap
Department of Earth Sciences
Indian Institute of Technology, Roorkee
Uttarakhand, India

Varun Khajuria
Department of Water Resources
Indian Institute of Remote Sensing
ISRO-Dehradun
Uttarakhand, India

Neloy Khare
Ministry of Earth Sciences
Prithvi Bhawan, Lodhi Road
New Delhi, India

Bahadur Singh Kotlia
Centre of Advanced Study in Geology
Kumaun University, Nainital
Uttarakhand, India

Girish Chandra Kothyari
Department of Petroleum Engineering
    and Earth Science,
University of Petroleum and Energy
    Studies, Dehradun, India

Manmohan Kukreti
Centre of Advanced Study in Geology
Kumaun University, Nainital
Uttarakhand, India

Anil V. Kulkarni
DST-Centre for Excellence in
    Climate Change
Divecha Centre for Climate
    Change, IISc
Bengaluru, India

Dhirendra Kumar
Department of Geology,
Central University of South Bihar,
    India

Kireet Kumar
G. B. Pant National Institute of
    Himalayan Environment and
    Sustainable Development
Kosi-Katarmal, Almora
Uttarakhand, India

M. Kumar
Defence Geoinformatics Research
    Establishment (DGRE), Him Parisar,
    Sector 37 A, Chandigarh, India

Pradeep Kumar
Polar Studies Division
Geological Survey of India
Faridabad, India

V. D. Mishra
Defence Geoinformatics Research
 Establishment (DGRE), Him Parisar,
 Sector 37 A, Chandigarh, India

S. M. Pednekar
National Centre for Polar and Ocean
 Research
Ministry of Earth Science
Vasco da Gama, Goa, India

R Phani
Indian Institute of Tropical
 Meteorology (IITM)
Ministry of Earth
Pune, India

Abid Farooq Rather
Department of Geography and Disaster
 Management
School of Earth and Environmental
 Sciences
University of Kashmir
Srinagar, India

Syed Mohammad Saalim
National Centre for Polar and Ocean
 Research
Vasco da Gama, Goa, India

M Sadiq
Polar Studies Division
Geological Survey of India
Faridabad, India

A.K. Sahai
Indian Institute of Tropical
 Meteorology (IITM)
Ministry of Earth Sciences
Pune, India

Ashim Sattar
DST-Centre for Excellence in
 Climate Change
Divecha Centre for Climate Change, IISc
Bengaluru, India

Savati Sharma
Department of Remote Sensing and GIS
University of Jammu
Jammu, India

Anoop Kumar Singh
Department of Geology
University of Lucknow
Lucknow, UP, India

Dhruv Sen Singh
Department of Geology
University of Lucknow
Lucknow, UP, India

K.K. Singh
Defence Geoinformatics Research
 Establishment (DGRE), Him Parisar,
 Sector 37 A, Chandigarh, India

Paramvir Singh
Defence Geoinformatics Research
 Establishment (DGRE), Him Parisar,
 Sector 37 A, Chandigarh, India

P.K. Srivastava
Defence Geoinformatics Research
 Establishment (DGRE), Him Parisar,
 Sector 37 A, Chandigarh, India

Ajay Kumar Taloor
Department of Remote Sensing and GIS
University of Jammu
Jammu, India

Abhishek Verma
Polar Studies Division
Geological Survey of India
Faridabad, India

Balkrishan Vishawakarma
Department of Geology
University of Lucknow
Lucknow, UP, India

Gowhar Farooq Wani
Department of Geography and Disaster
  Management
School of Earth and Environmental
  Sciences,
University of Kashmir
Srinagar, India

Acknowledgements

It is my great pleasure to express my gratitude and deep appreciation to all contributing authors. Without their valuable inputs on various facets of climate variability over the Antarctic and surrounding Southern Ocean region, the book would not have been possible. Various learned experts who have reviewed different chapters are graciously acknowledged for their timely, constructive and critical reviews.

I sincerely thank Dr M. Ravichandran, Secretary Ministry of Earth Sciences, Government of India, New Delhi (India) for their support and encouragement. Prof. Govardhan Mehta, FRS has always been a source of inspiration and is acknowledged for his kind support.

Prof. Anil Kumar Gupta, Indian Institute of Technology, Kharagpur (India) and Dr Rajiv Nigam, former Adviser at the National Institute of Oceanography, Goa (India) are deeply acknowledged for providing many valuable suggestions to this book. This book has received significant support from Akshat Khare and Ashmit Khare, who have helped me during the book preparation. Dr Rajni Khare has unconditionally supported enormously during various stages of this book. Shri Hari Dass Sharma from the Ministry of Earth Sciences, New Delhi (India) has helped immensely in formatting the text and figures of this book and bringing it to its present form. Valuable help rendered by Dr. Siddharth Dey is deeply acknowledged. Publishers (Taylor & Francis) have done a commendable job and are sincerely acknowledged.

Date: May 2023
Place: New Delhi
Neloy Khare

1 Cryosphere and Changing Climates
An Introduction

Neloy Khare

The geosphere and the biosphere are two primary components of the Earth system. The geosphere has four subcomponents: lithosphere (solid Earth), atmosphere (gaseous envelope), hydrosphere (liquid water), and cryosphere (frozen water). Similarly, the cryosphere is part of the Earth's climate system that includes solid precipitation, snow, sea ice, lake and river ice, icebergs, glaciers and ice caps, ice sheets, ice shelves, permafrost, and seasonally frozen ground. The term "cryosphere" traces its origins to the Greek word "kryos" for frost or ice cold. The cryosphere is an all-encompassing term for those portions of the Earth's surface where water is in solid form, including sea ice, lake ice, river ice, snow cover, glaciers, ice caps, ice sheets, and frozen ground (permafrost). The Earth's climate system is a complex system having five interacting components: the atmosphere (air), the hydrosphere (water), the cryosphere (ice and permafrost), the lithosphere (Earth's upper rocky layer), and the biosphere (living things) (Planton, 2013). The climate is the statistical characterization of the climate system, representing the average weather, typically over 30 years, and is determined by a combination of processes in the climate system, such as ocean currents and wind patterns (www.climatechange.environment.nsw.gov.au/global-climate-system; https://worldoceanreview.com/en/wor-1/climate-system/earth-climate-system/). Circulation in the atmosphere and oceans is primarily driven by solar radiation and transports heat from the tropical regions to regions that receive less energy from the Sun. The water cycle also moves energy throughout the climate system. In addition, different chemical elements, necessary for life, are constantly recycled between the different components.

The climate system can change due to internal variability and external forcings. These external forcings can be natural, such as variations in solar intensity and volcanic eruptions, or caused by humans. The accumulation of heat-trapping greenhouse gases, mainly emitted by people burning fossil fuels, is causing global warming. Human activity also releases cooling aerosols, but their net effect is far less than that of greenhouse gases (Planton, 2013). Changes can be amplified by feedback processes in the different climate system components.

Thus, there is a wide overlap with the hydrosphere. This sphere helps maintain the Earth's climate by reflecting incoming solar radiation into space. The frozen part of the Earth's hydrosphere is made of ice: glaciers, ice caps, and icebergs. The

DOI: 10.1201/9781003364115-1

frozen part of the hydrosphere is called the cryosphere. Water moves through the hydrosphere in a cycle. The conversion among solid water, liquid water, and vapour meets the law of mass balance, along with the coupling and transformation of energy. Thus, the coupled mass–energy processes provide the physical basis for the formation and changes of the cryosphere. Acting like a highly reflective shield, the cryosphere protects the Earth from getting too warm. Snow and ice reflect more sunlight than open water or bare ground. The presence or absence of snow and ice, therefore, affects heating and cooling over the Earth's surface, influencing the entire planet's energy balance. Thus, the cryosphere comprises all regions on and beneath the surface of the Earth and ocean where water is in solid form, including sea ice, lake ice, river ice, snow cover, glaciers and ice sheets, and frozen ground, including permafrost (Planton, 2013).

There is a wide overlap with the hydrosphere. The cryosphere is an integral part of the global climate system, with important linkages and feedback generated through its influence on surface energy and moisture fluxes, clouds, precipitation, hydrology, and atmospheric and oceanic circulation. Through these feedback processes, the cryosphere plays a significant role in the global climate and in the climate model response to global changes. Approximately 10% of the Earth's surface is covered by ice, but this is rapidly decreasing (Planton, 2013). The term *deglaciation* describes the retreat of cryospheric features. Cryology is the study of cryospheres.

The residence time of water in each of these cryospheric sub-systems varies widely. Snow cover and freshwater ice are essentially seasonal, and most sea ice, except for ice in the central Arctic, lasts only a few years if it is not seasonal. A given water particle in glaciers, ice sheets, or ground ice, however, may remain frozen for 10–100,000 years or longer, and deep ice in parts of East Antarctica may have an age approaching 1 million years.

Most of the world's ice volume is in Antarctica, principally in the East Antarctic Ice Sheet. In terms of areal extent, however, Northern Hemisphere winter snow and ice extent comprise the largest area, amounting to an average of 23% of the hemispheric surface area in January. The large areal extent and the important climatic roles of snow and ice, related to their unique physical properties, indicate that the ability to observe and model snow and ice-cover extent, thickness, and physical properties (radiative and thermal properties) is of particular significance for climate research.

Several fundamental physical properties of snow and ice modulate energy exchanges between the surface and the atmosphere. The most important properties are surface reflectance (albedo), the ability to transfer heat (thermal diffusivity), and the ability to change state (latent heat). These physical properties, together with surface roughness, emissivity, and dielectric characteristics, have important implications for observing snow and ice from space. For example, surface roughness is often the dominant factor determining the strength of radar backscatter (https://worldoceanrev iew.com/en/wor-1/climate-system/earth-climate-system/). Physical properties such as crystal structure, density, length, and liquid water content are important factors affecting the transfers of heat and water and the scattering of microwave energy.

The surface reflectance of incoming solar radiation is important for the surface energy balance (SEB). This is the ratio of reflected to incident solar radiation, commonly referred to as albedo. Climatologists are primarily interested in albedo integrated over the shortwave portion of the electromagnetic spectrum (~300 to 3500 nm), which coincides with the main solar energy input. Typically, albedo values for non-melting snow-covered surfaces are high (~80–90%) except in the case of forests. The higher albedos for snow and ice cause rapid shifts in surface reflectivity in autumn and spring in high latitudes, but the overall climatic significance of this increase is spatially and temporally modulated by cloud cover. (Planetary albedo is determined principally by cloud cover, and by the small amount of total solar radiation received in high latitudes during winter months.) Summer and autumn are times of high average cloudiness over the Arctic Ocean so the albedo feedback associated with the large seasonal changes in sea–ice extent is greatly reduced. Groisman et al. observed that snow cover exhibited the greatest influence on the Earth's radiative balance in the spring (April to May) period when incoming solar radiation was greatest over snow-covered areas (Barry and Hall-McKim, 2014). The thermal properties of cryospheric elements also have important climatic consequences. Snow and ice have much lower thermal diffusivities than air. Thermal diffusivity is a measure of the speed at which temperature waves can penetrate a substance. Snow and ice are many orders of magnitude less efficient at diffusing heat than air. Snow cover insulates the ground surface, and sea ice insulates the underlying ocean, decoupling the surface–atmosphere interface with respect to both heat and moisture fluxes. The flux of moisture from a water surface is eliminated by even a thin skin of ice, whereas the flux of heat through thin ice continues to be substantial until it attains a thickness of over 30–40 cm. However, even a small amount of snow on top of the ice will dramatically reduce the heat flux and slow down the rate of ice growth. The insulating effect of snow also has major implications for the hydrological cycle. In non-permafrost regions, the insulating effect of snow is such that only near-surface ground freezes and deep-water drainage is uninterrupted (Gettelman and Rood, 2016).

While snow and ice act to insulate the surface from large energy losses in winter, they also act to retard warming in the spring and summer because of the large amount of energy required to melt ice (the latent heat of fusion, 3.34×10^5 J/kg at 0°C). However, the strong static stability of the atmosphere over areas of extensive snow or ice tends to confine the immediate cooling effect to a relatively shallow layer, so associated atmospheric anomalies are usually short-lived and local to regional in scale (Gettelman and Rood, 2016). In some areas of the world such as Eurasia, however, the cooling associated with a heavy snowpack and moist spring soils is known to play a role in modulating the summer monsoon circulation (Kundzewicz, 2008). Gutzler and Preston (1997) recently presented evidence for a similar snow-summer circulation feedback over the southwestern United States (Goosse, 2015).

The role of snow cover in modulating the monsoon is just one example of a short-term cryosphere–climate feedback involving the land surface and the atmosphere. These operate over a wide range of spatial and temporal scales from local seasonal cooling of air temperatures to hemispheric-scale variations in ice sheets over time scales of thousands of years. The feedback mechanisms involved are often complex and incompletely understood. For example, Curry et al. (1995) showed that the so-called "simple" sea ice-albedo feedback involved complex interactions with lead fraction, melt ponds, ice thickness, snow cover, and sea–ice extent.

1.1 SNOW

Because of its close relationship with hemispheric air temperature, snow cover is an important indicator of climate change. Most of the Earth's snow-covered area is located in the Northern Hemisphere and varies seasonally from 46.5 million km^2 in January to 3.8 million km^2 in August (Gettlemen and Rood, 2016). North American winter snow cover increased during the 20th century (https://climate.nasa.gov/vital-signs/ocean-warming/; Desonie, 2008), largely in response to an increase in precipitation (Goosse, 2015). However, the IPCC Sixth Assessment Report found that Northern Hemisphere snow cover has been decreasing since 1978, along with snow depth (Goosse, 2015). Paleoclimate observations show that such changes are unprecedented over the last millennia in Western North America (Goosse, 2015; Houze, 2012). Snow cover is an extremely important storage component in the water balance, especially seasonal snow-packs in mountainous areas of the world. Though limited in extent, seasonal snow-packs in the Earth's mountain ranges account for the major source of the runoff for stream flow and groundwater recharge over wide areas of the midlatitudes. For example, over 85% of the annual runoff from the Colorado River Basin originates as snowmelt. Snowmelt runoff from the Earth's mountains fills the rivers and recharges the aquifers that over a billion people depend on for their water resources. Furthermore, over 40% of the world's protected areas are in mountains, attesting to their value both as unique ecosystems needing protection and as recreation areas for humans. Climate warming is expected to result in major changes to the partitioning of snow and rainfall and to the timing of snowmelt, which will have important implications for water use and management. These changes also involve potentially important decadal and longer time-scale feedback to the climate system through temporal and spatial changes in soil moisture and runoff to the oceans (Walsh, 1995). Freshwater fluxes from the snow cover into the marine environ-ment may be important, as the total flux is probably of the same magnitude as desalinated ridging and rubble areas of sea ice (Barry and Hall-McKim 2014). In addition, there is an associated pulse of precipitated pollutants which accumulate over the Arctic winter in snowfall and are released into the ocean upon ablation of the sea ice.

1.2 SEA ICE

Sea ice covers much of the polar oceans and forms by freezing seawater. Satellite data since the early 1970s reveal considerable seasonal, regional, and interannual variability in the sea ice covers of both hemispheres. Seasonally, sea-ice extent in the Southern Hemisphere varies by a factor of 5, from a minimum of 3–4 million km^2 in February to a maximum of 17–20 million km^2 in September (Gettelman and Rood, 2016; Haug and Keigwin, 2004). The seasonal variation is much less in the Northern Hemisphere where the confined nature and high latitudes of the Arctic Ocean result in a much larger perennial ice cover, and the surrounding land limits the equatorward extent of wintertime ice. Thus, the seasonal variability in Northern Hemisphere ice extent varies by only a factor of 2, from a minimum of 7–9 million km^2 in September to a maximum of 14–16 million km^2 in March (Haug and Keigwin, 2004; Gettelman and Rood, 2016). The ice cover exhibits much greater regional-scale interannual variability than it does hemispherically. For instance, in the region of the

Sea of Okhotsk and Japan, the maximum ice extent decreased from 1.3 million km^2 in 1983 to 0.85 million km^2 in 1984, a decrease of 35%, before rebounding the following year to 1.2 million km^2 (Haug and Keigwin, 2004). The regional fluctuations in both hemispheres are such that for any several-year period of the satellite record some regions exhibit decreasing ice coverage while others exhibit increasing ice cover (Goosse, 2015). The overall trend indicated in the passive microwave record from 1978 through mid-1995 shows that the extent of Arctic sea ice is decreasing by 2.7% per decade (Goosse, 2015). Subsequent work with the satellite passive-microwave data indicates that from late October 1978 through the end of 1996 the extent of Arctic sea ice decreased by 2.9% per decade, while the extent of Antarctic sea ice increased by 1.3% per decade (Smil, 2003). The Intergovernmental Panel on Climate Change publication *Climate Change 2013: The Physical Science Basis* stated that sea ice extent for the Northern Hemisphere showed a decrease of 3.8% ± 0.3% per decade from November 1978 to December 2012.

1.3 LAKE ICE AND RIVER ICE

Ice forms on rivers and lakes in response to seasonal cooling. The sizes of the ice bodies involved are too small to exert anything other than localized climatic effects. However, the freeze-up/break-up processes respond to large-scale and local weather factors, such that considerable interannual variability exists in the dates of appearance and disappearance of the ice. Long series of lake-ice observations can serve as a proxy climate record, and the monitoring of freeze-up and break-up trends may provide a convenient integrated and seasonally specific index of climatic perturbations. Information on river-ice conditions is less useful as a climatic proxy because ice formation is strongly dependent on the river-flow regime, which is affected by precipitation, snow melt, and watershed runoff as well as being subject to human interference that directly modifies channel flow, or that indirectly affects the runoff via land-use practices.

Lake freeze-up depends on the heat storage in the lake and therefore on its depth, the rate and temperature of any inflow, and water–air energy fluxes. Information on lake depth is often unavailable, although some indication of the depth of shallow lakes in the Arctic can be obtained from airborne radar imagery during late winter and spaceborne optical imagery during summer (Duguay and Lafleur, 1997). The timing of breakup is modified by snow depth on the ice as well as by ice thickness and freshwater inflow.

1.4 FROZEN GROUND AND PERMAFROST

Frozen ground (permafrost and seasonally frozen ground) occupies approximately 54 million km^2 of the exposed land areas of the Northern Hemisphere (Zhang et al., 2003) and therefore has the largest areal extent of any component of the cryosphere. Permafrost (perennially frozen ground) may occur where mean annual air temperatures (MAAT) are less than −1 or −2°C and is generally continuous where MAAT are less than −7°C. In addition, its extent and thickness are affected by ground moisture content, vegetation cover, winter snow depth, and aspect. The global extent of permafrost is still not completely known, but it underlies approximately 20% of

Northern Hemisphere land areas. Thicknesses exceed 600 m along the Arctic coast of north-eastern Siberia and Alaska, but, toward the margins, permafrost becomes thinner and horizontally discontinuous. The marginal zones will be more immediately subject to any melting caused by a warming trend. Most of the presently existing permafrost formed during previous colder conditions and is therefore relic. However, permafrost may form under present-day polar climates where glaciers retreat or land emergence exposes unfrozen ground. Washburn (1980) concluded that most continuous permafrost is in balance with the present climate at its upper surface, but changes at the base depend on the present climate and geothermal heat flow; in contrast, most discontinuous permafrost is probably unstable or "in such delicate equilibrium that the slightest climatic or surface change will have drastic disequilibrium effects" (Barry and Hall-McKim, 2014). Under warming conditions, the increasing depth of the summer active layer has significant impacts on the hydrologic and geomorphic regimes. Thawing and retreat of permafrost have been reported in the upper Mackenzie Valley and along the southern margin of its occurrence in Manitoba, but such observations are not readily quantified and generalized. Based on average latitudinal gradients of air temperature, an average northward displacement of the southern permafrost boundary by 50–150 km could be expected, under equilibrium conditions, for a 1°C warming. Only a fraction of the permafrost zone consists of actual ground ice. The remainder (dry permafrost) is simply soil or rock at subfreezing temperatures. The ice volume is generally greatest in the uppermost permafrost layers and mainly comprises pore and segregated ice in Earth material. Measurements of bore-hole temperatures in permafrost can be used as indicators of net changes in temperature regimes. Lachenbruch et al., (1982) infer a 2–4°C warming over 75–100 years at Cape Thompson, Alaska, where the upper 25% of the 400-m thick permafrost is unstable with respect to an equilibrium profile of temperature with depth (for the present mean annual surface temperature of −5°C). Maritime influences may have biased this estimate, however. At Prudhoe Bay, similar data imply a 1.8°C warming over the last 100 years (Lachenbruch et al., 1982). Further complications may be introduced by changes in snow-cover depths and the natural or artificial disturbance of the surface vegetation. The potential rates of permafrost thawing have been established by Osterkamp (2005) to be two centuries or less for 25-meter-thick permafrost in the discontinuous zone of interior Alaska, assuming warming from −0.4 to 0°C in 3–4 years, followed by a further 2.6°C rise. Although the response of permafrost (depth) to temperature change is typically a very slow process (Osterkamp, 2005), there is ample evidence for the fact that the active layer thickness quickly responds to a temperature change (Kane et al., 1991). Whether, under a warming or cooling scenario, global climate change will have a significant effect on the duration of frost-free periods in both regions with seasonally and perennially frozen ground.

1.5   GLACIERS AND ICE SHEETS

Ice sheets and glaciers are flowing ice masses that rest on solid land. They are controlled by snow accumulation, surface, and basal melt, calving into surrounding oceans or lakes, and internal dynamics. The latter results from gravity-driven creep flow ("glacial flow") within the ice body and sliding on the underlying land, which leads to thinning and horizontal spreading (Bridgman and Oliver, 2014). Any imbalance of this dynamic equilibrium between mass gain, loss, and transport due to flow results in either growing or shrinking ice bodies.

Ice sheets are the greatest potential source of global freshwater, holding approximately 77% of the global total. This corresponds to 80 m of world sea-level equivalent, with Antarctica accounting for 90% of this. Greenland accounts for most of the remaining 10%, with other ice bodies and glaciers accounting for less than 0.5%. Because of their size in relation to annual rates of snow accumulation and melt, the residence time of water in ice sheets can extend to 100,000 or 1 million years. Consequently, any climatic perturbations produce slow responses, occurring over glacial and interglacial periods. Valley glaciers respond rapidly to climatic fluctuations, with typical response times of 10–50 years (Bridgman and Oliver, 2014). However, the response of individual glaciers may be asynchronous to the same climatic forcing because of differences in glacier length, elevation, slope, and speed of motion. Oerlemans (1994) provided evidence of coherent global glacier retreat which could be explained by a linear warming trend of 0.66°C per 100 years (Barry and Hall-McKim, 2014). While glacier variations are likely to have minimal effects on global climate, their recession may have contributed one-third to one-half of the observed 20th-century rise in sea level (Meier, 1984). Furthermore, it is extremely likely that such extensive glacier recession as is currently observed in the Western Cordillera of North America (Barry and Hall-McKim, 2014), where runoff from glacierized basins is used for irrigation and hydropower, involves significant hydrological and ecosystem impacts. Effective water-resource planning and impact mitigation in such areas depend upon developing a sophisticated knowledge of the status of glacier ice and the mechanisms that cause it to change. Furthermore, a clear understanding of the mechanisms at work is crucial to interpreting the global change signals that are contained in the time series of glacier mass balance records. Combined glacier mass balance estimates of the large ice sheets carry an uncertainty of about 20%. Studies based on estimated snowfall and mass output tend to indicate that the ice sheets are near balance or taking some water out of the oceans (Gruza, 2009). Marine-based studies (Goosse, 2015) suggest sea level rise from the Antarctic or rapid ice-shelf basal melting. Some authors (Paterson, 1993; Alley, 2000) have suggested that the difference between the observed rate of sea-level rise (roughly 2 mm/y) and the explained rate of sea-level rise from melting of mountain glaciers, thermal expansion of the ocean, etc. (roughly 1 mm/y or less) is similar to the modelled imbalance in the Antarctic (roughly 1 mm/y of sea-level rise; Huybrechts, 1990), suggesting a contribution of sea-level rise from the Antarctic. Relationships between global climate and changes in ice extent are complex. The mass balance of land-based glaciers and ice sheets is determined by the accumulation of snow, mostly in winter, and warm-season ablation due primarily to net radiation and turbulent heat fluxes to melting ice and snow from warm-air advection (Goosse, 2015; Munro, 1990). However, most of Antarctica never experiences surface melting (www.metoffice.gov.uk/weather/learn-about/weather/how-weather-works/watercycle). Where ice masses terminate in the ocean, iceberg calving is the major contributor to mass loss. In this situation, the ice margin may extend out into deep water as a floating ice shelf, such as that in the Ross Sea. Despite the possibility that global warming could result in losses to the Greenland ice sheet being offset by gains to the Antarctic ice sheet (Bengtsson et al., 2014), there is major concern about the possibility of a West Antarctic Ice Sheet collapse. The West Antarctic Ice Sheet is grounded on bedrock below sea level, and its collapse has the potential of raising the world sea level by 6–7 m (20–23 ft) over a few hundred years. Most of the discharge of the West Antarctic Ice Sheet is via the five major ice streams (faster-flowing ice) entering the Ross Ice Shelf, the Rutford Ice Stream entering Filchner–Ronne Ice Shelf of the Weddell Sea, and the Thwaites Glacier and Pine Island Glacier entering the Amundsen Ice Shelf. Opinions differ as to the present mass balance of these systems (Bentley 1983), principally because of the limited data. The West Antarctic Ice Sheet is stable so long as the Ross Ice Shelf and Filchner-Ronne Ice Shelf are constrained by drag along their lateral boundaries and pinned by the local grounding of ice rises.

1.6   COMPONENTS OF THE CLIMATE SYSTEM

The *atmosphere* envelops the Earth and extends hundreds of kilometres from the surface. It consists mostly of inert nitrogen (78%), oxygen (21%), argon (0.9%), carbon dioxide, and trace gases (https://earthhow.com/earth-atmosphere-composition/). Some trace gases in the atmosphere, such as water vapour and carbon dioxide, are the gases most important for the workings of the climate system, as they are greenhouse gases that allow visible light from the Sun to penetrate to the surface, but block some of the infrared radiation the Earth's surface emits to balance the Sun's radiation. This causes surface temperatures to rise. The hydrological cycle is the movement of water through the atmosphere. Not only does the hydrological cycle determine patterns of precipitation, it also has an influence on the movement of energy throughout the climate system.

The *hydrosphere* contains all the liquid water on Earth, with most of it contained in the world's oceans. The oceans cover 71% of the Earth's surface to an average depth of nearly 4 kilometres (2.5 miles) (www.physicalgeography.net/fundamentals/8o.html), and ocean heat content is much larger than the heat held by the atmosphere (https://climate.nasa.gov/vital-signs/ocean-warming/). It contains seawater with a salt content of about 3.5% on average, but this varies spatially (Goosse, 2015). Brackish water is found in estuaries and some lakes, and most freshwater, 2.5% of all water, is held in ice and snow (Desonie, 2008).

The *cryosphere* contains all parts of the climate system where water is solid. This includes sea ice, ice sheets, permafrost, and snow cover. Because there is more land in the Northern Hemisphere compared to the Southern Hemisphere, a larger part of that hemisphere is covered in snow. Both hemispheres have about the same amount of sea ice. Most frozen water is contained in the ice sheets of Greenland and Antarctica, which average about 2 kilometres (1.2 miles) in height. These ice sheets slowly flow toward their margins. The *Earth's crust*, specifically mountains and valleys, shapes global wind patterns: vast mountain ranges form a barrier to winds and impact where and how much it rains. Land closer to the open ocean has a more moderate climate than land farther from the ocean. To model the climate, the land is often considered static as it changes very slowly compared to the other elements that make up the climate system. The position of the continents determines the geometry of the oceans and therefore influences patterns of ocean circulation. The locations of the seas are important in controlling the transfer of heat and moisture across the globe, and therefore, in determining global climate. Lastly, the *biosphere* also interacts with the rest of the climate system. Vegetation is often darker or lighter than the soil beneath, so more or less of the Sun's heat gets trapped in areas with vegetation. Vegetation is good at trapping water, which is then taken up by its roots. Without vegetation, this water would have run off to the closest rivers or other water bodies. Water taken up by plants instead evaporates, contributing to the hydrological cycle. Precipitation and temperature influence the distribution of different vegetation zones. Carbon assimilation from seawater by the growth of small phytoplankton is almost as much as land plants from the atmosphere. While humans are technically part of the biosphere, they are often treated as a separate component of the Earth's climate system, the *anthroposphere*, because of human's large impact on the planet.

1.7 POLAR REGION

The regions surrounding the geographical poles (the North and South Poles), lying within the polar circles, are known as the polar regions. These are also called the frigid zones, of the Earth. The unique characteristics of these regions are the floating sea ice around the Arctic Ocean. The Arctic is the geographical region north of the Arctic Circle (currently Epoch 2010 at 66°33'44" N), it can also be considered as the geographical region north of 60° north latitude or the region from the North Pole south to the timberline.

On the other hand, the Antarctic is the geographical region south of 60° south latitude, or the continent of Antarctica. The two polar regions possess distinguished climatic and biometric conditions of Earth. Polar regions are known to receive weaker solar radiation than the other parts of the Earth. This happens as the Sun's energy arrives at an oblique angle. It spreads over large areas and travels a longer distance. The Earth's atmosphere in which it gets scattered, absorbed, and reflected causes colder conditions. The axial tilt of the Earth plays a major governing role in polar climates. The polar regions are the farthest from the equator. The experience the weakest solar radiation and are therefore generally frigid year round due to the Earth's axial tilt of 23.5° not being enough to create a high maximum midday declination to sufficiently strengthen the Sun's rays even in summer, except for relatively brief periods in peripheral areas near the polar circles. A large amount of ice and snow also reflects a large part of what weak sunlight the polar regions receive, contributing to the cold. Polar regions are characterized by extremely cold temperatures, heavy glaciation wherever there is sufficient precipitation to form permanent ice, and extreme variations in daylight hours, with 24 hours of daylight in summer, and complete darkness in mid-winter.

1.8 POLAR ICE

Polar ice caps are dome-shaped sheets of ice found near the North and South Poles. They form because high-latitude polar regions receive less heat from the Sun than other areas on the Earth. As a result, average temperatures at the poles can be very cold. They consist of primarily water-ice with a few percent dusts. Frozen carbon dioxide makes up a small permanent portion of the Planum Australe or the South Polar Layered Deposits. In both hemispheres a seasonal carbon dioxide frost deposits in the winter and sublimates during the spring. Only 3.5% of all the water on the Earth is freshwater, while most of it is frozen in polar ice sheets. Polar ice caps are melting as global warming causes climate change. We lose Arctic sea ice at a rate of almost 13% per decade, and over the past 30 years, the oldest and thickest ice in the Arctic has declined by a stunning 95%. The ice sheets covering Greenland and Antarctica contain a vast amount of water. Antarctic ice alone would raise global sea levels by 187 feet if it all melted.

Sea ice extent is a measurement of the area of the ocean where there is at least some sea ice. Usually, scientists define a threshold of minimum concentration to mark the ice edge; the most common cutoff is 15 percent. Scientists use the 15 percent cutoff because it provides the most consistent agreement between satellite and ground observations.

1.9   THE ARCTIC

The Arctic is a polar region located in the northernmost part of the Earth. The Arctic consists of the Arctic Ocean, adjacent seas, and parts of Alaska (United States), Canada, Finland, Greenland (Denmark), Iceland, Norway, Russia, and Sweden. Land within the Arctic region has seasonally varying snow and ice cover, with predominantly treeless permafrost (permanently frozen underground ice) containing tundra. Arctic seas contain seasonal sea ice in many places. The Arctic region is a unique area among the Earth's ecosystems. The cultures in the region and the Arctic indigenous peoples have adapted to its cold and extreme conditions. Life in the Arctic includes zooplankton and phytoplankton, fish and marine mammals, birds, land animals, plants, and human societies (Krembs and Deming, 2008). Arctic land is bordered by the subarctic. The word Arctic comes from the Greek word ἀρκτικός (*arktikos*), "near the Bear, northern" and from the word ἄρκτος (*arktos*), meaning bear. The name refers either to the constellation Ursa Major, the "Great Bear", which is prominent in the northern portion of the celestial sphere, or to the constellation Ursa Minor, the "Little Bear", which contains the celestial north pole (currently very near Polaris, the current north Pole Star, or North Star). There are several definitions of what area is contained within the Arctic. The area can be defined as north of the Arctic Circle (about 66° 34'N), the approximate southern limit of the midnight sun and the polar night. Another definition of the Arctic, which is popular with ecologists, is the region in the Northern Hemisphere where the average temperature for the warmest month (July) is below 10°C (50°F); the northernmost tree line roughly follows the isotherm at the boundary of this region.

The Arctic is characterized by cold winters and cool summers. Its precipitation mostly comes in the form of snow and is low, with most of the area receiving less than 50 cm (20 in). High winds often stir up snow, creating the illusion of continuous snowfall. Average winter temperatures can go as low as −40°C (−40°F), and the coldest recorded temperature is approximately −68°C (−90°F). Coastal Arctic climates are moderated by oceanic influences, having generally warmer temperatures and heavier snowfalls than the colder and drier interior areas. The Arctic is affected by current global warming, leading to Arctic sea ice shrinkage, diminished ice in the Greenland ice sheet, and Arctic methane release as the permafrost thaws. The melting of Greenland's ice sheet is linked to polar amplification (Dormann and Woodin, 2002). Due to the poleward migration of the planet's isotherms (about 56 km (35 miles) per decade during the past 30 years as a consequence of global warming), the Arctic region (as defined by tree line and temperature) is currently shrinking (Tedesco et al., 2016). Perhaps the most alarming result of this is Arctic sea ice shrinkage. There is a large variance in predictions of Arctic sea ice loss, with models showing near-complete to complete loss in September from 2035 to sometime around 2067 (Hansen et al., 2006). Scientists tend to focus on Arctic sea ice extent more closely than other aspects of sea ice because satellites measure extent more accurately than they do other measurements, such as thickness. The Arctic sea ice minimum marks the day, each year, when the sea ice extent is at its lowest. The sea ice minimum occurs at the end of the summer melting season. The summer melt season usually begins in March and ends sometime during September. The sea ice minimum has been occurring later in recent years because of a longer melting season. However, ice growth and melt are local processes; sea ice in some areas will have already started growing before the date of the sea ice minimum, and ice in other areas will still shrink even after the date of the minimum. Changes in the timing of the sea ice minimum extent are especially important because more of the Sun's energy reaches the Earth's surface during the Arctic summer than during the Arctic winter. As explained above, sea ice reflects much of the Sun's radiation into space, whereas dark, ice-free ocean water absorbs more of the Sun's energy. Therefore, reduced sea ice during the sunnier summer months has a big impact on the Arctic's overall energy balance. The Arctic sea ice maximum marks the day of the year when Arctic sea ice reaches its largest extent. The sea ice maximum occurs at the end of the winter cold season. The Arctic cold season usually begins in September and ends in March. Monitoring winter sea ice is important to understanding the state of the sea ice. Scientists have found that Arctic sea ice has been recovering less in the winter, meaning the sea ice is already "weak" when the summer melting season arrives. A possible cause is that the underlying ocean is warmer.

1.10 THE ANTARCTIC

The Antarctic (/ænˈtɑːrtɪk/ or /ænˈtɑːrktɪk/, US English also /æntˈɑːrtɪk/ or /æntˈɑːrktɪk/; commonly /æˈnɑːrtɪk/) is a polar region around the Earth's South Pole, opposite the Arctic region around the North Pole. The Antarctic comprises the continent of Antarctica, the Kerguelen Plateau, and other island territories located on the Antarctic Plate or south of the Antarctic Convergence. The Antarctic region includes the ice shelves, waters, and all the island territories in the Southern Ocean situated south of the Antarctic Convergence, a zone approximately 32–48 km (20–30 miles) wide varying in latitude seasonally (https://web.archive.org/web/20131214094758/http://www.scar.org/). The region covers some 20 percent of the Southern Hemisphere, of which 5.5 percent (14 million km²) is the surface area of the Antarctica continent itself. All of the land and ice shelves south of 60°S latitude are administered under the Antarctic Treaty System. Biogeographically, the Antarctic realm is one of eight biogeographic realms of the Earth's land surface. As defined by the Antarctic Treaty System, the Antarctic region is everything south of the 60°S latitude. The Treaty area covers Antarctica and the archipelagos of the Balleny Islands, Peter I Island, Scott Island, the South Orkney Islands, and the South Shetland Islands (www.nsf.gov/geo/opp/antarct/anttrty.jsp). However, this area does not include the Antarctic Convergence, a transition zone where the cold waters of the Southern Ocean collide with the warmer waters of the north, forming a natural border to the region   (www.agriculture.gov.au/agriculture-land/fisheries/international/ccamlr). Because the Convergence changes seasonally, the Convention for the Conservation of Antarctic Marine Living Resources (CCAMLR) approximates the Convergence line by joining specified points along parallels of latitude and meridians of longitude. The implementation of the convention is managed through an international Commission headquartered in Hobart, Australia, by an efficient system of annual fishing quotas, licences, and international inspectors on fishing vessels, as well as satellite surveillance. The islands are situated between 60°S latitude parallel to the south and the Antarctic Convergence to the north, and their respective 200-nautical-mile (370 km) exclusive economic zones fall under the national jurisdiction of the countries that possess them: South Georgia and the South Sandwich Islands (United Kingdom), Bouvet Island (Norway), and Heard and McDonald Islands (Australia).

Kerguelen Islands (France; also an EU Overseas territory) are situated in the Antarctic Convergence area, while the Isla Grande de Tierra del Fuego, Falkland Islands, Isla de los Estados, Hornos Island with Cape Horn, Diego Ramírez Islands, Campbell Island, Macquarie Island, Amsterdam and Saint Paul Islands, Crozet Islands, Prince Edward Islands, and Gough Island and Tristan da Cunha group remain north of the Convergence and thus outside the Antarctic region. The oldest ice on Earth probably is hiding somewhere in Antarctica, because this frozen continent holds ice that's hundreds of thousands and even millions of years old. Sea ice near the Antarctic Peninsula, south of the tip of South America, has recently experienced a significant decline. The rest of Antarctica has experienced a small increase in Antarctic sea ice. Antarctica and the Arctic are reacting differently to climate change partly because of geographical differences. Antarctica is a continent surrounded by water, while the Arctic is an ocean surrounded by land. Wind and ocean currents around Antarctica isolate the continent from global weather patterns, keeping it cold. In contrast, the Arctic Ocean is intimately linked with the climate systems around it, making it more sensitive to changes in climate.

1.11   THE IPCC REPORT PERTAINING TO IMPACTS ON THE CRYOSPHERE REGIONS

Here we present excerpts from the Intergovernmental Panel on Climate Change (IPCC) report to highlight the significance of the cryosphere in the wake of ongoing global warming.

All people on Earth depend directly or indirectly on the oceans and cryosphere. The global oceans cover 71% of the Earth's surface and contain about 97% of the Earth's water. The cryosphere refers to frozen components of the Earth system. Around 10% of the Earth's land area is covered by glaciers or ice sheets. The oceans and cryosphere support unique habitats and are interconnected with other components of the climate system through the global exchange of water, energy, and carbon. The projected responses of the ocean and cryosphere to past and current human-induced greenhouse gas emissions and ongoing global warming include climate feedback, changes over decades to millennia that cannot be avoided, thresholds of abrupt change, and irreversibility. Human communities in close connection with coastal environments, small islands (including Small Island Developing States, SIDS), polar areas, and high mountains are particularly exposed to ocean and cryosphere change, such as sea level rise, extreme sea level, and shrinking cryosphere. Other communities further from the coast are also exposed to changes in the oceans, such as through extreme weather events. Today, around 4 million people live permanently in the Arctic region, of whom 10% are Indigenous. The low-lying coastal zone is currently home to around 680 million people (nearly 10% of the 2010 global population), projected to reach more than one billion by 2050. SIDS is home to 65 million people. Around 670 million people (nearly 10% of the 2010 global population), including Indigenous peoples, live in high mountain regions in all continents except Antarctica. In high mountain regions, the population is projected to reach between 740 and 840 million by 2050 (about 8.4–8.7% of the projected global population). In addition to their role within the climate system, such as the uptake and redistribution of natural and anthropogenic carbon dioxide (CO_2) and heat, as well as ecosystem support, services provided to people by the oceans and/or cryosphere include food and water supply, renewable energy, and benefits for health and well-being, cultural values, tourism, trade, and transport. The state of the oceans and cryosphere interacts with each aspect of sustainability reflected in the United Nations Sustainable Development Goals (SDGs).

Over the last few decades, global warming has led to a widespread shrinking of the cryosphere, with mass loss from ice sheets and glaciers (*very high confidence*), reductions in snow cover (*high confidence*), Arctic sea ice extent and thickness (*very high confidence*), and increased permafrost temperature (*very high confidence*). Ice sheets and glaciers worldwide have lost mass (*very high confidence*). Between 2006 and 2015, the Greenland Ice Sheet lost ice mass at an average rate of 278 ± 11 Gt yr^{-1} (equivalent to 0.77 ± 0.03 mm yr^{-1} of global sea level rise) mostly due to surface melting (*high confidence*). In 2006–2015, the Antarctic Ice Sheet lost mass at an average rate of 155 ± 19 Gt yr^{-1} (0.43 ± 0.05 mm yr^{-1}), mostly due to rapid thinning and retreat of major outlet glaciers draining the West Antarctic Ice Sheet (*very high confidence*). Glaciers worldwide outside Greenland and Antarctica lost mass at an average rate of 220 ± 30 Gt yr^{-1} (equivalent to 0.61 ± 0.08 mm yr^{-1} sea level rise) in 2006–2015.

Arctic June snow cover extent on land declined by $13.4 \pm 5.4\%$ per decade from 1967 to 2018, a total loss of approximately 2.5 million km^2, predominantly due to surface air temperature increase (*high confidence*). In nearly all high mountain areas, the depth, extent, and duration of snow cover have declined over recent decades, especially at lower elevations (*high confidence*).

Permafrost temperatures have increased to record high levels (1980s–present) (*very high confidence*) including the recent increase by $0.29°C \pm 0.12°C$ from 2007 to 2016 averaged across polar and high-mountain regions globally. Arctic and boreal permafrost contains 1460–1600 Gt organic carbon, almost twice the carbon in the atmosphere (*medium confidence*). There is *medium evidence* with *low agreement* on whether northern permafrost regions are currently releasing additional net methane and CO_2 due to thaw. Permafrost thaw and glacier retreat have decreased the stability of high-mountain slopes. Between 1979 and 2018, the Arctic sea ice extent has *very likely* decreased for all months of the year. September sea ice reductions are *very likely* 12.8 ± 2.3% per decade. These sea ice changes in September are *likely* unprecedented for at least 1000 years. Arctic sea ice has thinned, concurrent with a transition to younger ice: between 1979 and 2018, the areal proportion of multi-year ice at least 5 years old has declined by approximately 90% (*very high confidence*). Feedback from the loss of summer sea ice and spring snow cover on land has contributed to amplified warming in the Arctic (*high confidence*) where surface air temperature *likely* increased by more than double the global average over the last two decades. Changes in Arctic sea ice have the potential to influence mid-latitude weather, but there is *low confidence* in the detection of this influence for specific weather types. Antarctic sea ice extent overall has had no statistically significant trend (1979–2018) due to contrasting regional signals and large interannual variability.

REFERENCES

The National Snow and Ice Data Center Glossary. Archived 10 July 2009 at the Portuguese Web Archive.

"NSIDC Arctic Sea Ice News Fall 2007". nsidc.org. Retrieved 27 March 2008.

"State of the Cryosphere / Arctic and Antarctic Standardized Anomalies and Trends Jan 1979 – Jul 2009". National Snow and Ice Data Center. Retrieved 24 April 2010.

Alley, R.B., 2000. Continuity comes first: recent progress in understanding subglacial deformation. *Geological Society, London, Special Publications*, *176*(1): 171–179.

"Antarctic Sea Ice Reaches New Record Maximum". NASA Goddard Space Flight Center. 8 April 2015. Retrieved 10 May 2017.

"Arctic Sea Ice News & Analysis". National Snow and Ice Data Center. Retrieved 9 May 2010.

Barry, Roger G., and Eileen A. Hall-McKim. Essentials of the Earth's climate system. Cambridge University Press, 2014.

Bengtsson, Lennart. "Foreword: International Space Science Institute (ISSI) workshop on the earth's hydrological cycle." The Earth's Hydrological Cycle (2014): 485–488.

https://earthhow.com/earth-atmosphere-composition/

www.physicalgeography.net/fundamentals/8o.html

Bentley, C.R., 1983. West Antarctic ice sheet: Diagnosis and prognosis. In *Proceedings: Carbon Dioxide Research Conference: Carbon Dioxide, Science, and Consensus*. Washington, DC: Department of Energy.

Bridgman, Howard A., and John E. Oliver. The global climate system: patterns, processes, and teleconnections. Cambridge University Press, 2014.

Curry, J.A., Schramm, J.L. and Ebert, E.E., 1995. Sea ice-albedo climate feedback mechanism. *Journal of Climate*, *8*(2): 240–247.

Desonie, Dana. Climate: Causes and effects of climate change. Infobase Publishing, 2008.

Dormann, C.F. and Woodin, S.J., 2002. Climate change in the Arctic: using plant functional types in a meta-analysis of field experiments. *Functional Ecology*, *16*(1): 4–17.

Gettelman, Andrew, and Richard B. Rood. Demystifying climate models: A users guide to earth system models. Springer Nature, 2016.

Goosse, Hugues. Climate system dynamics and modeling. Cambridge University Press, 2015.

https://climate.nasa.gov/vital-signs/ocean-warming/

Gruza, George Vadimovich, ed. Environmental structure and function: climate system-volume II. EOLSS Publications, 2009.

www.metoffice.gov.uk/weather/learn-about/weather/how-weather-works/water-cycle

Grima, Cyril G.; Kofman, W.; Mouginot, J.; Phillips, R. J.; Herique, A.; Biccardi, D.; Seu, R.; Cutigni, M. (2009). "North polar deposits of Mars: Extreme purity of the water ice". Geophysical Research Letters. 36 (3): n/a. Bibcode:2009GeoRL..36.3203G. doi:10.1029/2008GL036326

Gutzler, D.S. and Preston, J.W., 1997. Evidence for a relationship between spring snow cover in North America and summer rainfall in New Mexico. *Geophysical Research Letters*, 24(17): 2207–2210.

Hansen, J., Sato, M., Ruedy, R., Lo, K., Lea, D.W. and Medina-Elizade, M., 2006. Global temperature change. *Proceedings of the National Academy of Sciences*, 103(39): 14288–14293.

Haug, Gerald H., Ralf Tiedemann, and Lloyd D. Keigwin. "How the Isthmus of Panama put ice in the Arctic." Oceanus 42 (2004): 94–97.

Houze Jr, Robert A. "Orographic effects on precipitating clouds." Reviews of Geophysics 50.1 (2012).

Huybrechts, P., 1990. A 3-D model for the Antarctic ice sheet: a sensitivity study on the glacial-interglacial contrast. *Climate Dynamics*, 5: 79–92.

Kane, D.L., Hinzman, L.D. and Zarling, J.P., 1991. Thermal response of the active layer to climatic warming in a permafrost environment. *Cold Regions Science and Technology*, 19(2): 111–122.

Krembs, Christopher, and Jody W. Deming. "The role of exopolymers in microbial adaptation to sea ice." Psychrophiles: from biodiversity to biotechnology (2008): 247–264.

Kundzewicz, Zbigniew W., et al. "The implications of projected climate change for freshwater resources and their management." Hydrological sciences journal 53.1 (2008): 3–10.

Lachenbruch, A.H., Sass, J.H., Marshall, B.V. and Moses Jr, T.H., 1982. Permafrost, heat flow, and the geothermal regime at Prudhoe Bay, Alaska. *Journal of Geophysical Research: Solid Earth*, 87(B11): 9301–9316.

Lafleur, P.M., Wurtele, A.B. and Duguay, C.R., 1997. Spatial and temporal variations in surface albedo of a subarctic landscape using surface-based measurements and remote sensing. *Arctic and Alpine Research*, 29(3): 261–269.

Langway, Chester (April 2008). "The history of early polar ice cores, Cold Regions Science and Technology". 52 (2): 101–117.

Meier, M.F., 1984. Contribution of small glaciers to global sea level. *Science*, 226(4681): 1418–1421

Munro, D.S., 1990. Comparison of melt energy computations and ablatometer measurements on melting ice and snow. *Arctic and Alpine Research*, 22(2): 153–162.

National Snow and Ice Data Center A real hole near the pole, 4 September 2012

Osterkamp, T.E., 2005. The recent warming of permafrost in Alaska. *Global and Planetary Change*, 49(3-4): 87–202.

Oerlemans, J., 1994. Quantifying global warming from the retreat of glaciers. *Science*, 264(5156): 243–245.

Paterson, W.S.B., 1993. World sea level and the present mass balance of the Antarctic ice sheet. In *Ice in the climate system* (pp. 131–140). Berlin, Heidelberg: Springer Berlin Heidelberg.

Parnell, Brid-Aine. "New Horizons Probe Snaps Possible Polar Ice Cap On Pluto". Forbes. Retrieved 20 May 2015.

Planton, S. (2013). "Annex III: Glossary" (PDF). In Stocker, T.F.; Qin, D.; Plattner, G.-K.; Tignor, M.; Allen, S.K.; Boschung, J.; Nauels, A.; Xia, Y.; Bex, V.; Midgley, P.M. (eds.). Climate Change 2013: The Physical Science Basis. Contribution of Working Group I to the Fifth Assessment Report of the Intergovernmental Panel on Climate Change. Cambridge University Press, Cambridge, United Kingdom and New York, NY, USA.

www.climatechange.environment.nsw.gov.au/global-climate-system

https://worldoceanreview.com/en/wor-1/climate-system/earth-climate-system/

https://worldoceanreview.com/en/wor-1/climate-system/earth-climate-system/

"Polar ice is melting more faster than predicted". The Watchers. Retrieved 18 May 2015.

Ravilious, Kate (28 February 2007). "Mars Melt Hints at Solar, Not Human, Cause for Warming, Scientist Says". National Geographic News. National Geographic Society. Retrieved 28 October 2008.

Smil, Vaclav. The Earth's biosphere: Evolution, dynamics, and change. Mit Press, 2003.

Taylor Redd, Nola. "Pluto Is Larger Than Thought, Has Ice Cap, NASA Probe Reveals". Space.com. Retrieved 10 September 2015.

Tedesco, M., Scalici, C., Vaccari, D., Cipollina, A., Tamburini, A. and Micale, G., 2016. Performance of the first reverse electrodialysis pilot plant for power production from saline waters and concentrated brines. *Journal of Membrane Science*, *500*: 33–45.

Thompson, Elvia. "Recent Warming of Arctic May Affect Worldwide Climate". NASA. Retrieved 2 October 2012.

Videl, John (19 September 2012). "Arctic Ice Shrinks 18% against Record, Sounding Climate Change Alarm Bells". The Guardian. London. Retrieved 3 October 2012.

Walsh, J.E., 1995. Long-term observations for monitoring of the cryosphere. *Climatic Change*, *31*(2–4): 369–394.

Washburn, A.L., 1980. Permafrost features as evidence of climatic change. *Earth-Science Reviews*, *15*(4): 327–402.

Zhang, J. and Rothrock, D.A., 2003. Modeling global sea ice with a thickness and enthalpy distribution model in generalized curvilinear coordinates. *Monthly Weather Review*, *131*(5): 845–861.

2 Assessment of the Recession Rate and Surface Velocity of the Gangotri Glacier Garhwal Himalaya India
A Geospatial Approach

Savati Sharma, Ajay Kumar Taloor, Harish Bisht,
Bahadur Singh Kotlia, Varun Khajuria,
Kireet Kumar, Girish Chandra Kothyari,
Rajkumar Kashyap, and Manmohan Kukreti

2.1 INTRODUCTION

According to the topographic features, the Himalayas is divided into four segments: Karakoram, Western Himalaya, Central Himalaya, and Eastern Himalaya (Figure 2.1). The Himalayan region consists of a large number of glaciers second only to the polar region and has been widely affected by climate changes over the years (Dyurgerov and Meier, 1997; Messer et al., 2004). As Himalayan glaciers are located near the Tropic of Cancer and therefore receive more heat and are sensitive to climate change, recent trends of global warming construe that most of the Himalayan glaciers are on the verge of extinction (Bahuguna, 2003; Negi et al., 2012; Dubey et al., 2017; Kothyari et al., 2017a, 2017b; Taloor et al., 2019; Kannaujiya et al., 2020; Sarkar et al., 2020; Singh et al., 2020). In the Indian Himalayan Region (IHR), glaciers are mainly situated in five states: Jammu and Kashmir, Himachal Pradesh, Uttarakhand, Sikkim, and Arunachal Pradesh. Of these five states, Jammu and Kashmir have the largest concentration with 3136 glaciers covering an area of about 32,000 km^2 (Figure 2.1) (Raina, 2005). The climate and regional hydrology of the Indian subcontinent are directly influenced by the Himalayan glaciers (Kothyari et al., 2020a). These Himalayan glaciers are a rich source of fresh water, agriculture, and electricity for around 800 million people residing in the lap of the Himalayan region (Bhambri et al., 2011; Khadka et al., 2014; Bisht et al., 2018; Bhat et al., 2019; Kothyari et al., 2019; Haque et al., 2020; Sood et al., 2020a). The glaciers

DOI: 10.1201/9781003364115-2

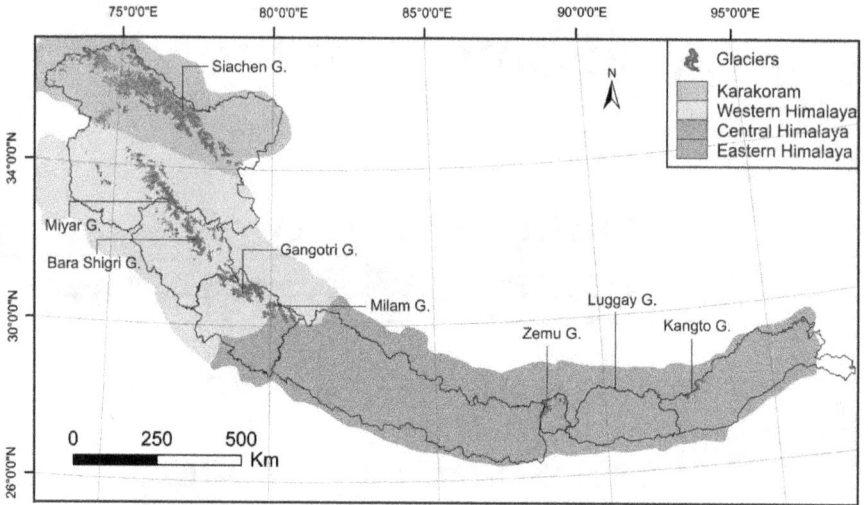

FIGURE 2.1 The four main segments of the Himalayan arc with major glaciers located in the Indian Himalayan Region (IHR).

are the main geomorphic agents of erosion and are a vital source of suspended sediment load in the meltwater streams (Bisht et al., 2020a; Kothyari et al., 2020b). Over the last few decades, the glaciers of the Himalayan region have undergone considerable changes which have influenced the biological, geological, hydrological, and ecological systems of the environment (Yan et al., 2015; Bhushan et al., 2017; Singh et al., 2017; Taloor et al., 2017; Khan et al., 2020; Sood et al., 2020b; Taloor et al., 2020). Remote sensing and geographical information systems (GIS) play a vital role in monitoring changes in glaciers. The fluctuations in the recession rate of Himalayan glaciers over recent years have initiated general discussions, mainly in the context of global warming (Dyurgerov and Meier, 1997; Bisht et al., 2020b; Kumar et al., 2020). The change in snout position varies for different glaciers (Dobhal et al., 2004), therefore it is essential to monitor these glaciers regularly. The extent of the snout of a glacier can be reconstructed from satellite images (Bisht et al., 2019). To monitor a change in snout position, area loss, volume loss, and equilibrium line altitude (ELA), remote sensing techniques can give helpful information (Negi et al., 2012). Therefore, over the past few decades, a remote sensing-based approach has been widely used for mapping glaciers utilizing various multitemporal and multispectral data. The recent glacier studies in the Himalayan region point to a wide-ranging inconsistency in the retreating rate and mass balance of the glacier (Dobhal et al., 2013). This is generally due to morpho-geometrical changes, the behaviour of winter and summer monsoons, local climatic changes, valley aspect, and debris cover on the glacier surface (Dumka et al., 2013; Kothyari and Luirei, 2016; Joshi et al., 2018; Bisht et al., 2020; Kumar et al., 2020; Kothyari et al., 2020a,2021).

Considerable changes in velocity estimates over the years indicate that changes in ice flow dynamics are intricately linked with the mass balance of glaciers (Heid and Kääb, 2012). Changes in the surface velocities of glaciers overtime can provide insights into the differing responses of mountain glaciers to climatic forcing by illustrating how glacier basal velocities vary during an advance/retreat cycle (Herman et al., 2011). The variations in surface velocity of glaciers over hours to a day have been observed on several glaciers (Iken, 1977; Sugiyama and Gudmundsson, 2004). To analyse such behaviour, it is important to acquire accurate information on the surface velocity of the glaciers (Riesen et al., 2011). Most of the surface velocity estimates of glaciers are based on remote sensing techniques using optical (Saraswat et al., 2013; Mishra et al., 2020; Kandregula et al., 2021 in press) and synthetic aperture radar (SAR) data (Copland et al., 2009; Holzer et al., 2015), while some studies are based on the kinematic GPS survey measurements (Kaser et al., 2003). A combination of both methods (remote sensing and kinematic GPS survey) provides significant insight into glacier dynamic studies with a better interpretation of results (Bisht et al., 2019, 2020b). The kinematic GPS survey is a unique technique to determine the crustal deformation as well as the retreating rate and surface velocity of a glacier (Bisht et al., 2020c). In the GPS survey method, estimation of velocity is carried out by measuring the displacement of stakes over a certain period, while in remote sensing it is accomplished by the feature tracking technique in which the displacement of visible features is tracked between two images over a certain time interval (Scherler et al., 2008; Copland et al., 2009). Both methods have their advantages and disadvantages, and it is generally difficult to derive velocity for regions with steeper slopes such as the Himalayas using InSAR methods (Saraswat et al., 2013). Also, it is very difficult to estimate the velocity of glaciers which are moving very fast due to the temporal decorrelation or loss of coherence between the pairs of images under consideration (Saraswat et al., 2013). Despite the limitations imposed by clouds and excessive snow cover, optical bands have been effectively used for glacier velocity estimation (Scherler et al., 2008; Saraswat et al., 2013). Due to the highly rugged terrain and inaccessibility of certain areas, satellite-obtained information is used to monitor the surface velocity and glacier retreat. In the present study, attempts have been made to estimate the retreating trend of the Gangotri glacier over a considerable continuous period from 1980 to 2018. In addition, an attempt has been made to estimate and validate the surface velocity of the Gangotri glacier by using the kinematic GPS survey method.

2.2 STUDY AREA

The Gangotri glacier is located in the higher Himalayan region and is the source of the most famous and sacred river in India. It lies in the latitudes of 30°43'10" to 30°55'50" N and longitudes of 79°04'55" to 79°17'18" E (Figure 2.2). It is a northwest-flowing valley-type glacier, which is about 30.2 km long and 0.5–2.5 km wide (Kaul, 1999). The surface elevation of the Gangotri glacier varies from~4000 to~7000 m, with a surface area of about 140 km^2 (Gantayat et al., 2014). Geologically, the area lies above the Main Central Thrust (MCT) and comprises quartzite, quartz-biotite schist, biotite, granite, and leucogranite (Metcalfe, 1993; Bisht et al., 2020a). The regional climate is mainly influenced by the Indian Summer Monsoon (ISM) and partly by the Indian Winter Monsoon (IWM), and

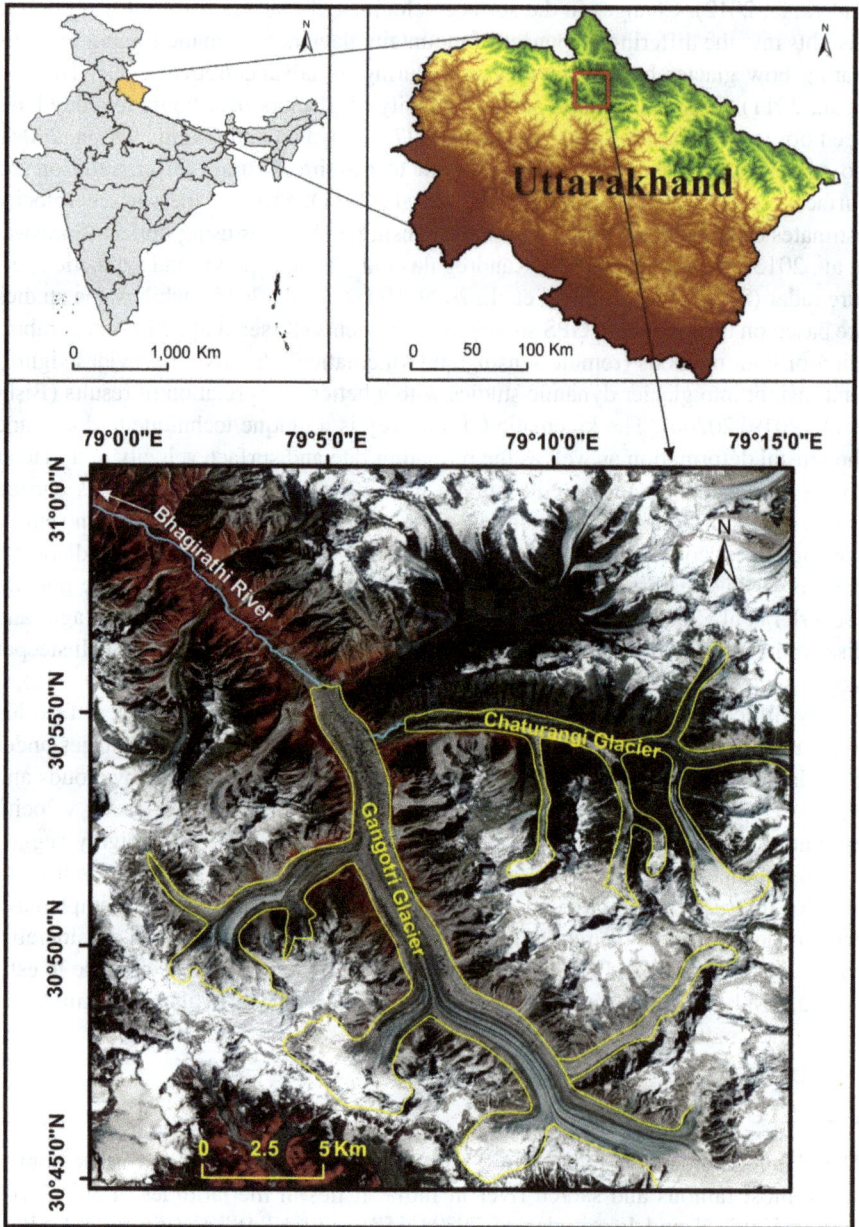

FIGURE 2.2   Landsat8(OLI) image showing Gangotri glacier.

the microclimate is affected by both the valley aspects and altitude (Naithani et al., 2001). Heavy snowfall is caused by winter disturbances from December to March, while melting generally extends from May to October (Dobhal et al., 2008). At present, the glacier comprises mainly four active tributary glaciers (KirtiBamak, SwachhandBamak, MaiandiBamak, and GhanohlmBamak) and two inactive tributary glaciers (RaktavarnaBamak and ChaturangiBamak).

2.3 METHODOLOGY AND DATA SETS

2.3.1 DIFFERENTIAL GLOBAL POSITIONING SYSTEM (DGPS) SURVEY METHOD FOR SURFACE VELOCITY MEASUREMENTS

Surveying the position of artificial velocity stakes on the surface of a glacier with DGPS is a simple but highly precise method of monitoring the surface velocity of the glacier. This method requires installing velocity stakes on the glacier surface and repeatedly surveying the position of the stakes using high-precision GPS. The change in distance of the location of the velocity stake and the known interval of time between surveys can be used to calculate the surface velocity of the glacier at the target location (Karpilo, 2009). A total of five velocity stakes were installed on the surface of the glacier with the help of an ice drilling machine (AR502) (Figure 2.3a). The stakes were arranged along a transverse line and their position was surveyed using DGPS in static mode during 1–2 September 2016. During the next month, from 1–2 October, the position of the velocity stakes was re-surveyed. The latitude and longitude of all the velocity stakes are given in Table 2.1. We also measured a location point of velocity stakes with the help of handheld GPS (Etrex 30) (Figure 2.3b). The DGPS survey was carried out using a pair of using Leica SR520 GPS receivers and AT502 antenna in rapid static and kinematic modes. In this method, one antenna was mounted on a solid bedrock with a 1 mm hole which was located off the glacier with one receiver (Figure 2.3c)and another as a rover with a moving antenna which made observations at intervals of 5 seconds lasting for 5 minutes to obtain precise position coordinates with centimetre-level accuracy. It was necessary to start the reference station 1 day/24 hours before the kinematic survey for instrument calibration to have as precise a location as possible for the reference point.

The roving antenna was fixed on the top of the velocity stake, which was nearly 1 m above the ground level (Figure 2.3d). The accuracy if results is also affected by the large geometric dilution of precession (GDOP), which could introduce a large value of error in GPS-derived positions. To minimize this error, all GPS observations with GDOP > 6 were not considered during the final analysis to attain a precession to the sub-centimetre level which maintained the high quality of data required for the present study. The raw data generated by the DGPS survey were processed in the laboratory using Leica SKI-PRO 3.0 software. Post-processing the GPS data mitigates the signal-degrading effects of the atmosphere and improves the data accuracy. The following processing parameters were used in the software: cut-off angle 15°, broadcast ephemeris with automatic selection of solution type and frequency. The coordinates of the location were presented in the form of a WGS 84

FIGURE 2.3 (a) Drilling on the surface of Gangotri glacier using an ice drill machine for inserting velocity stakes. (b) Measurement of the location of the ablation stake through handheld GPS. (c) Reference antenna mounted on hard bedrock. (d) Roving antenna mounted on a velocity stake for making observations to determine position coordinates through DGPS.

coordinate system. The position quality defined as RMS of standard deviations of X and Y coordinates was also calculated.

After the post-processing of raw data the reference displacements of stakes measured from GPS surveys were calculated using the following formula:

$$l = \sqrt{\Delta x^2 + \Delta y^2} \tag{2.1}$$

where Δx and Δy are the Easting and Northing displacements, respectively. The displacements were then converted into reference velocities (m/month) based on the time difference between the surveys (i.e., 30 days).

TABLE 2.1
The Latitude and Longitude of the Inserted Velocity Stakes into the Gangotri Glacier Measured with the Help of the Differential Global Positioning System (DGPS)

Stakes	Latitude	Longitude
VS 1	30°55′32.49″N	79°4′58.45″E
VS 2	30°55′30.68″N	79°4′57.04″E
VS 3	30°55′28.31″N	79°4′54.87″E
VS 4	30°55′26.15″N	79°4′52.36″E
VS 5	30°55′24.97″N	79°4′49.12″E

2.3.2 REMOTE SENSING METHOD FOR RETREAT MEASUREMENTS

The Landsat satellite series (MSS, TM, EMT+, and OLI 8) data sets from 1980 to 2018 were downloaded and the most suitable scene was selected. Cloud-free, as well as partly cloudy images that do not affect the snow-covered area (SCA) of the catchment, were selected by visual examination. Furthermore, these sets have a spatial resolution of 60m and 30m with the orthorectified data type. Mapping of the snow cover was accomplished by visual interpretation. The data for 2010 downloaded from the Landsat 7(ETM+) have a missing scan line error which was corrected by the Landsat tool in the ArcGIS 10.5 software. The details of different satellite data used in the present study are given in Table 2.2.

The uncertainties for terminus retreat and temporal changes in the areal extent of the glacier were measured using Equations(2.2) and (2.3), respectively (Hall et al., 2003)

$$e = \sqrt{a^2 + b^2} + E_{reg} \qquad (2.2)$$

where "a" is the spatial resolution of image 1, "b" is the spatial resolution of image 2, and "E_{reg}" is the image registration error.

$$U_{area} = 2eV \qquad (2.3)$$

where "e" is the terminus uncertainty and "V" is the spatial resolution of the satellite image.

In addition, to obtain a relationship between retreating trend and climatic conditions the meteorological data (temperature and precipitation) from the period of 1901–2016 were downloaded and processed using Climate Research Unit (CRU) Time Series (TS) Version 4.01 (CRUTS-4.01) dataset of 0.5°×0.5° grids interval.

TABLE 2.2
Comprehensive Details of the Satellite Data Sets Used in the Present Study

Year	Satellite/ sensor	ID	Spatial resolution (m)	Date of acquisition
1980	Landsat-3 (MSS)	LM03_L1TP_157038_19800915_ 20180417_01_T2	75	15/09/1980
		LM03_L1TP_157039_19801126_ 20180418_01_T2		26/11/1980
1990	Landsat-5 (TM)	LT05_L1TP_146038_19900802_ 20170130_01_T1	30	02/08/1990
		LT05_L1TP_146039_19900223_ 20170131_01_T1		23/02/1990
2000	Landsat-7 (EMT+)	LE07_L1TP_146038_20001211_ 20170208_01_T1	30	11/12/2000
		LE07_L1TP_146039_20001227_ 20170208_01_T1		27/12/2000
2010	Landsat-7 (EMT+)	LE07_L1TP_146038_20101121_ 20161212_01_T1	30	21/11/2010
		LE07_L1TP_146039_20101121_ 20161212_01_T1		21/11/2010
2018	Landsat-8 (OLI & TIRS)	LC08_L1TP_146038_20181119_ 20181129_01_T1	30	19/11/2018
		LC08_L1TP_146039_20181119_ 20181129_01_T1		19/11/2018

2.4   RESULTS AND DISCUSSION

2.4.1   RECESSION OF THE GANGOTRI GLACIER FROM 1980 TO 2018

The results obtained from the satellite images were used to determine the temporal changes in the snout position and area loss of the Gangotri glacier from 1980 to 2018 (Figures 2.4 and 2.5). The recession of the glacier was computed during four different periods, namely, 1980–1990, 1990–2000, 2000–2010, and 2010–2018 and was observed as 320.12 ± 44.3 m, 275.19 ± 37.2 m, 216.65±37.43 m, and 108.70±29.23 m, respectively (Table 2.3, Figure 2.4). The study revealed that the Gangotri glacier retreated 920.66±37.04 m, with an average rate of retreat of 24.23±0.97 m/yr from 1980 to 2018 (Table 2.3). The variable retreat rate was observed during the entire observational period for the glacier. The results obtained from the reconstruction of the past snout position suggested that the retreat rate was higher during 1980–1990 (32.01±4.43 m/yr) but then decreased consistently during 1990–2000 (27.52±3.72 m/yr), 2000–2010 (21.67±3.74 m/yr), and 2010–2018 (13.59±3.65 m/yr) (Table 2.3). The lowest recession rate of 13.59±3.65 m/yr was observed during 2010–2018, which showed that the recession rate for Gangotri glacier has declined during recent years. Bhattacharya et al. (2016) and Bisht et al. (2020b) also reported a decreased retreating

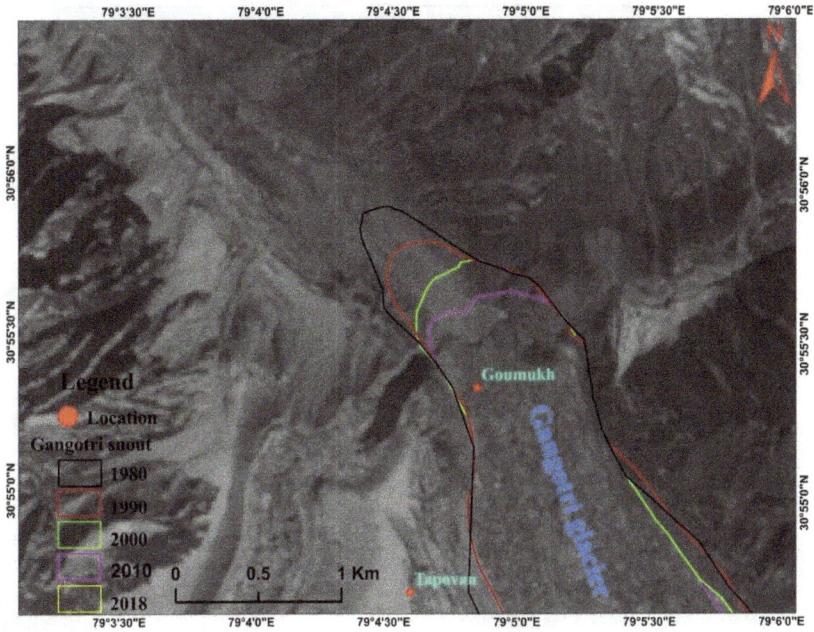

FIGURE 2.4 Temporal changes in the snout position of Gangotri glacier over the years from 1980 to 2018.

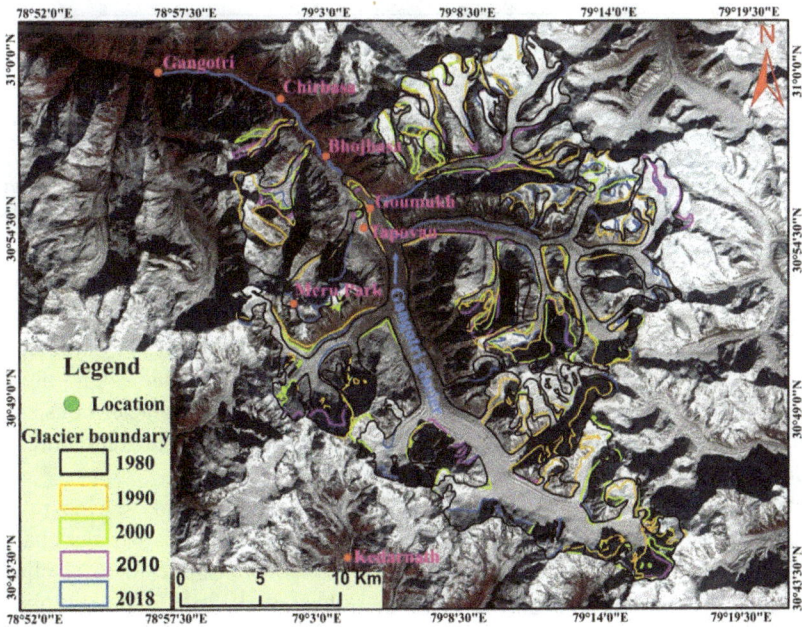

FIGURE 2.5 Spatiotemporal changes in the area of Gangotri glacier.

TABLE 2.3
Recession Rate of Gangotri Glacier Estimated Using Satellite Data from 1980 to 2018

Time period	Total recession (m)	The average rate of retreat (m/yr)
1980–1990	320.12 ± 44.3	32.01 ± 4.43
1990–2000	275.19 ± 37.2	27.52 ± 3.72
2000–2010	216.65 ± 37.43	21.67 ± 3.74
2010–2018	108.70 ± 29.23	13.59 ± 3.65
Total	920.66 ± 37.04	

TABLE 2.4
The Area Loss of Gangotri Glacier from 1980 to 2018

Year	Area (km²)	Time period	Area loss (km²)
1980	304.33 ± 23	–	–
1990	266.62 ± 25	1980–1990	37.71 ± 2.5
2000	254.42 ± 22	1990–2000	12.2 ± 2.2
2010	252.52 ± 19	2000–2010	1.9 ± 1.9
2018	251.06 ± 15	2010–2018	1.46 ± 1.8
Total			53.27 ± 2.1

rate for the Gangotri glacier in recent years. The study carried out by Ali et al. (2019) proved that the retreat rate of the Pindari glacier has slowed since 2010. Similarly, the Satopanth glacier receded at a rate of 22.86 m/yr prior to 2005 but slowed down to 6.5 m/yr in the following years (Nainwal et al., 2008). Considering all this, we believe that the recession rate of a few glaciers has declined in the Himalayan region. The area loss of Gangotri glacier was also computed during four different periods, namely, 1980–1990 (37.71±2.5 km²), 1990–2000 (12.2±2.2 km²), 2000–2010 (1.9±1.9 km²), and 2010–2018 (1.46±1.8 km²) (Table 2.4). The results indicate that the total area loss of Gangotri glacier from 1980 to 2018 was53.27±2.1 km² (Table 2.4, Figure 2.5). Similar to the measured retreat rate, the area loss of Gangotri glacier consistently decreased from 1980 to 2018, which also supports the decrease in the retreating trend of Gangotri glacier.

Earlier studies on the recession of the Gangotri glacier show varying rates of retreat in the last century (Table 2.5). The variability in retreat rate and mass balance of different glaciers in the Himalayan region are generally due to local climatic conditions and the topography of the region (Dobhal et al., 2013). To determine the retreating trend with local climatic conditions, the meteorological data (temperature and precipitation) which were derived from the CRU time series 0.5°×0.5° grid

TABLE 2.5
Retreating Rate of the Gangotri Glacier Estimated in Earlier Studies and in the Present Study

Time period	Annual snout retreat (m)	Reference
1935–1997	40	Mukherjee and Sangewar (2001)
1962–1982	40	Tangri (2002)
1962–1999	34	Naithani et al. (2001)
1975–1976	38	Puri (1984)
1976–1977	30	Puri (1984)
1977–1990	28.08	Puri (1991)
1990–1996	28.33	Sangewar (1997)
1990	37	Tangri (2002)
1999	25	Tangri (2004)
2004–2005	12.10	Kumar et al. (2008)
2005–2012	11.48	Singh et al. (2016)
2005–2015	10.26	Bisht et al. (2020b)
1980–1990	32.01	Present study
1990–2000	27.52	Present study
2000–2010	21.67	Present study
2010–2018	13.59	Present study

interval were also analysed. The meteorological data suggested that the temperature has been increasing consistently since the last century while the precipitation pattern is decreasing (Figure 2.6a–d). The result shows that even after the rising temperature pattern, the retreating rate of the Gangotri glacier has been declining in recent years, which might be due to glacier characteristics. Pudełko et al. (2018) and Venkatesh et al. (2012) also suggested that the variability in retreat rate with time not only depends on climatic factors but also on the length, slope, size, and type of glacier. The variability in retreat rate also could be due to various methodologies used by different researchers. The results also indicate that the recession rate of Gangotri glacier has been declining in recent years. The monitoring of the snout using field data and satellite data indicates that since 1980 the rate of retreat has been decreasing and has come down to 13.59 ± 3.65 m/yr, which is not in agreement with human-induced global warming. The recession of Gangotri glacier documented by various institutions by several methods illustrates that the rate of retreat is decreasing with time (Table 2.5). Furthermore, if global warming is the only reason for the retreat rate of Gangotri glacier then the rate of retreat for all the glaciers situated in a similar climate regime should be identical (Singh et al., 2016). However, different retreat rates have been observed in different glaciers and some glaciers have advanced too, which supports that local factors and glacier characteristics play a vital role in varying retreating rates, apart from global warming (Singh and Mishra, 2001; Singh et al., 2016). The thick debris cover and supraglacial lakes have also been recognized as important agents that affect the recession rate (Fujita et al.,2009; Bisht et al., 2020b). In the Himalayan region, monitoring of 2018 glacier snouts using satellite data in 2000 and 2011

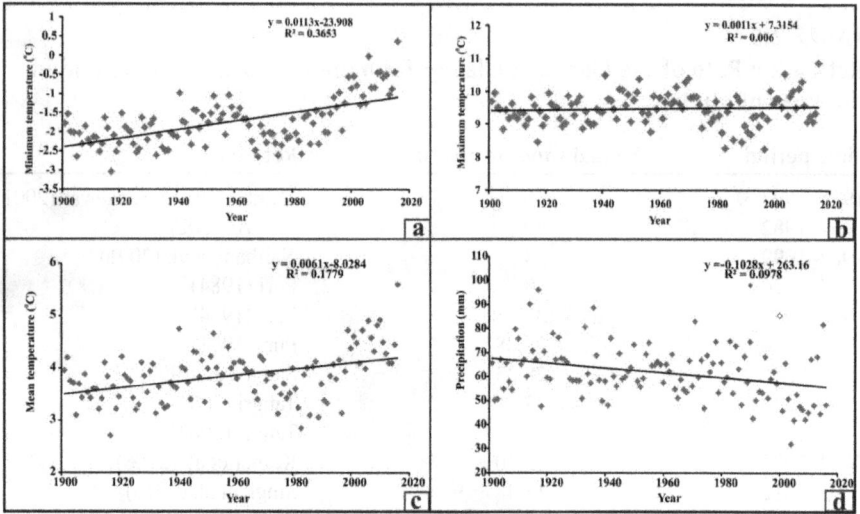

FIGURE 2.6 (a) Average annual minimum temperature. (b) Average annual maximum temperature. (c) Average annual mean temperature. (d) Average annual precipitation from 1901 to 2016 (data derived from CRUTS 4.1).

shows that 1752 glaciers have negligible changes in the snout positions, whereas 248 glaciers are receding at a considerable rate, and 18 are advancing (Bahuguna et al., 2014). Therefore, it is still debatable whether the recession of the Himalayan glaciers is only in response to human-induced global warming (Bali et al., 2011). Finally, we have suggested that different rates of glacier retreat are quite often advanced in a few cases, showing that different glaciers respond differently to regional and global climatic changes. This refutes the general concept that global warming has a direct impact on glaciers, and demonstrates that glacier dynamics is not only controlled by climatic changes but also governed by debris cover, glacier characteristics, and local aspects.

2.4.2 SURFACE VELOCITY OF THE GANGOTRI GLACIER

The surface velocity of the glacier is a measure of how fast the surface ice is moving towards the terminus of the glacier. The flow can be slow or fast, depending on how much the glacier is melting. Fast-moving glaciers bring more ice towards the terminus for melting, which in turn is one of the important factors leading mass balance of the glacier (Tiwari et al., 2014). Monitoring of glacier movement provides an important link between fluctuations in mass balance and the consequent changes in glacier geometry (Fountain et al., 1997). The rate of glacier movement determines how quickly mass is redistributed and is an important factor in determining whether the glacier is retreating or advancing (Hooke, 2005). Glacier movement is seasonal, with the highest flow velocities observed during summer, with the lowest during winter. A few glaciers exhibit summer velocities that are twice as fast as their winter rates

(Singh and Singh, 2001). The surface velocity of glaciers is affected by increases in climatic variables such as temperature and precipitation. An increase in air temperature increases the ice temperature; which correspondingly reduces the ice viscosity and increases glacier velocity. In addition, air temperature also increases the volume of basal meltwater, which reduces basal friction and increases glacier velocity. If precipitation increases, the thickness of the ice also increases, which can result in increased glacier velocity (Karpilo, 2009). The results derived through DGPS based on 1-month data from September to October 2016 show that the surface velocity of the glacier is around 2.28±0.03 m/month (Table 2.6). Thus the average velocity for the whole year is around 27.36±0.03 m/year, which is approximately equal to the previous studies done by Satyabala (2016) for the winter season. According to Satyabala (2016), the surface velocity of the Gangotri glacier in peak summer was 63.1±5.4 m/yr in 1992, 66.6±6.0 m/yr in 1999, 58.2±4.5 m/yr in 2004, and 42.8±2 m/yr in 2007, whereas winter speed was relatively stable (25–30 m/yr) during the same period by using synthetic aperture radar (SAR) data. When we compare our results with those of Satyabala (2016) we conclude that our results are similar to those results which were observed in the winter period. Meanwhile, the average annual results of the previous studies are much higher than for the present study. The present results also indicate that the resultant surface velocity of the Gangotri glacier is higher in the northern part (3.64±0.02 m/month) than in the southern part (1.11±0.01 m/month) of the glacier (Table 2.6, Figure 2.7). Based on these monthly data results we estimate that the average annual surface velocity of the glacier in the northern, central, and southern parts is around 43.68±0.02 m/yr, 36.72±0.09 m/year, and 13.32±0.01 m/year, respectively. The calculated velocity is also presented on a map as a vector representing the magnitude and direction of the measurement (Figure 2.7). The results also reveal that the central part of the glacier is also moving at a higher rate as compared to its

---

TABLE 2.6
Kinematic GPS-derived Position Coordinates of Stakes in 2016 for Determining the Monthly Average Surface Velocity of Gangotri Glacier

Months/ points	X (North) (m)	Y (East) (m)	Position quality (cm)	Change in N (dX) (m)	Change in E (dY) (m)	Resultant Velocity (m/month)
Sept/2016-1	1037790.63	5380500.10	1.72	3.29	−1.57	3.64±0.02
Oct/2016-1	1037793.91	5380498.53	1.26			
Sept/2016-2	1037792.09	5380503.35	1.19	2.22	0.16	2.22±0.02
Oct/2016-2	1037794.30	5380503.52	1.84			
Sept/2016-3	1037790.89	5380505.75	1.9	2.67	−1.51	3.06±0.09
Oct/2016-3	1037793.56	5380504.24	0.24			
Sept/2016-4	1037790.48	5380508.49	1.59	1.35	−0.29	1.38±0.01
Oct/2016-4	1037791.82	5380508.20	0.5			
Sept/2016-5	1037790.16	5380507.19	1.32	1.09	−0.21	1.11±0.01
Oct/2016-5	1037791.25	5380506.98	0.87			
Mean				2.12	−0.74	2.28±0.03

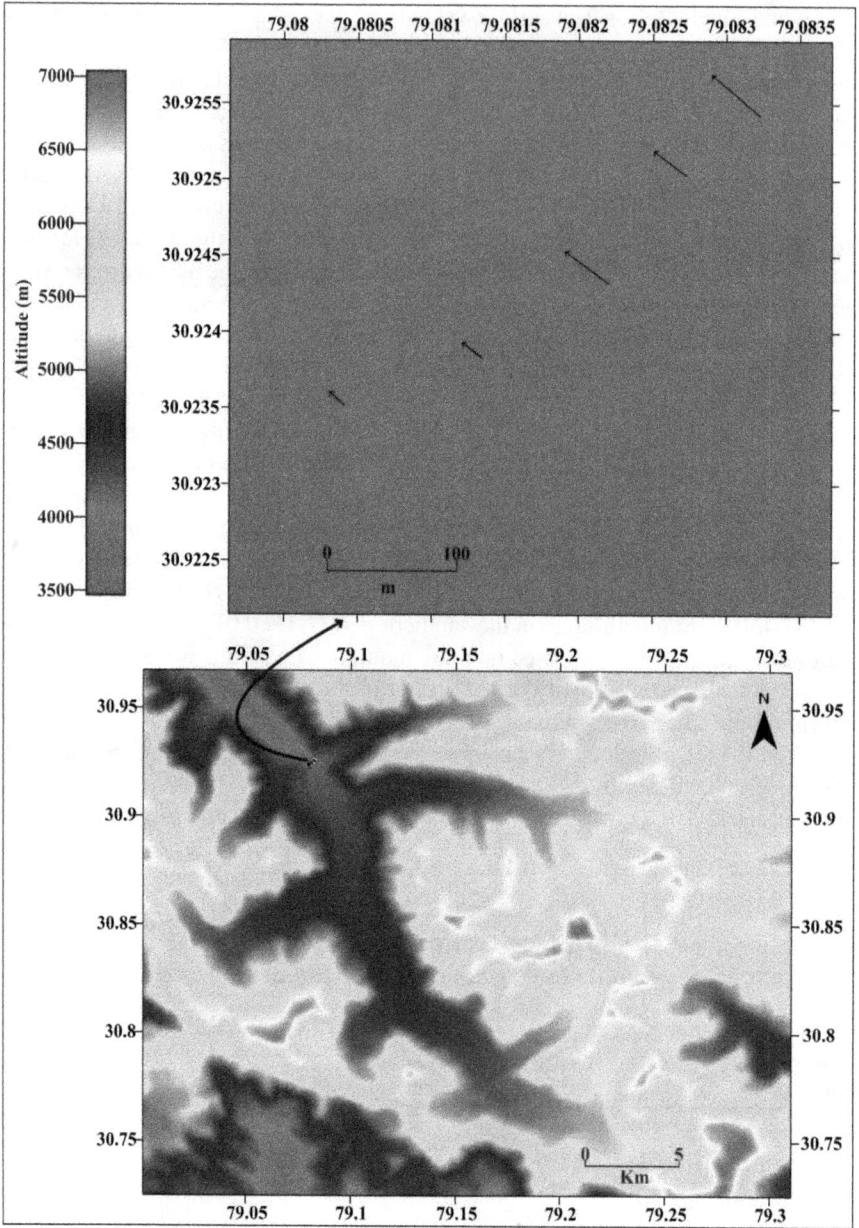

FIGURE 2.7 The average annual resultant surface velocity of Gangotri glacier was measured through a kinematic GPS survey, where the velocity vector represents the magnitude and direction of the flow.

marginal part. However, the marginal part of the northern side moving at a higher rate might be due to less overburden over the surface in the northern flank. Because of the unequal surface velocity of the glacier in the central and its marginal part, several longitudinal and transverse crevasses are formed on the surface of the glacier.

2.5 CONCLUSIONS

In view of the general concern about global warming and glacier recession, it is important to monitor these glaciers regularly. We successfully measured the surface ice motion and recession rate of Gangotri glacier using DGPS survey and satellite data, respectively. The results derived from the kinematic GPS survey suggested that the average annual surface velocity of Gangotri glacier is 27.36±0.03 m/year. The results concluded that the monthly surface velocity measurement of the glacier is much more relevant than the annual measurement because during the annual measurement the velocity stakes could be washed out due to higher debris cover in the Himalayan valley glaciers. In many cases, continuous position logging may not be practical; therefore, weekly, monthly, or annual survey visits may be appropriate. The velocity estimate obtained from the current study concluded that the surface velocity of the glacier has decreased compared with previous studies. In addition, the results derived from the satellite data interpretation concluded that the total retreat and area loss of Gangotri glacier from 1980 to 2018 are 920.66±37.04 m (24.23±0.97 m/yr) and 53.27±2.1 km², respectively. The results also suggested that from 1980 to 1990 the recession rate was higher, at 32.01±4.43 m/yr, while during the more recent years of 2010 to 2018 a lower recession rate of 13.59±3.65 m/yr has been observed. Finally, the results concluded that the Gangotri glacier is dynamic throughout its length as well as its surface velocity and recession rate having slowed down during recent years. The decrease in retreating rate in recent years is probably due to high debris thickness on the glacier surface near the snout, which prevents the ice from melting. The decrease in the retreating rate supports the view that apart from global warming, the retreat rate of the glacier is also governed by the debris cover, local aspects, and glacier characteristics. The present study also emphasizes the velocity estimation and recession rate of all the tributary glaciers to assess the overall changes in the Gangotri glacier system.

ACKNOWLEDGEMENTS

We thank the Department of Science and Technology, New Delhi, for partial financial support to carry out this research. We also thank the Director, G.B. Pant National Institute of Himalayan Environment and the Head, Department of Remote Sensing and GIS, University of Jammu, Jammu for providing us with the necessary working facilities.

FUNDING

Some of this work has been carried out with the help of funds provided by the Department of Science and Technology (Grant No. SR/DGH/58/2013).

DATA AVAILABILITY STATEMENT

Due to the nature of this research, the authors did not agree for their data to be shared publicly, therefore supporting data are not available. However, the intermediate results supporting the findings of this study are available from the corresponding author upon reasonable request.

REFERENCES

Ali, S.N., Singh, R., Pandey, P. (2019) Estimation of the frontal retreat rate of the Pindari glacier, Central Himalaya using remote sensing technique. *Earth Science India* 12: 146–157.

Bahuguna, I.M. (2003). *Satellite Stereo Data Analysis in Snow and Glaciated Region. Training Document: Course on Remote Sensing for Glaciological Studies*. Manali, India, SAC/RESA/MWRG/ ESHD/TR/13/2003, 85–99.

Bahuguna, I.M., Rathore, B.P., et al. (2014) Are the Himalayan glaciers retreating? *Current Science* 106: 1008–1013.

Bali, R., Agarwal, K.K., Ali, S.N., Srivastava, P. (2011) Is the recessional pattern of Himalayan glaciers suggestive of anthropogenically induced global warming? *Arabian Journal of Geosciences* 4: 1087–1093.

Bhambri, R., Bolch, T., Chaujar, R.K. (2011) The frontal recession of Gangotri glacier, Garhwal Himalaya, from 1965 to 2006, was measured through high-resolution data. *Current Science* 102: 489–494.

Bhat, M.S., Alam, A., Ahmad, B., Kotlia, B.S., Farooq, H., Taloor, A.K., Ahmad, S. (2019) Flood frequency analysis of river Jhelum in Kashmir basin. *Quaternary International* 507: 288–294.

Bhattacharya, A., Bolch. T., Mukherjee. K., Pieczonka, T., Kropacek. J., Buchroithner, M.F. (2016) Overall recession and mass budget of Gangotri glacier, Garhwal Himalayas, from 1965 to 2015 using remote sensing data. *Journal of Glaciology* 62: 1115–1133.

Bisht, H., Arya, P.C., Kumar, K. (2018) Hydro-chemical analysis and ionic flux of meltwater runoff from Khangri Glacier, West Kameng, Arunachal Himalaya, India. *Environmental Earth Sciences* 77: 1–16.

Bisht, H., Kotlia, B.S., Kumar, K., Dumka, R.K., Taloor, A.K., Upadhyay, R. (2020c) GPS-derived crustal velocity, tectonic deformation and strain in the Indian Himalayan arc. *Quaternary International* 575: 141–152.

Bisht, H., Kotlia, B.S., Kumar, K., Arya, P.C., Sah, S.K., Kukreti, M., Chand, P. (2020a) Estimation of suspended sediment concentration and meltwater discharge draining from the Chaturangi glacier, Garhwal Himalaya. *Arabian Journal of Geosciences* 13: 4–12.

Bisht, H., Kotlia, B.S., Kumar, K., Joshi, L.M., Sah. S.K., Kukreti, M. (2020b) Estimation of the recession rate of Gangotri glacier, Garhwal Himalaya (India) through kinematic GPS survey and satellite data. *Environmental Earth Sciences* 79: 1–14.

Bisht, H., Rani, M., Kumar, K., Sah, S., Arya, P.C. (2019) Retreating rate of Chaturangi glacier, Garhwal Himalaya, India derived from kinematic GPS survey and satellite data. *Current Science* 116: 304–311.

Bhushan, S., Syed, T.H., Kulkarni, A.V., Gantayat, P., Agarwal, V. (2017) Quantifying changes in the Gangotri glacier of Central Himalaya: Evidence for increasing mass loss and decreasing velocity. *Journal of Applied Earth Observation and Remote Sensing* 10: 5295–5306.

Copland, L., Pope, S., et al. (2009) Glacier velocities across the central Karakoram. *Annals of Glaciology* 50: 41–49.

Dobhal, D.P., Gergan, J.T., Thayyen, R.J. (2008) Mass balance studies of the Dokriani Glacier from 1992 to 2000, Garhwal Himalaya, India. *Bulletin of Glaciological Research Japanese Society Snow Ice* 25: 9–17.

Dobhal, D.P., Gergan, J.T., Thayyen, R.J. (2004) Recession and morphometrical changes of Dokriani glacier (1962–1995) Garhwal Himalaya, India. *Current Science* 86: 692–696.

Dobhal, D.P., Mehta, M., Srivastava, D. (2013) Influence of debris cover on terminus retreat and mass changes of Chorabari Glacier, Garhwal region, central Himalaya, India. *Journal of Glaciology* 59: 961–971.

Dubey, R.K., Dar, J.A., Kothyari, G.C. (2017) Evaluation of relative tectonic perturbations of the Kashmir Basin, Northwest Himalaya, India: An integrated morphological approach. *Journal of Asian Earth Sciences* 148: 153–172.

Dumka, R.K., Kotlia, B.S., Miral, M.S., Joshi, L.M., Kumar, K., Sharma, A.K. (2013) First GPS-derived recession rate in Milam glacier, higher central Himalaya, India. *International Journal of Engineering and Science* 2: 58–63.

Dyurgerov, M.B., Meier, M.F. (1997) Mass balance of mountain and sub-polar glaciers: a new global assessment for 1961–1990. *Arctic Alpine Research* 29: 379–391.

Fountain, A.G., Krimmel, R.M., Trabant, D.C. (1997) A strategy for monitoring glaciers. *U.S. Geological Survey Circular* 1132: 19.

Fujita, K., Suzuki, R., Nuimura, T., Yamaguchi, S., Sharma, R.R. (2009) Recent changes in Imja Glacial Lake and its damming moraine in the Nepal Himalayas were revealed by in situ surveys and multi-temporal ASTER imagery. *Environmental Research Letters* 4: 45–50.

Gantayat, P., Kulkarni, A.V., Srinivasan, J. (2014) Estimation of ice thickness using surface velocities and slope: a case study at Gangotri Glacier, India. *Journal of Glaciology* 60: 277–282.

Hall, D.K., Bahr. K.J., Shoener, W., Bindschadler, R.A., Chien, J.Y.L. (2003) Consideration of the errors inherent in mapping historical glacier positions in Austria from the ground and space. *Remote Sensing Environment* 86: 566–577.

Haque, S., Kannaujiya, S., Taloor, A.K., Keshri, D., Bhunia, R.K., Ray, P.K.C., Chauhan, P. (2020) Identification of groundwater resource zone in the active tectonic region of the Himalayas through earth observatory techniques. *Groundwater for Sustainable Development* 10: 100337.

Heid, T., Kääb, A. (2012) Repeat optical satellite images reveal widespread and long-term decreases in land-terminating glacier speeds. *Cryosphere* 6: 467–478.

Herman, F., Anderson, B., Leprince, S. (2011) Mountain glacier velocity variation during a retreat/advance cycle quantified using sub-pixel analysis of ASTER images. *Journal of Glaciology* 57: 197–207.

Holzer, N., Vijay, S., Yao, T., Xu, B., Buchroithner, M., Bolch, T. (2015) Four decades of glacier variations at Muztagh Ata (eastern Pamir): A multisensory study including Hexagon KH-9 and Pleiades data. *Cryosphere* 9: 2071–2088.

Hooke, R.L. (2005) *Principles of Glacier Mechanics*, 429 pp. Cambridge, United Kingdom, Cambridge University Press.

Iken, A. (1977) Variations of surface velocities of some Alpine glaciers measured at intervals of a few hours: comparison with Arctic glaciers. *Z. Gletscherkd. Glazialgeol* 13: 23–35.

Joshi, N., Singh, S., Pant, P.D., Puniya, M.K., Kothyari, G.C. (2018) Polyphase or time-dependent kinematics and quaternary reactivation of thrust bounding Baijnath Klippe: western Kumaun Himalaya, India. *International Journal of Earth Sciences*.

Kandregula, R.S., Kothyari, G.C., Swamy, K.V., Taloor, A.K., Lakhote, A., Chauhan, G., Thakkar, M.G., Pathak, V., Malik, K. (2021) Estimation of regional surface deformation post the 2001 Bhuj earthquake in the kachchh region, western India using RADAR inter-ferometry. *Geocarto International* https://doi.org/10.1080/10106049.2021.1899299

Kannaujiya, S., Gautam, P.K.R., Chauhan, P., Roy, P.N.S., Pal, S.K., Taloor, A.K. (2020) Contribution of seasonal hydrological loading in the variation of seismicity and geo-detic deformation in Garhwal region of Northwest Himalaya. *Quaternary International* 575: 62–71.

Karpilo, R.D. (2009) Glacier monitoring techniques. In: Young, R., Norby, L. (Eds.) *Geological Monitoring*, pp. 141–162. Boulder, Colorado: Geological Society of America.

Kaser, G., Fountain, A., Jansson, P. (2003) A manual for monitoring the mass balance of moun-tain glaciers. In: *Proceedings of the IHP-VI Technical Documents Hydrology*, 55 pp. Paris, France.

Kaul, M.K. (1999) Inventory of Himalayan glaciers. *Geological Survey of India, Special Publication* 34: 136–137.

Khadka, D., Babel, M.S., Shrestha, S., Tripathi, N.K. (2014) Climate change impact glacier and snow melt and runoff in the Tamakoshi basin in the Hindu Kush Himalayan (HKH) region. *Journal of Hydrology* 511: 49–60.

Khan, A., Govil, H., Taloor, A.K., Kumar, G. (2020) Identification of artificial groundwater recharge sites in parts of Yamuna river basin India based on remote sensing and geograph-ical information system. *Groundwater for Sustainable Development* 11: 100415.

Kothyari, G.C., Juyal, N. (2013) Implications of fossil valleys and associated epigenetic gorges in parts of Central Himalaya. *Current Science* 105: 383–388.

Kothyari, G.C., Luirei, K. (2016) Late Quaternary tectonic landforms and fluvial aggradation in the Saryu River valley: Central Kumaun Himalaya. *Geomorphology* 268: 159–176.

Kothyari, G.C., Sharma, A.D., Juyal, N. (2017a) Reconstruction of Late Quaternary climate and seismicity using fluvial landforms in Pindar River valley, Central Himalaya, Uttarakhand, India. *Quaternary International* 443: 248–264.

Kothyari, G.C., Kandregula, A., Luirei, K. (2017b) Morphotectonic records of neotectonic activity in the vicinity of North Almora Thrust Zone, Central Kumaun Himalaya. *Geomorphology* 285: 272–286.

Kothyari, G.C., Joshi, N., Taloor, A.K., Kandregula, R.S., Kotlia, B.S., Pant, C.C. (2019) Landscape evolution and deduction of surface deformation in the Soan Dun, NW Himalaya, India. *Quaternary International* 507: 302–323.

Kothyari, G.C., Kotlia, B.S., Talukdar, R., Pant, C.C., Joshi, M. (2020a) Evidence of neotectonic activity along Goriganga River, Higher Central Kumaun Himalaya, India. Geological *Journal* 55: 6123–6146.

Kothyari, G.C., Pant, P.D., Talukdar, R., Taloor, A.K., Kandregula, R.S., Rawat, S. (2020b) Lateral variations in sedimentation records along the strike length of North Almora Thrust: Central Kumaun Himalaya. *Quaternary Science Advances* 2: 100009.

Kothyari, G.C., Joshi, N., Thakur, M., Taloor, A.K., Pathak, V. (2021) Reanalyzing the geo-morphic developments along tectonically active Soan Thrust, NW Himalaya, India. *Quaternary Science Advances* 3: 100017.

Kumar, K., Dumka, R.K., Miral, M.S., Satyal, G.S., Pant, M. (2008) Estimation of retreat rate of Gangotri glacier using rapid static and kinematic GPS survey. *Current Science* 94: 258–262.

Kumar, D., Singh, A.K., Taloor, A.K., Singh, D.S. (2020) The recessional pattern of Thelu and Swetvarn glaciers between 1968 and 2019, Bhagirathi basin, Garhwal Himalaya, India. *Quaternary International* 575: 227–235.

Messer, B., Viviroli, D., Wiengartner, R. (2004) Mountains of the world: Vulnerable water towers for the 21st century. *Ambio* 33: 29–34.

Metcalfe, R.P. (1993) Pressure, temperature and time constraints on metamorphism across the MCT zone of the High Himalaya slab in the Garhwal Himalaya. In: Treloar, R.J., Searle, M.P. (Eds.) *Himalaya Tectonics. Geological Society of London, Special Publication* 74: 495–509.

Mishra, A., Agrawal, K.K., Kothyari, G.C., Joshi, G. (2020) Quantitative geomorphic approach for identifying active deformation in the foreland region of central Indo-Nepal Himalaya. *Geotectonics* 54: 543–562.

Mukherjee, B.P., Sangewar, C.V. (2001) The recession of the Gangotri glacier through the 20th century. *Geological Survey of India, Special Publication* 65: 1–3.

Nainwal, H.C., Negi, B.D.S., Chaudhary, M., Sajwan, K.S., Gaurav, A. (2008) Temporal changes in the rate of recession: evidence from Satopanth and Bhagirath Kharak Glacier, Uttarakhand, using total station Survey. *Current Science* 94: 653–660.

Naithani, A.K., Nainwal, H.C., Sati, K.K., Prasad, C.P. (2001) Geomorphological evidence of retreat of Gangotri glacier and its characteristics. *Current Science* 80: 87–94.

Negi, H.S., Thakur, N.K., Ganju, A. (2012) Monitoring of Gangotri glacier using remote sensing and ground observations. *Journal of Earth System Science* 121: 855–866.

Pudełko, R., Angiel, P.J., Potocki, M., Jedrejek, A., Kozak, M. (2018) Fluctuation of glacial retreat rates in the Eastern part of Warszawa icefield, King George Island, Antarctica, 1979–2018. *Remote Sensing* 10: 2–25.

Puri, V.M.K. (1984) *Gangotri Glacier-report on the Interdepartmental Expedition-1975.* (Progress Report) Rep. Geological Survey of India (Unpublished).

Puri, V.M.K., Shukla, S.P. (1991) Tongue fluctuations studies on Gangotri glacier, Uttarkashi district, Uttar Pradesh. *Geological Survey of India, Special Publication* 21: 289–291.

Raina, V.K. (2005) Status of the glacier in India. *Himalayan Geology* 26: 285–293.

Riesen, P., Strozzi, T., Bauder, A., Wiesmann, A., Funk, M. (2011) Short-term surface ice motion variations were measured with a ground-based portable real aperture radar interferometer. *Journal of Glaciology* 57: 53–60.

Sangewar, C.V. (1997) *Report on glacier front fluctuation in parts of H.P. and U.P.* Geological Survey of India (Unpublished).

Saraswat, P., Syed, T.H., Famiglietti, J.S., Fielding, E.J., Crippen, R., Gupta, N. (2013) Recent changes in the snout position and surface velocity of Gangotri glacier observed from space. International *Journal of Remote Sensing* 34: 8653–8668.

Sarkar, T., Kannaujiya, S., Taloor, A.K., Ray, P.K.C., Chauhan, P. (2020) Integrated study of GRACE data derived interannual groundwater storage variability over water-stressed Indian regions. *Groundwater for Sustainable Development* 10: 100376.

Satyabala, S.P. (2016) Spatiotemporal variations in surface velocity of the Gangotri glacier, Garhwal Himalaya, India: Study using synthetic aperture radar data. *Remote Sensing Environment* 181: 151–161.

Scherler, D., Leprince, S., Strecker, M.R. (2008) Glacier-surface velocities in alpine terrain from optical satellite imagery: Accuracy improvement and quality assessment. *Remote Sensing Environment* 112: 3806–3819.

Singh, A.K., Jasrotia, A.S., et al. (2017) Estimation of quantitative measures of total water storage variation from GRACE and GLDAS-NOAH satellites using geospatial technology. *Quaternary International* 444: 191–200.

Singh, D.S., Mishra, A. (2001) Gangotri glacier characteristics, retreat and processes of sedimentation in the Bhagirathi valley. *Geological Survey of India, Special Publication* 65: 17–20.

Singh, D.S., Tangri, A.K., Kumar, D., Dubey, C.A., Bali, R. (2016) The pattern of retreat and related morphological zones of Gangotri glacier, Garhwal Himalaya India. *Quaternary International* 144: 172–181.

Singh, P., Singh, V.P. (2001) *Snow and Glacier Hydrology*, 742 pp. Dordrecht, Netherlands: Kluwer Academic Publishers.

Singh, S., Sood, V., Taloor, A.K., Prashar, S., Kaur, R. (2020) Qualitative and quantitative analysis of topographically derived CVA algorithms using MODIS and Landsat-8 data over Western Himalayas, India. Quaternary International 575: 85–95.

Sood, V., Gusain, H.S., Gupta, S., Taloor, A.K., Singh, S. (2020a) Detection of snow/ice cover changes using subpixel-based change detection approach over Chhota-Shigri glacier, Western Himalaya, India. *Quaternary International* 575: 204–212.

Sood, V., Singh, S., Taloor, A.K., Prasher, S., Kaur, R. (2020b) Monitoring and mapping of snow cover variability using topographically derived NDSI model over north Indian Himalayas during the period 2008–19. *Applied Computing and Geosciences* 8: 100040.

Sugiyama, S., Gudmundsson, G.H. (2004) Short-term variations in glacier flow controlled by subglacial water pressure at Lauteraargletscher, Bernese Alps, Switzerland. *Journal of Glaciology* 50: 353–362.

Taloor, A.K., Joshi, L.M., et al. (2020) Tectonic imprints of landscape evolution in the Bhilangana and Mandakini basin, Garhwal Himalaya, India: A geospatial approach. *Quaternary International* 575: 21–36.

Taloor, A.K., Kotlia, B.S., et al. (2019) Tectono-climatic influence on landscape changes in the glaciated During Drung basin, Zanskar Himalaya, India: A geospatial approach. *Quaternary International* 507: 262–273.

Taloor, A.K., Ray, P.K.C., et al. (2017) Active tectonic deformation along reactivated faults in Binta basin in Kumaun Himalaya of north India: Inferences from the tectono-geomorphic valuation. *Zeitschrift für Geomorphologie* 61: 159–180.

Tangri, A.K. (2002) Shrinking glaciers of Uttaranchal, a cause of concern and hope for the future. In: Pant, C.C., Sharma, A.K. (Eds.) *Aspects of Geology and Environment of the Himalaya*, pp. 359–367. Gyanoday Prakashan, Nainital.

Tangri, A.K., Chandra, R., Yadav, S.K.S. (2004) Temporal monitoring of the snout, equilibrium line and ablation zone of Gangotri Glacier through remote sensing and GIS techniques – an attempt at deciphering the climatic variability. *Geological Survey of India, Special Publication* 80: 145–153.

Tiwari, R.K., Gupta, R.P., Arora, M.K. (2014) Estimation of surface ice velocity of Chhota-Shigri glacier using sub-pixel ASTER image correlation. *Current Science* 106: 853–859.

Venkatesh, T.N., Kulkarni, A.V., Srinivasan, J. (2012) The relative effect of slope and equilibrium line altitude on the retreat of Himalayan glaciers. *Cryosphere* 6: 301–311.

Yan, S., Liu, G., Wang, Y., Ruan, Z. (2015) Accurate determination of glacier surface velocity fields with a DEM-assisted pixel-tracking technique from SAR imagery. *Remote Sensing* 7: 10898–10916.

3 Modelling a Glacial Lake Outburst Flood of Sona-Sar Lake in the Jhelum Basin of Kashmir Himalayas, India, Using HEC-RAS

Rayees Ahmed, Gowhar Farooq Wani,
Abid Farooq Rather, Sumaira Javaid,
Syed Towseef Ahmad, Savati Sharma, and
Ajay Kumar Taloor

3.1 INTRODUCTION

The Indian Himalayan region is a vital source of freshwater for the highly populated areas downstream (Singh et al., 2016). The temperatures in the entire Indian Himalayan region are significantly increasing as is evident from earlier studies. The changing precipitation regimes and increasing temperatures over the Himalayan region are a major concern for the snow cover and glacier health in the region (Negi et al., 2021). Thus, they pose a serious threat to India's water supply. According to current water demand and usage estimates, only 50% of water demands will be met by 2030, and by 2050, the water scarcity is predicted to reach 68 percent (Mir, 2021). With the ever-increasing population, industrialization and urbanization, the availability of water, both quantitively and qualitatively, will be affected. Based on satellite data, Kulkarni et al. (2013) reported a glacier loss of about 1868 glaciers of 11 basins of the Indian Himalayas during the years 1962 to 2002. They reported an overall reduction in glacier area of 16% ranging from 2.7% to 20% among different basins. The Indian Himalayan region is home to numerous potentially dangerous glacial lakes which may result in high-intensity GLOFs due to sudden dam breaches.

Glacial lake outburst floods (GLOFs), also known as "jökulhlaups", refer to a sudden release of an enormous quantity of water and debris from a lake that may cause damage to infrastructure and populations downstream (Watson et al., 2015). Glacial lakes exacerbate the rate of mass loss from their mother glacier due to the

DOI: 10.1201/9781003364115-3

mechanical calving process and sub-aqueous melting (Benn et al., 2007; Truffer and Motyka, 2016). Currently, in the Himalayan region there are more than 700 proglacial lakes that are capable of directly determining the behaviour of their feeding glacier (Nie et al., 2017; Zhang et al., 2015). Glacial lakes are increasing in number and rapidly expanding in size across the world as a result of de-glaciation (Carrivick and Tweed, 2013; Shugar et al., 2020; Zheng et al., 2021). However, the rate of glacial lake expansion differs from region to region and watershed to watershed. For instance, the glacial lakes in the central Himalayas have shown a 57% increase in area from 1994 to 2017 (Pandey et al., 2021). Begum et al. (2019) reported a 43.6% expansion in the area of glacial lakes in Central and Eastern Himalaya over a period of 25 years. Similarly, in Nepal's Himalayas, glacial lakes expanded in surface area by 25% from 1987 to 2017, as documented by Khadka et al. (2018). Proglacial lakes fed by the host glacier expanded in area by 50% between 1990 and 2015 in the Himalayan region, as reported by Nie et al. (2017).

The preparation of inventory, identification of potentially dangerous glacial lakes, and assessing their level of hazard are some of the key steps that are crucial to timely guiding of the appropriate remedial works, including further analysis and the development of early warning systems (Worni et al., 2013; Ahmed et al., 2022). Climate change over the past few decades had a significant impact on the hydrological cycle (glaciers, glacial lakes, permafrost, snow, etc.) in the Himalayan mountain ranges (Singh and Kumar, 1997; Singh and Bengtsson, 2004; Hunt et al., 2020; Qazi et al., 2020; Taloor et al., 2021; Ahmad et al., 2022), consequently, glaciers are retreating and forming a number of glacial lakes. These glacial lakes are formed as a result of glacier melt water accumulation behind dams (moraine dammed or glacier dammed), within the glacier body (englacial lake), on the glacier bed (sub-glacial lake), on the surface of the glacier (supra-glacial lake), behind the bed rock (bed-rock dammed lake), blocked by landslides (landslide dammed) in glacier retreated depressions (cirque lakes), etc. The Himalayan glacier bearing catchments are home to a large number of such glacial lakes which are directly or indirectly fed by the melt waters of glaciers. The rapid expansion of the glacial lakes due to the melting of glaciers under the influence of climate change has enhanced the susceptibility to GLOFs in the Himalayan region. A recent study on the hazard assessment of the glacial lakes in the Indian Himalayan region carried out by Dubey and Goyal (2020) reported 73 glacial lakes as high- or very-high-risk lakes. Hence, understanding the hazard and analysing the risks associated with these glacial lakes through the modelling approaches becomes imperative to provide first-hand information for policy making and hazard mitigation strategies. The potential GLOF impacts of the glacial lakes with high hazard levels can be evaluated using hydrodynamic models, such as HEC-RAS, MIKE-11, DAMBREAK, HEC-HMS, RAMFLOW, etc. These models have been extensively used in past studies for hydraulic GLOF routing and to analyse the hydraulic properties of potential floods at various locations in the downstream regions.

The key objective of this study was to analyse the changes in the Sona-Sar Lake and to assess its GLOF susceptibility using important parameters. Furthermore, hydro-dynamic modelling was performed for the lake using HEC-RAS software to evaluate the potential GLOF impacts at various important locations in the downstream region.

Finally, inundation maps were prepared and then overlaid on the land use land cover (LULC) map to analyse the area potentially inundated due to the GLOF event.

3.2   STUDY AREA

This study has focused on the Sona-Sar glacial lake located in the Lidder catchment of Jhelum basin, western Himalayan region. The lake is a proglacial lake by nature (Ahmed et al., 2022), fed by the melt water supply of its mother glacier. The water from the Sona-Sar Lake is directly drained into the Lidder river through a small stream with a length of 2.45 km. The lake is located between the 34 04 02.23 N and 75 28 31.53 E coordinates. The distance between the lake and the mother glacier is estimated at 1.02 km with an average slope of 25.5 degrees. The nearest settlement, i.e., Chandanwari, is situated a distance of 6.9 km² from the lake outlet. The lake has an area of 0.19 km² according to recent estimates (2022). Sona Sar lake has been identified as a potentially dangerous glacial lake by Ahmed et al. (2022). Mal et al. (2021) categorized Sona-Sar Lake as very high dangerous (VHD) lake, whereas Dubey and Goyal (2020) classified it as a very-high-risk (VHR) lake.

The nearest meteorological station to the lake is located at a distance of 12.3 km, as shown in Figure 3.1. The analysis of climatic variables (1981–2020) revealed that the maximum temperature has increased significantly from 1981 to 2020 to the tune of 2.4°C, from 18.2°C to 15.8°C with an R^2 value of 0.33 (Figure 3.2). The minimum temperature shows a significant increasing trend and increased from 2.0°C

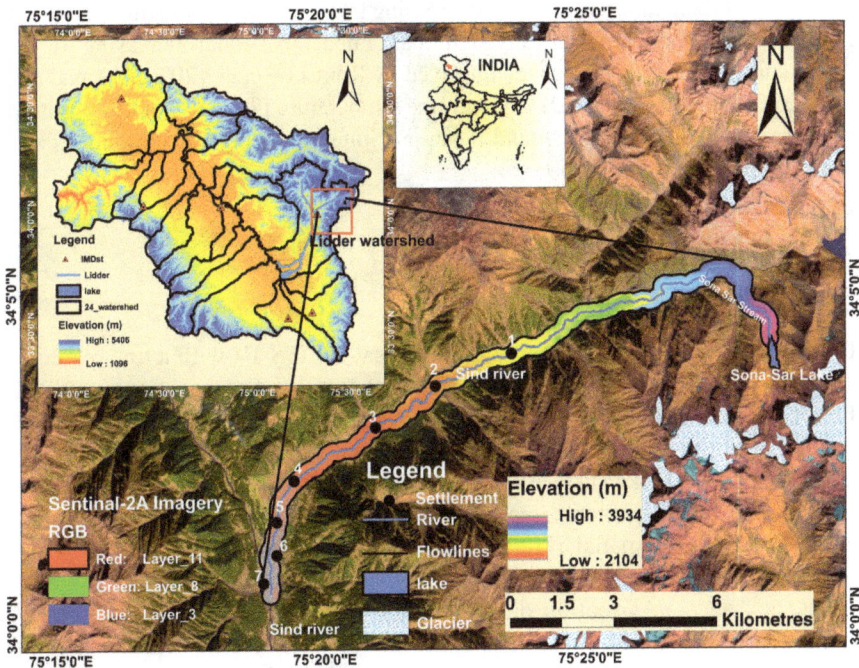

FIGURE 3.1   Location map of the study area.

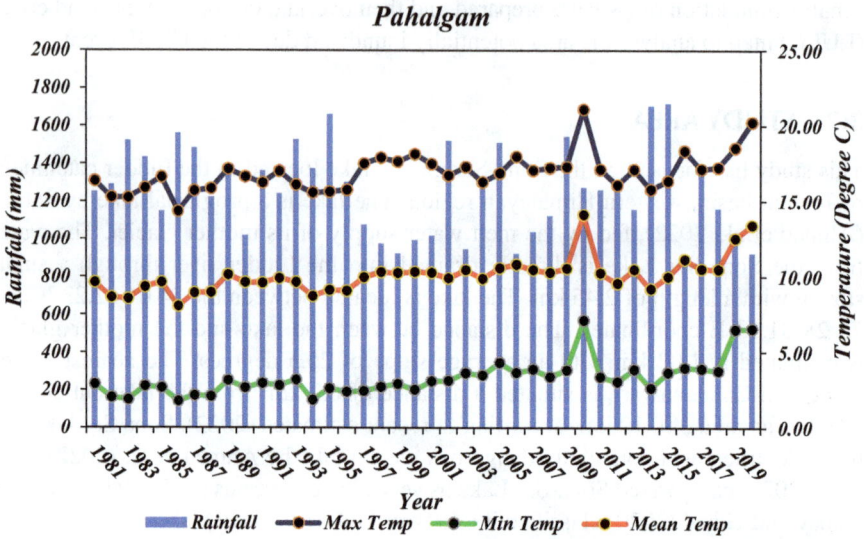

FIGURE 3.2 Temperature (T_{max}, T_{mean} and T_{min}) and annual precipitation trend at Pahalgam meteorological station.

to 4.50°C to the extent of 2.5°C over the last 40 years. For mean temperature, a significant increase of 2.6°C for the last 40 years was found, with an R^2 value of 0.43 (Figure 3.2). Precipitation followed a decreasing trend with respect to time series analysis; however, it failed to show any significant impact. A similar increasing trend of temperatures in the region has been observed by Romshoo et al. (2020), Ahmad et al. (2021) and Ahmed et al. (2022). Overall, the temperature (T_{max}, T_{mean} and T_{min}) shows an increasing trend, whereas annual mean precipitation shows a decreasing trend at Pahalgam station (Figure 3.2).

3.3 DATABASE AND METHODOLOGY

3.3.1 DATA SETS

High-resolution satellite images of Resourcesat-2 LISS-IV with a fine resolution of 5.8 m procured from the India's National Remote Sensing Centre (NRSC) Hyderabad web portal (source) were used to map the glacial lake outlines for the year 2020. Multi-date Landsat imagery was used to analyse the spatio-temporal trend in the glacial lake surface area. Advanced Land Observing Satellite Phased Array Type L-band Synthetic Aperture Radar (ALOS-PALSAR) Digital Elevation Model (DEM) with spatial resolution 12.5 m was downloaded from the USGS earth explorer to obtain topographic parameters such as elevation, slope and aspect, etc. and cross-section data. The Planet data available on a daily level with a spatial resolution of 3 m were used to obtain the breach parameters and to validate the glacial lake extents. Freely available high-resolution Google earth imagery was also used for the validation and cross checking. Necessary geometric

TABLE 3.1
Satellite Data Used in the Study

S. No	Data	Spatial resolution (m)	Source/website
1	Landsat, 5, 6, 7, 8 and 9	60, 30, 30, 30, 15	https://earthexplorer.usgs.gov/
2	Sentinnel-2A	10	https://earthexplorer.usgs.gov/
3	ALOS-PALSAR DEM	12.5	https://asf.alaska.edu/data-sets
4	Planet cube-sat	3	www.planet.com/products
5	Resourcesat-2	5.8	https://bhuvan.nrsc.gov.in/
6	Google Earth images	-	https://earth.google.com/

and radiometric correction was done using Arc GIS software. Landsat data with a spatial resolution of 30 m were used for the spatio-temporal analysis of glacial lakes because of the non-availability of high-resolution LISS-IV imagery. Resampling technique (nearest neighbour analysis) was used to match the spatial resolution of 30 m for Landsat imagery for all the years. All the satellite imagery scenes downloaded for the present study were preferred to have low cloud cover or be cloud free and were procured for the months of September, October and November, because during these months glacial lakes remain relatively stable. The details of the data sets used in this study are provided in Table 3.1.

3.3.2 METHODOLOGY

Identification and mapping of glacial lakes is essential to evaluate the hazards and risks associated with a glacial lake. The glacial lake boundaries can be extracted through various image enhancement techniques (Khadka et al., 2018; Allen et al., 2019; Aggarwal et al., 2019; Jain et al., 2019; Mir, 2018; Ahmed et al., 2021; Ahmed et al., 2022; Gupta et al., 2021; Islam and Patel 2022; Wang et al., 2021). In this study, we have applied the Normalized Difference Water Index (NDWI) technique to extract glacial lakes using normalization between two bands (green and NIR). Since we have used Landsat imagery, band 3 was subtracted from band 8 to get the vector layer of glacial lakes. However, the NDWI technique misclassifies shadow areas, cloud cover and built-up noise as water pixels. In order to remove such errors, DEM-based slope, aspect and hill shade maps were generated and subsequently overlaid on the vector layer to correct the lake extents. Finally, the vector layer was cross checked and validated from the high-resolution Google Earth imagery. The NDWI technique was propounded by MC-Feeteres in 1996 and later developed and applied by Huggel et al. (2002) to automatically extract the lake extents. The equation developed by Huggel et al. (2002) is as follows:

$$NDWI = \frac{Green - NIR}{Green + NIR} \tag{3.1}$$

where the green and NIR bands in Equation (3.1) represent the spectral reflectance.

3.4   GLACIAL LAKE OUTBURST FLOOD SIMULATION

After the identification of potentially dangerous glacial lakes and their hazard level assessment, the next step was to model the GLOF occurrence by simulating a dam breach and mapping the resultant flood in order to evaluate the potential downstream impacts. Glacial lake outbursts flood through any natural landscape, resulting due to a sudden dam breach, can be simulated using various kinds of mathematical and hydrodynamic models (Allen et al., 2009; Prakash and Nagarajan, 2018; Mergili and Schneider, 2011; Wang et al., 2015; Petrakov et al., 2012; Worni et al., 2013; Westoby et al., 2014). Hydraulic models offer an opportunity to provide a picture of the probable consequences of partial or complete dam failure, which is crucial to carrying out emergency planning and conducting reliable risk assessment. In the present study, Hydrologic Engineering Centre's River Analysis System (HEC-RAS) version 5.0.7 was employed to simulate a potential GLOF of Sona-Sar Lake. HEC-RAS is one of the most popular and extensively used open-source hydrodynamic models utilized for glacier hazard studies (Carling et al., 2011; Klimeš et al., 2014; Hussain et al., 2020). It is a reliable model to investigate the hydraulic behaviour of steady and unsteady flows due to its user-friendly interface and capacity for executing sophisticated dynamic simulations (Brunner, 2002). HEC-RAS was developed by the United States Army Corps Engineering Hydrologic Engineering Center, and has been designed to perform both one- and two-dimensional hydraulic routing. It is based on the St. Venant equations that are used to simulate flood scenarios generated by a glacial lake outburst. The equations used (continuity and momentum) are as follow:

$$\text{Continuity equation } (\partial Q/\partial X) + \partial (A + A_0) / \partial t - q = 0 \qquad (3.2)$$

$$\text{Momentum equation } (\partial Q/\partial t) + \{\partial (Q^2/A)/\partial X\} + g A ((\partial h/\partial X) +$$
$$S_f + S_c) = 0 \qquad (3.3)$$

where Q in Equations (3.1) and (3.2) denotes the discharge, A is active flow area, A_0 represents the inactive storage area, h is water surface elevation, q represents lateral flow, X is the distance along the waterway, t represents the time, S_f denotes friction slope, S_c is the expansion contraction slope and g is gravitational acceleration. HEC-Geo RAS, a plug-in of Arc-GIS, was used to create the geometry data that were subsequently imported to the HEC-RAS environment. HEC-RAS uses several input parameters for GLOF routing through a flow channel.

3.4.1   GLOF INPUT PARAMETERS

Lake outburst floods rise to peak flow and then, either gradually or rapidly, subside to normal levels after the water supply is drained. The peak flow is governed by the lake volume, width and height of the dam, composition of dam material, downstream terrain, failure mechanism and availability of sediment. The breaching of moraine dams of such glacial lakes also must be taken into consideration along with channel routing in order to obtain the highest GLOF peak at any site. Dam breach modelling provides various hydraulic properties of the flood wave generated from the GLOF.

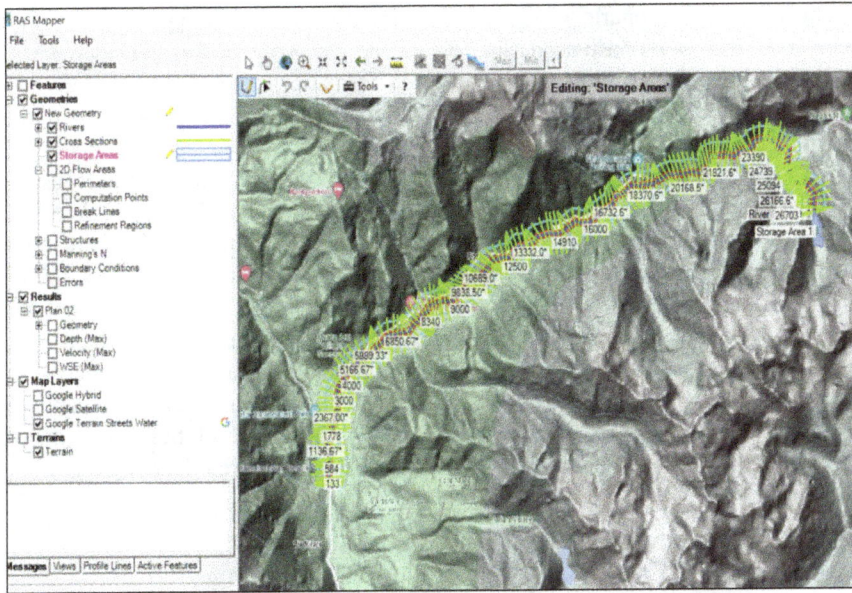

FIGURE 3.3   Cross-section profile.

The flood wave parameters include water surface elevation, flood depth, flood velocity and peak flood discharge at different sites in the downstream region of the glacial lake. In the HEC-RAS model, two types of data sets are required as an input to run the computations (steady and unsteady flow).

Geometry data: Geometry data include cross-sections, centre lines, flow lines and bank lines. The geometry data were created in HEC-Geo RAS plugin in Arc GIS interface and then imported into the HEC-RAS model. The cross-sections were created at intervals of 100 m, as depicted in Figure 3.3. The elevation on both sides of the cross-section was generated from the DEM.

Flow data: A breach flood hydrograph was generated using 1-D HEC-RAS with volume, breach depth, breach width and breach formation time as the input data. The breach hydrograph was used as an inflow data for the unsteady flow analysis. The volume of the lake was estimated through the improved volume area scaling equation provided by Qi et al. (2022). Breach width and breach formation time were estimated using Equation (3.4) developed by Froelich in 1995 (Froehlich 1995). This equation is as follows:

$$b_f = 0.1803 \, V_w^{\,0.32} \, h_b^{\,0.19} \tag{3.4}$$

$$t_f = 0.00254 \, V_w^{\,0.53} h_b^{\,-0.90} \tag{3.5}$$

where V_w is the volume of the storage area in m^3 and h_b is the height of the water above the breach invert level.

The worst-case scenario was considered for the GLOF routing along the flood channel. In this worst-case scenario, 75% of the lake volume was considered with a 20 m breach width and 30 minutes breach formation time. The failure event is represented as a breaching of the inline structure (dam), with the normal depth, determined by the slope along the channel taken as the downstream boundary condition. The normal depth of 0.06 m was obtained from the bed slope profile measured between the upper- and lower-most points of the flood channel. Manning's coefficient value of "n" = 0.06 was considered for the GLOF routing of Sona-Sar Lake. A Manning coefficient value of between 0.04–0.06 is considered as best for the Himalayan terrain (Chow, 1959) especially for hilly regions with a steep slope and low or no vegetation cover, gravels, cobbles, boulders and bushes on the banks, etc. The generated breach hydrograph was routed through the flood channel to analyse the potential flood impacts at various locations in the downstream areas. Subsequently, a freely available Sentinel-2A-based LULC map with 10 m resolution was extracted for the area in order to analyse the area inundated under different land use classes.

3.5　RESULTS AND DISCUSSIONS

3.5.1　Characteristics of a Glacial Lake

According to recent estimates from the Landsat 9 imagery, the area of the glacial lake was calculated as 0.19 km². This lake is classified as a proglacial lake located at a distance of 1.02 km from the feeding glacier snout. The average slope between the lake and its feeding glacier was estimated as 25.55 degrees, which is greater than the threshold (>10 degree) that can trigger potential mass movements into a lake. It is dammed by a moraine with no visible outlet, having an average height of 11 m as observed on the Google Earth imagery. However, seepage is visible in the bottom of the moraine dam which signifies the presence of loose material inside the dam which allows water to pass through the dam. The depth of the lake was estimated as 17.14 m calculated using the empirical formula provided by Huggel et al. (2002). The total volume of the lake was calculated as 9.49×10^6 m³ obtained through the improved volume area scaling equation provided by Qi et al. (2022). Lakes having more than 10 million cubic metres of water are generally considered as highly prone to cause a GLOF (Fujita et al., 2013; Hazra and Krishna, 2022). Meanwhile, the total peak discharge of the lake was estimated as 9602 m³. The terrain profile from the glacier to the downstream region of the moraine dam is illustrated in Figure 3.4. From Figure 3.4, the steep slope between the lake and moraine dam downstream can be clearly observed. However, no past GLOF event has been reported in the region, although cloud bursts and landslides have been frequently reported in the adjoining areas of the glacial lakes[1]. In the case that the lake is hit by a cloud burst or any mass movement, an outburst flood may result.

[1]　www.tribuneindia.com/news/j-k/13-amarnath-yatris-dead-as-cloudburst-hits-shrine-area-410650

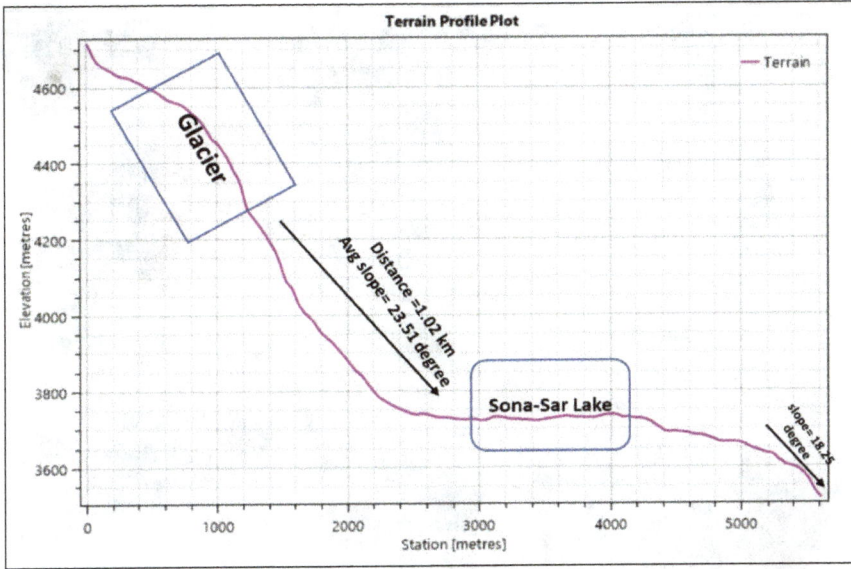

FIGURE 3.4 Terrain profile of the glacial lake and its feeding glacier and downstream area.

3.5.2 GLACIAL LAKE CHANGES

The multi-temporal satellite data show the remarkable expansion of Sona-Sar Lake over a period of 50 years from 1972 to 2022. The lake has expanded from 0.08 km² in 1972 to 0.19 km² in 2022, showing a growth rate of 58% at a rate of 0.002 km²/year (Figure 3.5). The average percentage change in the lake area per decade was observed as 14.5%. The lake showed a maximum increase in areal extent during the years 1972 (0.08 km²) to 1992 (0.16 km²). The perimeter of the lake has increased by 993 m from 1218 m in 1972 to 2251 m in 2022. The average length and width of the lake was also estimated during the various time points from 1972 to 2022. The analysis shows that the length of the lake has increased from 494 in 1972 to 951 in 2022, whereas width of the lake increased from 212 m in 1972 to 298 m in 2022 (Table 3.2). From Table 3.2, it can be concluded that the lake area has shown a gradual increase, but other parameters (perimeter, length and width) have revealed a fluctuation over this period of time. As parameters like glacial lake depth, volume and peak discharge were derived from regression equations which are dependent on the lake area, these also showed a gradual increase over the period of 50 years.

The glacial lake is fed by two small glaciers with areas of 0.42 km² (G1) and 0.18 km² (G2) located at distances of 1.01 km and 1.02 km, respectively. These glaciers have shown a substantial decrease in areal extents over the last 50 years from 1972 to 2022. The areas of the glaciers were 0.59 km² and 0.26 km² in 1972 and reduced to 0.45 km² and 0.18 km² in 2022, respectively (Figure 3.6). Thus, a total area of 0.22 km² (G1 and G2) has been lost over the past 50 years. The snouts of the glaciers have retreated by 236 m and 242 m during the study period, with annual retreat rates of 4.72 m/year and 4.84 m/year, respectively.

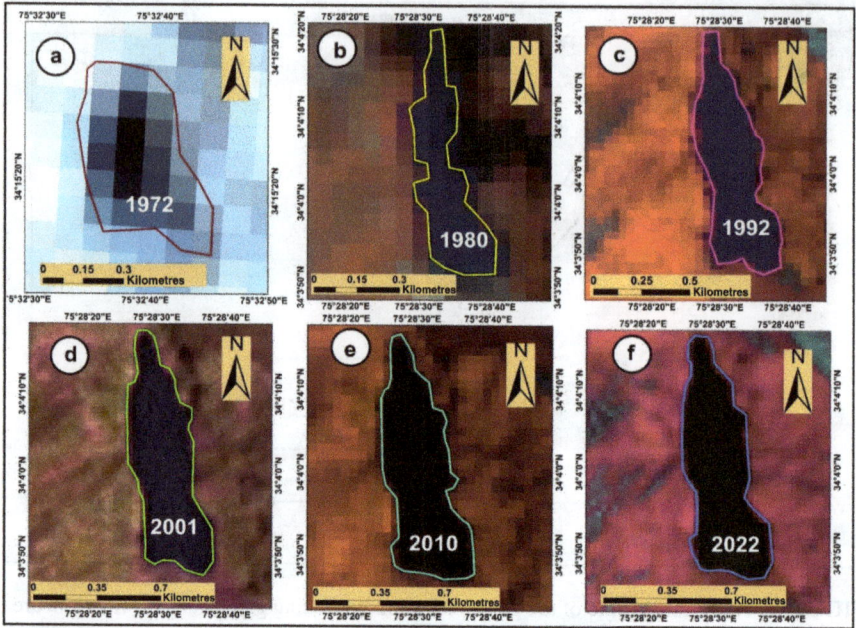

FIGURE 3.5 Glacial lake area change, a) 0.08 km² in 1972, b) 0.12 km² in 1980, c) 0.16 km² in 1992, d) 0.17 km² in 2001, e) 0.18 km² in 2010, and f) 0.19 km² in 2022.

TABLE 3.2
Changes in Lake Area, Perimeter, Length and Width from 1972 to 2022

Year	Area	Perimeter (m)	Length (m)	Width (m)
1972	0.08	1218	494	212
1980	0.12	2239	892	189
1992	0.16	2263	927	247
2001	0.175	2216	948	245
2010	0.181	2256	926	292
2022	0.19	2251	951	298

3.5.3 GLOF SUSCEPTIBILITY

Sona-Sar Lake was selected for the GLOF simulation based on its high GLOF susceptibility. In past studies, most GLOF assessment studies have been carried out on glacial lakes which have been classified as critical/high-hazard lakes (Sattar et al., 2020; Jain et al., 2019; Hazara and Krishna, 2021; Rawat et al., 2022; Ahmed et al., 2022). Sona-Sar Lake has been considered as a very high dangerous (VHD) lake by Mal et al. (2021), a high-risk lake by Dubey and Goyal (2020), and a less dangerous lake by Ives et al. (2010), Fujita et al. (2013) and Worni et al. (2013). In the recent study by Ahmed et al. (2022) Sona-Sar Lake was categorized as a lake with a high hazard level and it was recommended for further assessment (GLOF simulation and

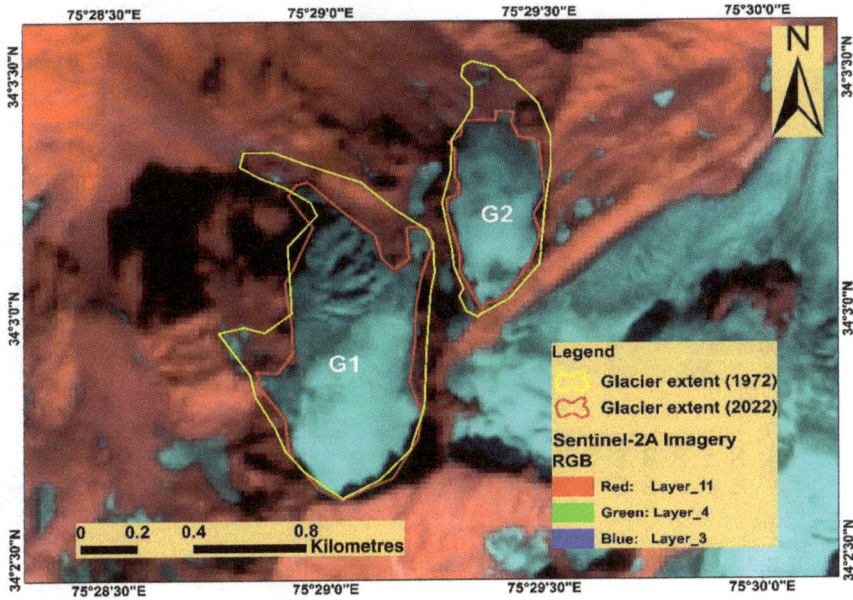

FIGURE 3.6   Glacier area change from 1972 to 2022.

risk assessment) using integrated approaches of remote sensing, GIS and dam breach modelling.

3.5.4   WORST-CASE SCENARIO GLOF

In the worst-case scenario, 75% of the volume of the lake was considered for the breach hydrograph, with a breach width of 20 m and breach formation time of 30 minutes. a One-dimensional HEC-RAS model was used to create the breach hydrograph of Sona-Sar Lake, as depicted in Figure 3.7. The generated breach hydrograph was routed through the flood channel up to a distance of 23 km from the lake outlet using HEC-RAS 2D computations. The model output provides some important flood wave parameters such as water surface elevation, flood depth, flood velocity and potential flood discharge at various sites in the downstream region. The generated breach hydrograph resulted in A peak discharge of 2024.16 m³/s which was attained 1 hour and 10 minutes after the breach started at the lake site (dam).

3.6   GLOF HAZARD ASSESSMENT

The breach hydrograph with a peak flood discharge of 2024.16 m³/s was routed through the channel to evaluate the potential impacts of GLOF at various locations/ sites in the downstream region. The derived hydraulic parameters of the potential flood were evaluated at six sites located at distances of 7.12 km (Chandanwari), 15.6 km (Betab valley), 19.9 km (Laripora), 21.2 km (Upper Pahalgam), 22.6 km (Lower Pahalgam) and 23.4 km (Lidder View Park). The flood inundation at these locations is

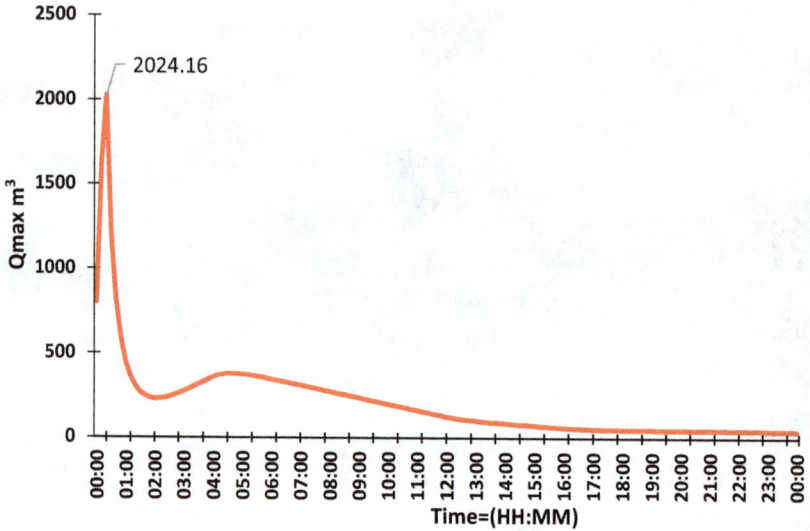

FIGURE 3.7   Breach hydrograph at lake site (inline structure).

depicted in Figure 3.7. Analysis of the GLOF hydrograph revealed that the flood wave would reach the nearest settlement located at a distance of 7.2 km (Chandanwari) within 1 hour and 50 minutes after the breach initiation with a peak discharge of 1366 m^3 s^{-1} with a maximum flood depth of 6.4 m and a flood velocity of 6 m/s. Further downstream, the GLOF hydrograph was evaluated at a famous tourist site located at a distance of 15.6 km from the lake outlet.

At this site, the routed GLOF hydrograph revealed a peak discharge of 1230 m^3 s^{-1} with a flood depth, flood velocity and water surface elevation estimated as 5.3 m, 6.4 m/s and 3534 m, respectively. The modelled GLOF event at Laripora generated a peak flood discharge of 1120 m^3 s^{-1} as illustrated in Figure 3.8. At the upper Pehalgam site, the peak flood was attained 3 hours and 5 minutes after the dam breach. The maximum peak discharge of 1109 m^3 s^{-1} was observed at this location with a maximum flood depth of 4.3 m and maximum flood velocity of 5.7 m/s. The evaluation of the potential flood downstream shows that the intensity of flood wave parameters decreases as it advances towards the low-lying areas as a result of increasing channel width.

The potential flood evaluated at site 5 (Lower Pahalgam) resulted in a peak discharge of 1017 m^3 s^{-1} with a flood height and flood velocity of 2.7 m and 3.6 m/s, respectively. Finally, the routed GLOF was evaluated at the last selected location, i.e., Lidder View Park, located a distance of 23.4 km from the lake outlet. The model-generated flood wave arrived at this location 3 hours and 15 minutes after the breach started at the lake (moraine dam). The peak flood of 989 m^3 s^{-1} was observed at the Lidder View Park with a flood depth of 2.3 m and flood velocity of 3 m/s. The routed GLOF hydrographs at each location are shown in Figure 3.8. Overall, the flood depth varies between 2.1 m to 25 m, with an average depth of 5.6 m. Meanwhile, flood velocity varies between 2.5 to 35 with a mean flood velocity of 7 m/s observed across the routed flood channel.

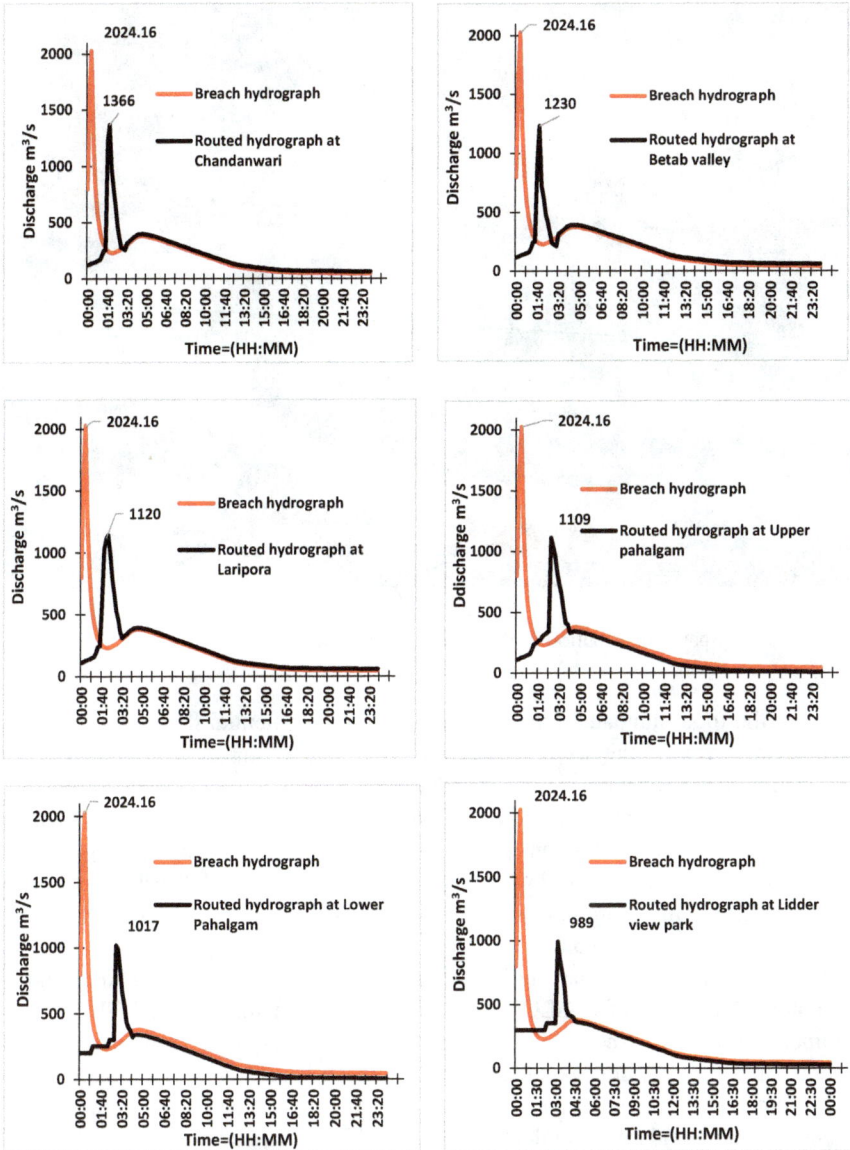

FIGURE 3.8 Routed GLOF hydrographs at various sites downstream.

3.6.1 GLOF INUNDATION

The potential GLOF impacts on the downstream region were also evaluated by analysis of temporal and spatial 2-D characteristics of the flood wave across the routed channel. The hydraulic properties of the flood wave (flood depth, flood velocity, inundation area, water surface elevation and peak discharge) from the lake site to the last selected location vary greatly, as observed from the model output. In the worst-case

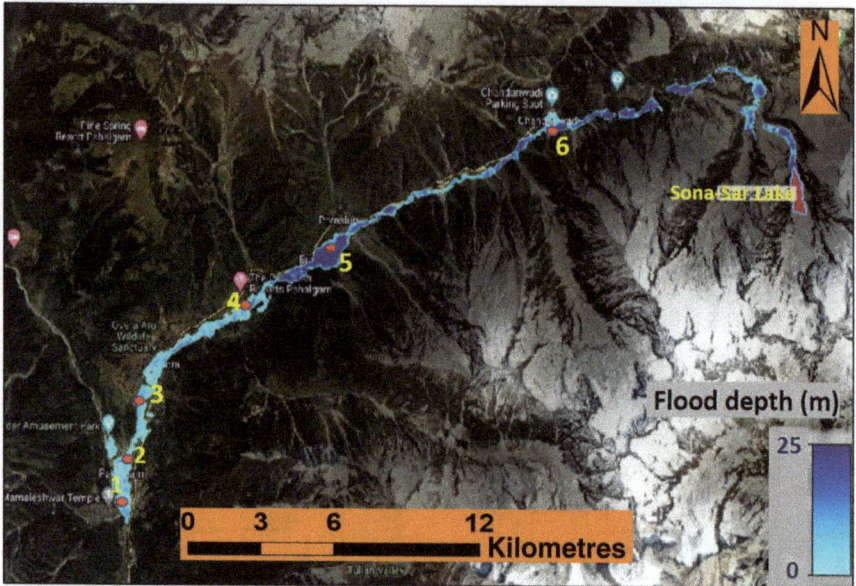

FIGURE 3.9   Spatial distribution of flood inundation depth.

scenario, the total inundated area resulting due to the potential flood was estimated as 4.10 km². Among the six evaluated sites, the maximum area was inundated at site 2 (Betab valley) with a total inundated area of 0.57 km², whereas the minimum inundated area (0.056 km²) was estimated at site 1 (Chandanwari). The flood inundated a total of nine locations (including houses, agricultural land, bridges, roads and forest land, etc) across the 23.4-km river channel. The spatial distribution of flood depth is depicted in Figure 3.9, with the flood velocity in Figure 3.10 and the water surface elevation in Figure 3.11.

Analysing the various characteristics of the flood wave is crucial to understanding the potential impacts of a GLOF event. The output in the form of GLOF wave parameters obtained can be affected by the various uncertainties associated with several input data parameters such as DEM, breach height, breach formation time, lake volume, etc. The variation in the spatial resolution of the DEMs influences the extent of flood inundation and depths (Hazra and Krishna, 2021). The volume of the lake derived from the empirical equation is also directly dependent upon the lake area. The volume of the lake is one of the most important parameters used as an input in the breach hydrograph which is routed through the river channel to obtain the various GLOF wave parameters. Therefore, model results may be affected by the lake volume calculated through empirical equations which sometimes yield overestimated or underestimated values. However, in the absence of ground data and high-resolution DEM data, such kinds of studies provide first-hand information about the probability of a GLOF event and its potential impacts on the various important locations in the downstream region. The input parameters derived through in-depth field investigations and the use of high-resolution DEMs will greatly improve the

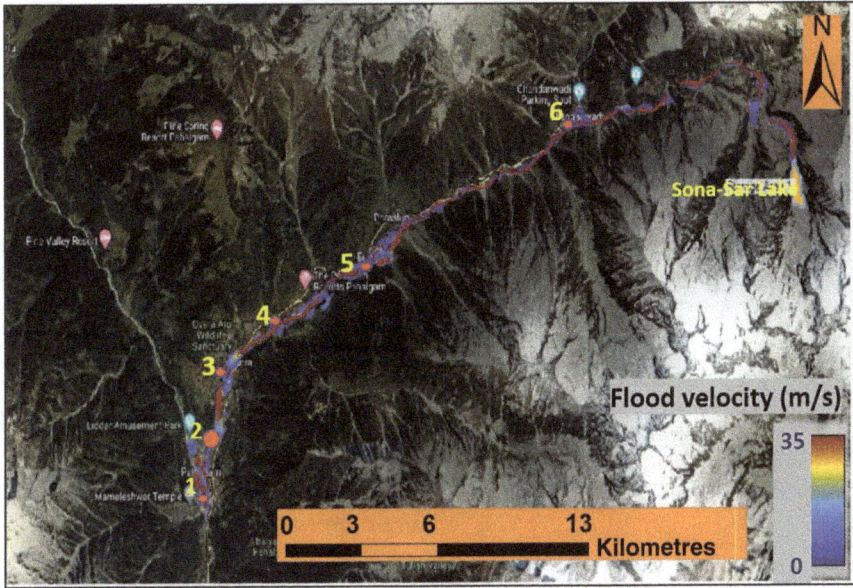

FIGURE 3.10   Spatial distribution of flood velocity.

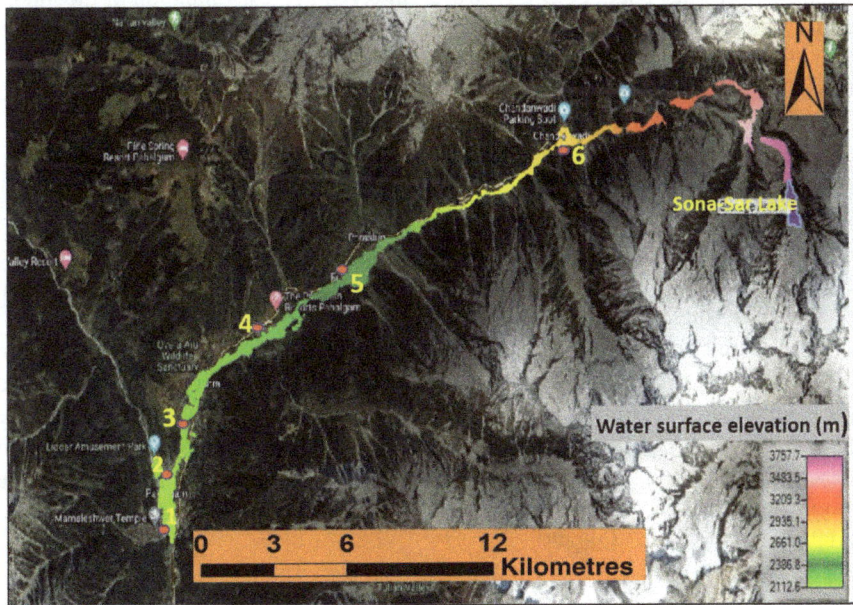

FIGURE 3.11   Spatial distribution of water surface elevation.

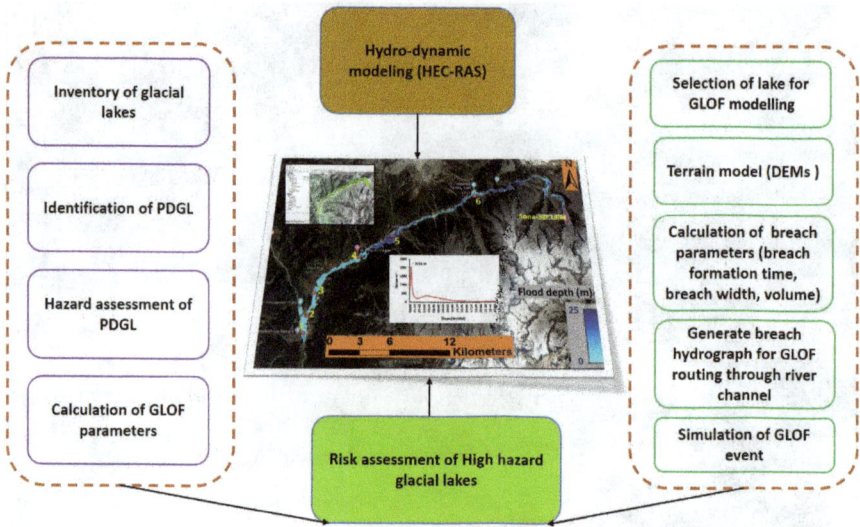

FIGURE 3.12   A comprehensive framework for GLOF hazard and risk assessment.

accuracy of model results. Furthermore, in several past studies model results have been validated using data from the previous GLOF events (Majeed et al., 2021; Sattar et al., 2020; Klimeš et al., 2021). However, in certain cases, validation of the model results was not done because of no previous history of GLOF events (Gilany and Iqbal, 2020; Hazra and Krishna, 2022; Rawat et al., 2022; Taloor et al., 2022b). This is also the case with our study where no previous GLOF event has been witnessed or reported from the lake or in the region.

In this study, we have modelled clear water simulation, but in reality GLOF can evolve into debris flow. A GLOF carrying debris may result in severe damage, whereas clear water can travel a longer distance and will inundate large areas. However, a clear water flow scenario is simple to use and may produce first-order results that may be helpful to other researchers to learn about the GLOF hazard and risk assessment in detail. Monitoring and mitigation of GLOF events has now become the key focus of GLOF research considering the destructive impacts of GLOFs on the downstream regions. In order to comprehend GLOF propagation, timely and effective hazard and risk assessment mapping is of prime importance. Such types of studies aid in the development of mitigation measures and landuse/urban planning in the downstream areas. The hazard and risk assessment of the glacial lake is a multi-stage process. The overall framework for GLOF hazard and risk assessment is illustrated in Figure 3.12 and the key steps are as follows:

1.   Selection of region/area for GLOF study and precreation of glacial lake database (inventory).
2.   Identification of potentially dangerous glacial lakes.
3.   Hazard assessment of potentially dangerous glacial lakes using a multi-criteria decision-making method.

4. Estimation of GLOF parameters through field survey or remote sensing-based technique.
5. Risk assessment of glacial lakes with high hazard level through hydro-dynamic modelling.

3.7 CONCLUSIONS

The increasing frequency of GLOF events in the Himalayan region due to climate-warming-induced glacier recession has enhanced the threat GLOF risk to mountain communities. In this study, we have evaluated the potential GLOF of Sona-Sar Lake in the Jhelum basin of Kashmir Himalayas which has been categorized as a high-hazard glacial lake by earlier studies. The changes in the lake and its feeding glacier area were also assessed using multi-temporal satellite data. The lake area has shown a substantial increase (56%), whereas the area of the glacier has reduced by 25.6% over a period of 50 years. The worst-case scenario GLOF event for the lake was also modelled considering 75% of the lake volume, 20 m breach width and 30 minutes breach formation time. The routed breach hydrograph generated flood would arrive at the nearest settlement (Chandanwari) located at a distance of 7.12 km within 1 hour and 50 minutes after the breach initiation with a peak discharge of 1366 m^3 s^{-1} and a maximum flood depth of 6.4 m and flood velocity of 6 m/s. At the last selected location (Lidder View Park), at a distance of 23.4 km, it would arrive at this location 3 hours and 15 minutes after the breach started at the lake (moraine dam). The peak flood discharge of 989 m^3 s^{-1} was observed at this location with a flood depth of 2.3 m and flood velocity of 3 m/s. The output in the form of GLOF wave parameters obtained through the HEC-RAS model can be affected by the various uncertainties associated with the several input data parameters such as DEM, breach height, breach formation time, lake volume, etc. Besides the afore-mentioned drawbacks, such kinds of studies produce first-order information about the potential GLOFs that may be helpful to other researchers for learning in detail about GLOF hazard and risk assessment. The difficulties of attempting to predict GLOF hazard and risk were highlighted despite the fact that it is not possible to do so based on currently available knowledge. The use of high-resolution DEMs and accurate breach parameters and bathymetry data of the lake will increase the accuracy of the HEC-RAS model output.

REFERENCES

Aggarwal, A., Jain, S.K., Lohani, A.K., Jain, N. (2016) Glacial lake outburst flood risk assessment using combined approaches of remote sensing, GIS and dam break modelling. *Geomatics, Natural Hazards and Risk* 7(1): 18–36.

Ahmad, S.T., Ahmed, R., Wani, G.F., Sharma, P., Ahmed, P., Mir, R.A., Alam, J.B. (2022). Assessing the status of glaciers in Upper Jhelum Basin of Kashmir Himalayas using multi-temporal satellite data. *Earth Systems and Environment* 6(2): 375–389.

Ahmed, R., Ahmad, S.T., Wani, G.F., Mir, R.A., Ahmed, P., Jain, S.K. (2022) High resolution inventory and hazard assessment of potentially dangerous glacial Lakes in Upper Jhelum Basin, Kashmir Himalaya, India. *Geocarto International* 1–32.

Ahmed, R., Wani, G.F., Ahmad, S.T., Sahana, M., Singh, H., Ahmed, P. (2021) A review of glacial lake expansion and associated glacial lake outburst floods in the Himalayan region. *Earth Systems and Environment* 5(3): 695–708.

Allen, S.K., Schneider, D., Owens, I.F. (2009) First approaches towards modelling glacial hazards in the Mount Cook region of New Zealand's Southern Alps. *Natural Hazards and Earth System Sciences* 9(2): 481–499.

Allen, S.K., Zhang, G., Wang, W., Yao, T., Bolch, T. (2019) Potentially dangerous glacial lakes across the Tibetan Plateau revealed using a large-scale automated assessment approach. *Science Bulletin* 64(7): 435–445.

Begam, S., Sen, D. (2019) Mapping of moraine dammed glacial lakes and assessment of their areal changes in the central and eastern Himalayas using satellite data. *Journal of Mountain Science* 16(1): 77–94.

Benn, D.I., Warren, C.R., Mottram, R.H. (2007) Calving processes and the dynamics of calving glaciers. *Earth-Science Reviews* 82(3–4): 143–179.

Brunner, G. W. (2002) Hec-ras (river analysis system). In: *North American Water and Environment Congress & Destructive Water*, pp. 3782–3787. ASCE.

Carling, P.A., Knaapen, M., Borodavko, P., Herget, J., Koptev, I., Huggenberger, P., Parnachev, S. (2011) Palaeoshorelines of glacial Lake Kuray–Chuja, south-central Siberia: form, sediments and process. *Geological Society, London, Special Publications* 354(1): 111–128.

Carrivick, J.L., Tweed, F.S. (2013) Proglacial lakes: character, behaviour and geological importance. *Quaternary Science Reviews* 78: 34–52.

Chow, V.T. (1959) *Open-Channel Hydraulics*; McGraw-Hill: New York, NY, USA; p. 680.

Dubey, S., Goyal, M.K. (2020). Glacial lake outburst flood hazard, downstream impact, and risk over the Indian Himalayas. *Water Resources Research* 56(4); e2019WR026533.

Froehlich, D.C. (1995) Peak outflow from breached embankment dam. *Journal of Water Resources Planning and Management* 121(1): 90–97.

Froehlich, D.C. (2008). Embankment dam breach parameters and their uncertainties. *Journal of Hydraulic Engineering* 134(12): 1708–1721.

Fujita, K., Sakai, A., Takenaka, S., Nuimura, T., Surazakov, A.B., Sawagaki, T., Yamanokuchi, T. (2013) Potential flood volume of Himalayan glacial lakes. *Natural Hazards and Earth System Sciences* 13(7): 1827–1839.

Gilany, N., Iqbal, J. (2020). Geospatial analysis and simulation of glacial lake outburst flood hazard in Shyok Basin of Pakistan. *Environmental Earth Sciences* 79(6): 139.

Gupta, A., Maheshwari, R., Guru, N., Sweta, Rao, B.S., Raju, P.V., Rao, V.V. (2021) Quantitative prioritization of potentially critical glacial Lakes in the Indus River basin using satellite derived parameters. *Geocarto International* 1–23.

Hazra, P., Krishna, A. P. (2022) Assessment of proglacial lakes in Sikkim Himalaya, India for glacial lake outburst flood (GLOF) risk analysis using HEC-RAS and geospatial techniques. *Journal of the Geological Society of India* 98(3): 344–352.

Huggel, C., Kääb, A., Haeberli, W., Teysseire, P., Paul, F. (2002) Remote sensing-based assessment of hazards from glacier lake outbursts: a case study in the Swiss Alps. *Canadian Geotechnical Journal* 39(2): 316–330.

Hunt, K.M., Turner, A.G., Shaffrey, L.C. (2020) The impacts of climate change on the winter water cycle of the western Himalaya. *Climate Dynamics* 55(7): 2287–2307.

Hussain, A., Nasab, N., Bano, D., Karim, D., Anwar, W., Hussain, K., Uddin, N. (2020) Glacier lake outburst flood modeling of Khurdopin glacier lake using HEC-RAS and GIS. In: *Селевые потоки: катастрофы, риск, прогноз, защита*, pp. 208–220.

Islam, N., Patel, P.P. (2022) Inventory and GLOF hazard assessment of glacial lakes in the Sikkim Himalayas, India. *Geocarto International* 37(13): 3840–3876.

Ives, J.D., Shrestha, R.B., Mool, P.K. (2010). *Formation of Glacial Lakes in the Hindu Kush-Himalayas and GLOF Risk Assessment*, pp. 10–11. Kathmandu: ICIMOD.

Jain, S.K., Mir, R.A. (2019) Glacier and glacial lake classification for change detection studies using satellite data: a case study from Baspa basin, western Himalaya. *Geocarto International* 34(4): 391–414.

Khadka, N., Zhang, G., Thakuri, S. (2018) Glacial lakes in the Nepal Himalaya: Inventory and decadal dynamics (1977–2017). *Remote Sensing* 10(12): 1913

Klimeš, J., Benešová, M., Vilímek, V., Bouška, P., Cochachin Rapre, A. (2014) The reconstruction of a glacial lake outburst flood using HEC-RAS and its significance for future hazard assessments: an example from Lake 513 in the Cordillera Blanca, Peru. *Natural Hazards* 71(3): 1617–1638.

Klimeš, J., Novotný, J., Rapre, A.C., Balek, J., Zahradníček, P., Strozzi, T., et al. (2021) Paraglacial rock slope stability under changing environmental conditions, Safuna Lakes, Cordillera Blanca Peru. *Frontiers in Earth Science* 9: 142

Kulkarni, A., Patwardhan, S., Kumar, K.K., Ashok, K., Krishnan, R. (2013) Projected climate change in the Hindu Kush–Himalayan region by using the high-resolution regional climate model PRECIS. *Mountain Research and Development* 33(2): 142–151.

Majeed, U., Rashid, I., Sattar, A., Allen, S., Stoffel, M., Nüsser, M., Schmidt, S. (2021) Recession of Gya Glacier and the 2014 glacial lake outburst flood in the Trans-Himalayan region of Ladakh, India. *Science of the Total Environment* 756: 144008.

Mal, S., Allen, S.K., Frey, H., Huggel, C., Dimri, A.P. (2021) Sectorwise assessment of glacial lake outburst flood danger in the Indian Himalayan region. *Mountain Research and Development* 41(1): R1.

Maskey, S., Kayastha, R.B., Kayastha, R. (2020) Glacial lakes outburst floods (GLOFs) modelling of Thulagi and lower Barun glacial lakes of Nepalese Himalaya. *Progress in Disaster Science* 7: 100106.

Mergili, M., Schneider, J.F. (2011) Regional-scale analysis of lake outburst hazards in the southwestern Pamir, Tajikistan, based on remote sensing and GIS. *Natural Hazards and Earth System Sciences* 11(5): 1447–1462.

Mir, R.A. (2018) Recent changes of two parts of Kolahoi Glacier and its controlling factors in Kashmir basin, western Himalaya. *Remote Sensing Applications: Society and Environment* 11: 265–281.

Mir, R.A. (2021) Remote sensing based assessment of glacier resources in parts of Ladakh Mountain Range, a Trans-Himalayan Region. In: *Water, Cryosphere, and Climate Change in the Himalayas: A Geospatial Approach*, pp. 85–100. Cham: Springer International Publishing.

Negi, H.S., Kumar, A., Kanda, N., Thakur, N.K., Singh, K.K. (2021) Status of glaciers and climate change of East Karakoram in early twenty-first century. *Science of the Total Environment* 753: 141914.

Nie, Y., Liu, Q., Wang, J., Zhang, Y., Sheng, Y., Liu, S. (2018) An inventory of historical glacial lake outburst floods in the Himalayas based on remote sensing observations and geomorphological analysis. *Geomorphology* 308: 91–106

Nie, Y., Sheng, Y., Liu, Q., Liu, L., Liu, S., Zhang, Y., Song, C. (2017) A regional-scale assessment of Himalayan glacial lake changes using satellite observations from 1990 to 2015. *Remote Sensing of Environment* 189: 1–13.

Pandey, P., Ali, S.N., Champati Ray, P.K. (2021) Glacier-glacial lake interactions and glacial lake development in the central Himalaya, India (1994–2017). *Journal of Earth Science* 32(6): 1563–1574.

Petrakov, D.A., Tutubalina, O.V., Aleinikov, A.A., Chernomorets, S.S., Evans, S.G., Kidyaeva, V.M., ... Seynova, I.B. (2012) Monitoring of Bashkara Glacier lakes (Central Caucasus, Russia) and modelling of their potential outburst. *Natural Hazards* 61(3): 1293–1316.

Prakash, C., Nagarajan, R. (2018) Glacial lake changes and outburst flood hazard in Chandra basin, North-Western Indian Himalayas. *Geomatics, Natural Hazards and Risk* 9(1): 337–355

Qazi, N.Q., Jain, S.K., Thayyen, R.J., Patil, P.R., Singh, M.K. (2020) Hydrology of the Himalayas. In: *Himalayan Weather and Climate and their Impact on the Environment*, pp. 419–450. Springer, Cham.

Qi, M., Liu, S., Wu, K., Zhu, Y., Xie, F., Jin, H., ... Yao, X. (2022) Improving the accuracy of glacial lake volume estimation: a case study in the Poiqu basin, central Himalayas. *Journal of Hydrology* 127973.

Rawat, M., Pandey, A., Gupta, P.K. (2022) Geospatial analysis of glacial lake outburst flood (GLOF). In: *Geospatial Technologies for Resources Planning and Management*, pp. 141–160. Springer, Cham.

Romshoo, S.A., Fayaz, M., Meraj, G., Bahuguna, I.M. (2020) Satellite-observed glacier recession in the Kashmir Himalaya, India, from 1980 to 2018. *Environmental Monitoring and Assessment* 192(9): 1–17.

Sattar, A., Haritashya, U.K., Kargel, J.S., Leonard, G.J., Shugar, D.H., Chase, D.V. (2021) Modeling lake outburst and downstream hazard assessment of the Lower Barun Glacial Lake, Nepal Himalaya. *Journal of Hydrology* 598: 126208.

Sattar, A., Goswami, A., Kulkarni, A.V., Emmer, A. (2020) Lake evolution, hydrodynamic outburst flood modeling and sensitivity analysis in the Central Himalaya: A case study. *Water* 12: 237.

Shugar, D.H., Burr, A., Haritashya, U.K., Kargel, J.S., Watson, C.S., Kennedy, M.C., ... Strattman, K. (2020) Rapid worldwide growth of glacial lakes since 1990. *Nature Climate Change* 10(10): 939–945.

Singh, P., Bengtsson, L. (2004) Hydrological sensitivity of a large Himalayan basin to climate change. *Hydrological Processes* 18(13): 2363–2385.

Singh, P., Kumar, N. (1997). Impact assessment of climate change on the hydrological response of a snow and glacier melt runoff dominated Himalayan River. *Journal of Hydrology* 193(1–4): 316–350.

Singh, S., Kumar, R., Bhardwaj, A., Sam, L., Shekhar, M., Singh, A., Gupta, A. (2016) Changing climate and glacio-hydrology in Indian Himalayan Region: a review. *Wiley Interdisciplinary Reviews: Climate Change* 7(3): 393–410.

Taloor, A.K., Kothyari, G.C., Manhas, D.S., Bisht, H., Mehta, P., Sharma, M., Mahajan, S., Roy, S., Singh, A.K., Ali, S., (2021) Spatio-temporal changes in the Machoi glacier Zanskar Himalaya India using geospatial technology. *Quaternary Science Advances* 4: 100031.

Taloor, A.K., Thakur, P.K., Jakariya, M. (2022a) Remote sensing and GIS applications in water science. *Groundwater for Sustainable Development* 100817.

Taloor, A.K., Thapliyal, A., Kimothi, S., Kothyari, G.C., Gupta, S. (2022b) Geospatial technology-based monitoring of HAGL in the context of flash flood: A case study of Rishi Ganga Basin, India. *Geosystems and Geoenvironment* 1(3): 100049.

Thapliyal, A., Kimothi, S., Taloor, A.K., Bisht, M.P.S., Mehta, P. Kothyari, G.C. (2022). Glacier retreat analysis in the context of climate change impact over the Satopanth (SPG) and Bhagirathi-Kharak (BKG) glaciers in the Mana Basin of the Central Himalaya, India: A geospatial approach. Geosystems and Geoenvironment 100128.

Truffer, M., Motyka, R.J. (2016) Where glaciers meet water: Subaqueous melt and its relevance to glaciers in various settings. *Reviews of Geophysics* 54(1): 220–239.

Wang, S., Yang, Y., Gong, W., Che, Y., Ma, X., Xie, J. (2021) Reason analysis of the Jiwenco glacial lake outburst flood (GLOF) and potential hazard on the Qinghai-Tibetan Plateau. *Remote Sensing* 13(16): 3114.

Wang, W., Xiang, Y., Gao, Y., Lu, A., Yao, T. (2015) Rapid expansion of glacial lakes caused by climate and glacier retreat in the Central Himalayas. *Hydrological Processes* 29(6): 859–874.

Watson, C.S., Carrivick, J., Quincey, D. (2015) An improved method to represent DEM uncertainty in glacial lake outburst flood propagation using stochastic simulations. *Journal of Hydrology* 529: 1373–1389.

Westoby, M.J., Glasser, N.F., Brasington, J., Hambrey, M.J., Quincey, D.J., Reynolds, J.M. (2014) Modelling outburst floods from moraine-dammed glacial lakes. *Earth-Science Reviews* 134: 137–159.

Worni, R., Huggel, C., Stoffel, M. (2013) Glacial lakes in the Indian Himalayas—From an area-wide glacial lake inventory to on-site and modeling-based risk assessment of critical glacial lakes. *Science of the Total Environment* 468: S71–S84.

Zhang, G., Zheng, G., Gao, Y., Xiang, Y., Lei, Y., Li, J. (2017) Automated water classification in the Tibetan plateau using Chinese GF-1 WFV data. *Photogrammatic Engineering and Remote Sensing* 83(7): 509–519.

Zhang, G., Yao, T., Xie, H., Wang, W., Yang, W. (2015) An inventory of glacial lakes in the Third Pole region and their changes in response to global warming. *Global and Planetary Change* 131: 148–157.

Zheng, G., Allen, S. K., Bao, A., Ballesteros-Cánovas, J. A., Huss, M., Zhang, G., ... Stoffel, M. (2021) Increasing risk of glacial lake outburst floods from future Third Pole deglaciation. *Nature Climate Change* 11(5): 41–417.

4 Hindukush Karakoram Himalayan Glaciers
A Cryospheric Asset

Anil V. Kulkarni, Roja Asharaf, and Ashim Sattar

4.1 INTRODUCTION

The third pole cryosphere, Hindukush Karakoram Himalaya (HKH), is a part of High Mountain Asia (HMA), known for having the greatest ice and snow reservoir outside the Arctic and Antarctic regions[1,2]. The "Water tower of Asia", as it is described, is due to the vast network of rivers including ten major river basins that support fresh water supply, food, energy, ecosystems, and livelihoods to ~1.9 billion people[3-5]. In the three major river basins in HKH, namely Ganga, Brahmaputra and Indus, glacier-fed rivers have great socio-economic importance, where over 800 million people[6] are dependent on the freshwater resources in the basins[7]. The meltwaters from the glaciers and snow, groundwater and rainfall-dominated runoff are among the major sources of freshwater in these basins[6,8]. The region has the world's largest irrigated area and highest installed hydropower capacity[8]. Glaciers are sensitive to changes in temperature and precipitation, and therefore are considered to be visible indicators of climate change[3]. Since the 1980s, climate change has significantly affected these glaciers, disturbing their water resource equilibrium and leading to widespread glacier mass loss[9]. With time, the impact of climate change on glaciers, snow and monsoons will subsequently alter the freshwater supply over these glacierized basins. This will, in turn, affect the runoff characteristics including annual and seasonal discharge[9]. Further, it will also influence the water management mechanisms, water governance levels and transboundary concerns[10]. Previous studies have reported that the glaciers over the HKH region are continuously shrinking[11]. These reported retreat and thinning rates have questioned the sustainability of the water supply, as they show considerable variability on temporal and spatial scales. Furthermore, glacier recession is causing sea level rise and frequent flooding events, and may result in a higher frequency of these events in the future,[11] thereby potentially altering the overall hazard situation. Thus, the distribution of HKH glaciers and the variability in glacier characteristics have received much attention from the scientific community. However, a robust relationship describing the glacier's sensitivity to climatic parameters like temperature and precipitation is still poorly constrained. Hence, studying the HKH glaciers has become a significant research area to focus on.

DOI: 10.1201/9781003364115-4

This study focuses on the three major basins in the HKH including the Indus, Ganga and Brahmaputra (Figure 4.1), spreading across six major mountain ranges in countries including China, Afghanistan, Pakistan, India, Nepal, Tibet and Bhutan. The region consists of a total of 39,822 glaciers covering an area of 46,312.07 km^2, accounting for ~2% of the total basin area. Most of these glaciers are maritime or temperate glaciers[11]. Some of the crucial parameters controlling the regional glacier dynamics are heterogeneity in the climatic conditions and topography[5,12]. The mountain ranges of the region, due to their location and high altitude, act as an orographic barrier to southerly and westerly flows, resulting in complex climatic conditions over the HKH region[10]. Two circulation systems influence the region's climate, namely the western disturbances and the Indian summer monsoons. The western disturbances influence the Karakoram and Western Himalayan region leading to winter precipitation, while the Central Himalayas and Eastern Himalayas are influenced by the Indian summer monsoon[3,6,10]. The Eastern Himalaya receives the most extensive annual precipitation of about 2800 mm, of which 89% is in the form of rainfall. In comparison, the Karakoram region acquires 81% of its annual precipitation (584 mm) in the form of snowfall in the winter season[6,10]. The basins can be divided into upper and lower parts based on altitude and varied hydrological processes. The region lying above

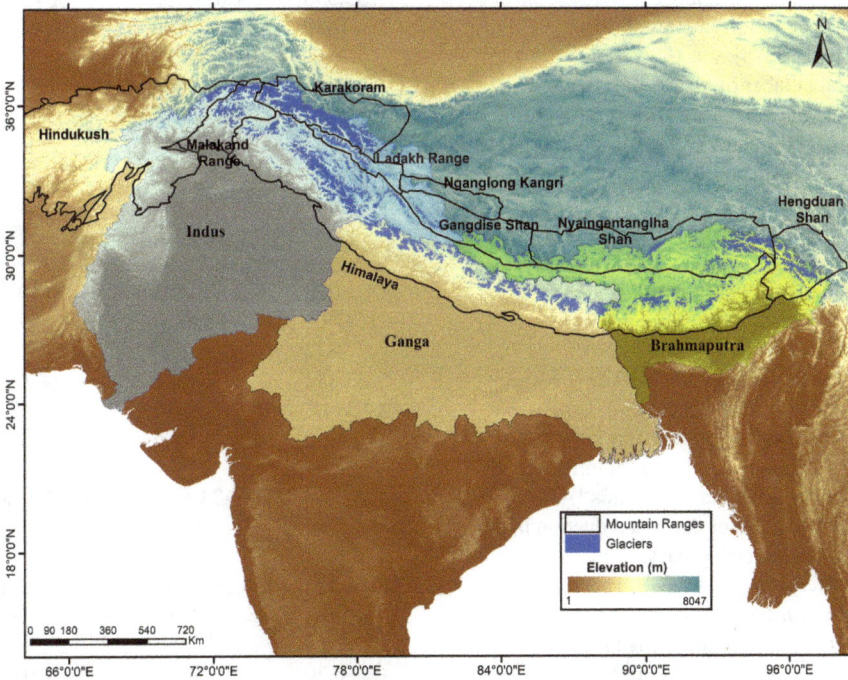

FIGURE 4.1 Map of the HKH region showing the Indus, Ganga and Brahmaputra basins; the glacier boundaries are acquired from RGI (V 5.0) datasets and the outlines of the mountain ranges from the GMBA (Global Mountain Biodiversity Assessment) mountain inventory v1.2 are used; background: ASTER Digital Elevation Model (DEM).

an elevation of 2000 m above sea level (a.s.l.) is considered the upper basin, and below 2000 m a.s.l. is considered to be the lower basin[6,13]. The upper basin areas consist of glaciers and snow, and the lower basin marks the major population, infrastructure and socio-economic activities[6]. The characteristics of these three basins are represented in Table 4.1. The Indus and Brahmaputra basins are marked by extensive upper basin areas of 33% and 65.2%, respectively, compared to the Ganga Basin (12.0%)[6,13]. The Indus basin has the largest glaciated area (26,548.5 km²) followed by the Brahmaputra (11,391.14 km²) and Ganga basins (8372.46 km²). The Ganges is the most densely populated basin, with approximately 637 million people[6,10,13]. The Ganga and Brahmaputra basins are wetter than the Indus basin, while the Indus and Ganga basins support large-scale irrigation systems and have the highest net irrigation water demand[13]. The contribution of snow and glacier melt to the Indus River is highest (44.7%), followed by the Brahmaputra (15%) and Ganga (13.4%)[6]. These

TABLE 4.1
Main Features and Characteristics of the Indus, Ganga and Brahmaputra Basins

Characteristics	Indus	Ganga	Brahmaputra
Total basin Area (km²)	1,155,550	1,079,160	564,389
Mean elevation (metres above sea level) [6]	1699	789	3137
Total number of glaciers	21,724	6606	11,492
Total glacierized area (km²)	26,548.47	8372.46	11,391.14
Upper basin area (%) [6]	33	12	65.2
Lower basin area (%) [6]	67	88	34.8
Mean snow cover area (%) [6]	15.1	4.8	20.3
Average annual runoff (upper basin) (mm/annum) [6]	577	1293	1575
Annual precipitation (mm/annum) [6]	423	1035	1071
Irrigation area (km²) [6]	144,900	15,6300	5989
Total population (in millions) [6]	312	637	71
Hydropower potential (MW) [6]	33,832	20,711	66,065
Installed hydropower (MW) [6]	14,294.3	5317.2	3974
Population density (persons/km²) [10]	279	586	131
Population impact index (in millions) [10]	31	2	1
Annual excess discharge (m³s⁻¹) [10] (additional water due to a reduction in the water stored by glaciers)	−125 ± 63	−24 ± 24	−163 ± 66
Glacier stored water (Gt)	2772.8 ± 1226	528.39 ± 212	701.71 ± 276
Glacier mass loss [6] (Gt/year)	3.5 ± 1	3.2 ± 0.6	4.9 ± 1
Average retreat rate [6] (m/annum)	2.7	15.5	20.2
Area loss rate (%/annum) [6]	−0.30	−0.33	−0.66
Snow and glacier melt contribution on runoff (%) [6]	44.7	13.4	15
Net irrigation water demand (mm)[13]	908	716	480

findings represent the heterogeneity of the basins subjected to changes in the future due to the ongoing climate change[14].

The HKH region is portrayed as a climate change hotspot, as these regions are experiencing greater warming than the global average[9,15]. This higher temperature anomaly is due to the elevation-dependent warming phenomenon, which intensifies the rate of change in mountain ecosystems and cryospheric systems[16]. An average temperature rise of 0.44°C per decade was observed over the HMA region from 1979 to 2020, whereas the rise in global average temperature was recorded as 0.19°C per decade[9]. The glaciers have experienced considerable mass loss and retreat during the past few decades due to climate change. It is presumed from the glacier feedback that future climate change is likely to alter the HKH river runoffs. With the growing population and development in the HKH, there has been an increase in the overall water demand. Approximately 129 million farmers in these basins are dependent on glacier melt water for their livelihoods and irrigation purposes. The Indus river contributes to 96% and 26% of Pakistan's and India's crop production, respectively[6]. The Ganga river supports about 37% of India's agriculture[6]. The emerging water stress and climate sensitivity due to the growing population, irrigation and socio-economic developments are making these river basins more vulnerable to climate change. The imbalance in these water towers could lead to serious problems in future irrigation needs, and also transboundary conflicts and threats[9]. Thus, the current knowledge of hydrological, ecological and socio-economic factors needs thorough evaluation from time to time to understand this imbalance. Hence, a complete understanding of HKH glaciers and the different hydrological processes is required to implement better management of water resources.

This chapter reviews the status of glaciers in the major HKH basins including the Indus, Ganga and Brahmaputra. Here we rely on the available long-term in situ and satellite observations along with model-based estimates. The study presents a complete database of glacier inventory, ice thickness, glacier stored water and mass balance. Furthermore, the study also emphasizes recent advancements in field-based mass balance and ice thickness measurements. Also, future projections in glacier area and mass and the response to climate change are considered. In addition, the chapter discusses the research gaps and future directions for HKH glacier research.

4.2 GLACIER INVENTORY

Remote sensing techniques and GIS have been extensively used in glacier studies and mappings[17-21]. A glacier inventory is a collection of mapped glacier outlines and is a vital dataset in glacier studies, especially in regions like the HKH where glaciers are located in remote inaccessible locations. Glacier inventories are important in not only understanding the evolution of glaciers from time to time but also evaluating aspects of freshwater sustainability (for irrigation and livelihood), potential catastrophes, estimating ice volumes, the contribution of glacier melting to sea-level rise, and the functioning of hydroelectric stations[22]. Moreover, it also helps evaluate future climate impacts based on records of the past glacier inventories[22-24]. The key principle of glacier delineation is the identification of glacier features. These include accumulation

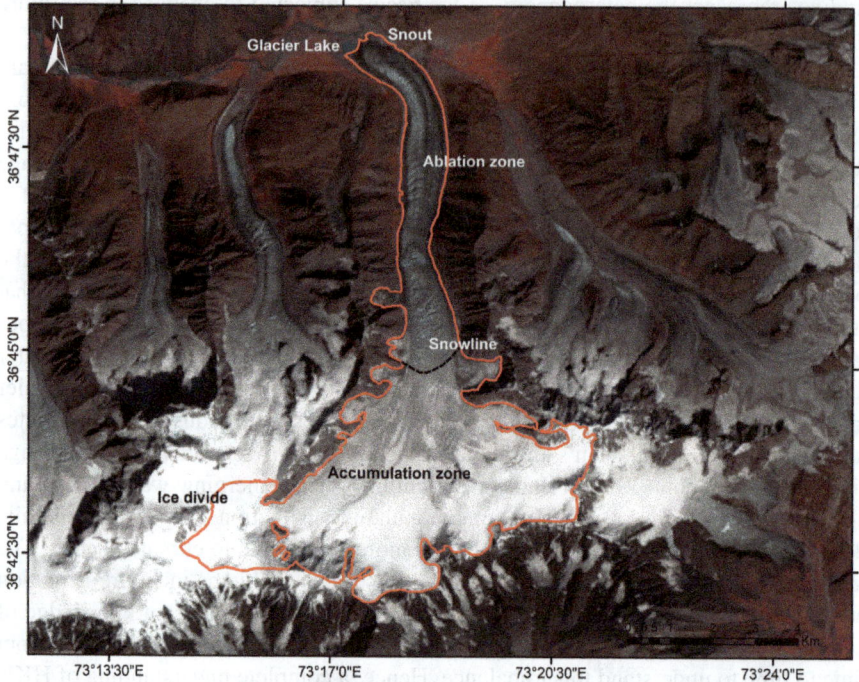

FIGURE 4.2 Mapped glacier boundary of a Kabul glacier with glacier features; background: Landsat 8 composite image acquired on 9 September 2017.

zones, ablation zones, snow or equilibrium lines, crevasses, snouts, moraines and ice divides. Figure 4.2 shows the different glacier features in a mapped glacier.

Earlier, the Himalayan glaciers were mapped using manual methods with the help of toposheets, field expeditions and aerial photographs[25]. These methods have great limitations due to the harsh climatic conditions and complex terrain conditions. The Space Application Centre (SAC, ISRO), for the first time in 1991, estimated the glacier areal extent of Indian Himalayan glaciers to be 23,314 km² using satellite images from the year 1987[25,26]. Further, the GSI (Geological Survey of India, 2009) reports the total glaciated area of the Indian Himalayas to be 26,775 km² using toposheets and aerial photographs captured in 1962–1963[25,27]. The International Centre for Integrated Mountain Development (ICIMOD) (2011) mapped the glaciers of the entire HKH region using Landsat images and SRTM DEM (Shuttle Radar Topography Mission) datasets and reported 54,252 glaciers spanning an area of 60,054 km²[28]. Further, SAC ISRO (1991) created a glacier inventory for the major basins, Indus, Ganga and Brahmaputra, using LISS-III images (Indian Remote Sensing Satellite [IRSS]) and mapped 32,392 glaciers covering an area of 71,182 sq. km[1,6,29,30]. The Randolph Glacier Inventory (RGI) (2012) is a freely available global database of glacier outlines which was developed after the Fifth Assessment Report of the Intergovernmental Panel on Climate Change (IPCC AR5)[31,32]. Here we evaluate the glacier distribution in the Indus, Ganga and Brahamaputra basins based on the

RGI glacier datasets. These basins consist of a total of 39,822 glaciers, spanning a total area of 46,312.07 km^2. Furthermore, the "Glacier Area Mapping for Discharge from the Asian Mountains" (GAMDAM) inventories (2015) provide a complete glacier inventory for High Mountain Asia, where glaciers are manually delineated using Landsat images, DEM and high-resolution Google Earth images. There are a total of 87,084 glaciers covering an area of 91,263 ± 13,689 km^2, in which a total glaciated area of 41,008 km^2 is seen in the Indus, Ganga and Brahmaputra basins[33]. In the HKH region, the GAMDAM glacial areal extent delineated is comparable to the ICIMOD inventory, with 93% similarity[33]. While it shows a 31% decrease in the glacier-covered area compared to RGI datasets, there is a total glacier-covered area of 54,058 km^2 [33]. The glacier-covered areas in Karakoram, Western Himalaya, Central Himalaya and Eastern Himalaya are reported as 17,385 km^2, 8402 km^2, 8221 km^2 and 2836 km^2, respectively[33]. These estimates report an 11% decrease in glacier area when compared to the inventory of Bolch et al. (2012)[33,34]. Considering the disparities, an updated GAMDAM glacier inventory was developed by considering summertime satellite images, steep snow-covered slopes and shaded areas that mapped a total of 134,770 glaciers over the HMA covering a total area of 100,693 ± 11,790 km^2 [24]. Further, the International Water Management Institute (IWMI) report shows that the HKH region has a glaciated area of 56,416 km^2 [6,35]. Here we compare the seven different glacier inventory datasets including ICIMOD, GAMDAM, IWMI, SAC, RGI 6.0, RGI 5.0 and RGI 4.0 for the Indus, Ganga and Brahmaputra basins. The intercomparison of the different glacier inventory datasets is shown in Figure 4.3. This study shows that the overall glacier-covered area in the Indus, Ganga and Brahmaputra basins vary from 41,008 km^2 to 71,182 km^2 (ICIMOD, GAMDAM, IWMI, SAC, RGI). Furthermore, it reveals that the Indus basin holds the largest number of glaciers and has the largest glacier-covered area. In contrast, the Ganga basin marks the least glaciated area with the lowest number of glaciers. The study also shows that the average glaciated areas obtained from the ensemble of all the inventories of the Indus, Ganga and Brahmaputra basins are 26,253 ± 3746 km^2, 11,110 ± 4038 km^2 and 14,399 ± 3782 km^2, respectively.

The difference in the mapping approaches of the debris-covered glaciers and the non-mapping of glacierets brings significant discrepancies between the SAC and GSI inventories[25]. The areal extents of glaciers obtained from the GAMDAM glacier inventory and ICIMOD inventory are approximately 12–14% and 3–18% less than for the RGI inventories, respectively, over the Indus, Ganga and Brahmaputra basins. Meanwhile the glacial areal extents obtained from SAC and IWMI inventories are approximately 31–54% and 4–22% higher than RGI inventories, respectively. The RGI and GAMDAM inventories have uncertainties of 5% and 15%, respectively, in total glacier areal extent[31,33]. Owing to the uncertainty in the glacier boundaries, these datasets often require manual corrections, for example, the ICIMOD glacier inventory of the Baspa basin (a basin in Himachal Pradesh, Western Himalaya) shows a glaciated area of 163 ± 12 km^2, whereas it is 193 ± 16 km^2 based on RGI 4.0 datasets[25]. The glacier outlines from these inventories are modified manually using satellite images (2013–2014), and the total glacier extent of the Baspa basin is estimated as 194 ± 14 km^2 [25]. Similar basin-wise glacier delineations are carried out in various regions of HKH. The modified inventory of the Bhaga basin (a basin in Himachal

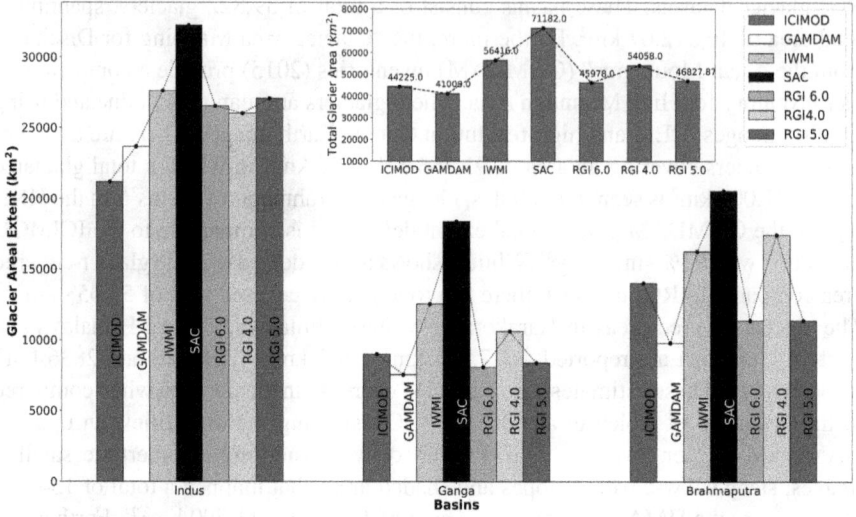

FIGURE 4.3   The intercomparison of the glacier-covered area of HKH basins (Indus, Ganga and Brahmaputra) based on different glacier inventories (ICIMOD, GAMDAM, IWMI, SAC, RGI 6.0, RGI 5.0 and RGI 4.0); the datasets are obtained from Kulkarni et al. (2021)[6]; the inset figure shows the total glacier-covered area for all three basins combined; SAC reports the highest glacier-covered area, and the GAMDAM reports the lowest glacier-covered area.

Pradesh, Western Himalaya) showed 319 glaciers covering an area of 320 ± 16 km^2 [36]. The Spiti basin (also a basin in Himachal Pradesh, Western Himalaya) showed 786 glaciers covering an area of 550.5 km^2 [37]. Furthermore, a study in the Kabul basin estimated 3097 glaciers covering an area of 2078.69 ± 104 km^2 [38]. A study in the Satluj basin (Western Himalaya) showed a total of 2026 glaciers covering an area of 1426 km^2 [39]. A study in the Chandra basin (Western Himalaya) showed 251 glaciers (area > 0.5 km^2) covering an area of 973 ± 70 km^2 [23].

These inventories are used to investigate the glaciers' retreat and area loss. The glaciers of the HKH region are retreating, and this rate has increased over recent decades[6,34,40,41]. The snout fluctuations of about 285 glaciers spread over the Indus, Ganga and Brahmaputra basins are reported by Kulkarni et al. (2021)[6] and this study estimated a mean retreat rate of -14.9 ± 15.1 meters per annum for the period 1842 to 2018. The basin-wise studies over the HKH region reported the retreat rates of the Indus, Ganga and Brahmaputra basins as -12.7 ± 13.2 m/a, -15.5 ± 14.4 m/a and -20.2 ± 19.7 m/a, respectively[6]. Furthermore, in the Karakoram it is observed as -1.37 ± 22.8 m/a, representing a stable or advancing condition[6]. The area change studies over the HKH basin have been analysed using 149 records, and the average rate of area change (1906–2018) was observed as $-0.36 \pm 0.38\%$ a^{-1} [6]. The area shrinkage rates of the Indus, Ganga and Brahmaputra basins for the period 1906 to 2018 were estimated as $-0.30 \pm 0.25\%$ a^{-1}, $-0.33 \pm 0.27\%$ a^{-1} and $-0.66 \pm 0.72\%$ a^{-1}, respectively[6]. In addition, studies over the Himalayan and HMA region reported average area loss rates of -0.36% a^{-1} [40] and -0.4% a^{-1} [42] respectively, for the period 1960–2010.

The higher variabilities in area change and retreat are due to the influence of topography, glacier extent, debris cover and climatic drivers[6].

All glacier datasets vary from each other and have notable uncertainties in their glacier areal extents. These uncertainties arise due to differences in the datasets, mapping methodologies, time period of satellite data acquisition, glacier classifications, scale and manual interpretations used to delineate glacier boundaries[6,23]. These disparities put forth the need to improve the methodological approaches and for regular updates of the glacier outlines.

4.3   GLACIER VOLUME

Detailed and precise information on the glacier volume is an essential prerequisite for understanding glaciohydrology[8]. Glacier ice thickness is a crucial parameter that aids in estimating glacier volume or stored water[43]. The knowledge about accurate glacier volume in the Himalaya is very limited due to inaccessible and difficult terrain[44]. Furthermore, the region's debris cover, crevasses and harsh climate limit field studies[6]. Radio echo sounding, ground-penetrating radar (GPR) and borehole measurements are the conventional field measurements used to estimate glacier ice thickness[36,45-47]. Currently, to our knowledge, field measurements of about 21 glaciers are available in the HKH (Table 4.2). Alternatively, empirical and numerical modelling approaches have been developed for the estimation of ice thickness and total glacier volume. This includes the volume–area and volume–length scaling method[45,48,49], models based on basal shear stress and perfect plasticity[50-52], slope-dependent approaches[53-56], laminar flow[57,58], flow dynamics, and mass conservation[59-61]. The HIGTHIM[36,62], GlabTop (Glacier bed Topography), GlabTop2[54,63,64] and HF models[65] are among the available established numerical/physical models to estimate glacier ice thickness.

The ice volume estimate of the whole HKH region has been computed by Farinotti et al. (2019)[46] using an ensemble of ice thickness models. The results showed that Central Asia, South Asia West and South Asia East (RGI subregions) hold $3.27 \pm 0.85 \times 10^3$ km^3, $2.87 \pm 0.74 \times 10^3$ km^3 and $0.88 \pm 0.23 \times 10^3$ km^3 ice volume, respectively. Further, Millan et al. (2022)[80] used an ice velocity and slope-based method to estimate the volume of the whole Himalayan region, which is calculated to be $9.6 \pm 3.7 \times 10^3$ km^3. The glacier volume obtained from physical-based models can be further used to develop regional volume–area (V-A) scaling equations, which can contribute to the estimation of basin-wide glacier-stored water. The various existing scaling methods and the regionally developed scaling equations are listed in Table 4.3. In the present study, the glacier-stored mass of the Indus, Ganga and Brahmaputra basins using different V-A scaling methods is calculated (Table 4.3; Figure 4.4). This analysis shows that the whole HKH region's glacier-stored mass ranges from 2377 Gt to 6715.96 Gt. The plot indicates that the Arendt et al. (2006)[81] scaling equations estimate the highest values and the Cuffey and Paterson (2010)[53] equation estimates the lowest stored mass. The combined average from the equations shows that the Indus basin holds the largest stored glacier mass of 2772.77 ± 1226 Gt, and the lowest is in the Ganga basin of 528.39 ± 212 Gt. The Brahmaputra basin holds a total glacier-stored mass of 701.71 ± 276 Gt. Further, Farinotti et al. (2019)[46] estimated the ice volume of the Indus, Ganga and Brahmaputra basins as 2327 km^3, 473 km^3 and 622 km^3,

TABLE 4.2
List of Field Measurements on Glacier Ice Thickness Conducted in the Study Area

Glacier	Locations	Time of investigation	Measured glacier ice thickness (m)	References
Barpu glacier	Upper Indus Basin	2015–2018	55.29 ± 245.46	Zou et al. (2021)[66]
Batura glacier	Upper Indus Basin	2015–2018	40.91 ± 201.52	Zou et al. (2021)[66]
Changri Nup glacier	Nepal, Central Himalaya (27.964 N, 86.792 E)	2011	150	Vincent et al. (2016)[67]
Chhungphar glacier	Upper Indus Basin	2015–2018	54.45 ± 150.81	Zou et al. (2021)[66]
Chhota Shigri glacier	Western Himalaya (32.231 N, 77.515 E)	2009	124–270	Azam et al. (2012)[68]
		2010	120–150	Singh et al. (2012)[69]
Dokriani glacier	Bhagirathi River Basin (30.852 N, 78.823 E)	1995	15–120	Gergan et al. (1999)[70]
Gharko glacier	Upper Indus Basin	2015–2018	84.89 ± 230.25	Zou et al. (2021)[66]
Hamtah glacier	Chandra Basin (32.241 N, 77.369 E)	–	35–95	Swain et al. (2018)[71]
Kangwure glacier	Central Himalaya (28.468 N, 85.814 E)	2008	47.3–65.7	Ma et al. (2010)[72]
Lirung glacier	Nepal	2015	100	Pritchard et al. (2020)[73]
Langtang glacier	Nepal	2015	155	Pritchard et al. (2020)[73]
Menthose glacier	Lahaul region, North-Western Himalaya	2016, 2017	24–55	Prakash et al. (2021)[74]
Mera glacier	Central Himalaya (27.725 N, 86.887 E)	2009 and 2012	55–130	Wagnon et al. (2013)[75]
Ngozumpa glacier	Nepal	2016	270	Pritchard et al. (2020)[73]
Parang glacier	Lahaul and Spiti (32.453 N, 78.054 E)	–	40–140	Swain et al. (2018)[71]

Glacier	Basin	Year	Value	Reference
Pasu glacier	Upper Indus Basin	2015–2018	96.72 ± 206.59	Zou et al. (2021)[66]
Patseo glacier	Bhaga Basin (32.761 N, 77.332 E)	2004	40	Singh et al. (2010)[76]
		2013	38.3	Singh et al. (2018)[77]
Sachen glacier	Upper Indus Basin	2015–2018	74.74 ± 228.56	Zou et al. (2021)[66]
Samudra Tapu glacier	Chenab Basin (32.505 N, 77.455 E)	2004	Values could not be estimated due to instrument limitations	Singh et al. (2010)[76]
Satopanth glacier	Alaknanda Basin (30.753 N, 79.362 E)	2016	38 to 112	Mishra et al. (2018)[78]
Yala glacier	Langtang River Basin (28.238 N, 85.619 E)	2009	36	Sugiyama et al. (2013)[79]

TABLE 4.3
Details of the Existing Scaling Equations and the Regionally Developed V-A Scaling Equations of HKH Region

Sl. no.	Scaling equations	References
1.	$V (m^3) = 0.191 * (A (m^2))^{1.375}$	Bahr (1997)[45]
2.	$V (m^3) = 0.2055 * (A(m^2))^{1.35}$	Chen and Ohmura (1990)[49]
3.	$V (m^3) = 0.28 * (A (m^2))^{1.375}$	Arendt et al. (2006)[81]
4.	$D (m) = 100,000/ (917 * 9.8 * slope)$	Cuffey and Paterson (2010)[53]

Regional scaling equations developed

Sl. no.	Scaling equations	Study region	References
5.	$V (km^3) = 0.0289*(A (km^2))^{1.3285}$	55 glaciers of Sikkim Himalaya (Brahmaputra)	Srinivasalu et al. (2018)[18]
6.	$V (km^3) = 0.0323*(A (km^2))^{1.289}$	298 glaciers of Satluj and Beas (Indus)	Prasad et al. (2019)[39]
7.	$V (km^3) = 0.030*(A (km^2))^{1.34}$	15 glaciers of Dhauliganga basin (Ganga)	Sattar et al. (2019)[82]
8.	$V (km^3) = 0.033* (A(km^2))^{1.29}$	59 glaciers of Bhaga basin & 298 glaciers of Satluj and Beas (Indus)	Gopika et al (2020)[36]
9.	$V (km^3) = 0.0344* (A(km^2))^{1.2579}$	290 glaciers of Kabul Basin (Indus)	Asharaf et al. (under review)[38]
10.	$V (km^3) = 0.0335* (A(km^2))^{1.3079}$	223 glaciers of Chenab Basin (Indus)	Gopika et al (under review)[85]

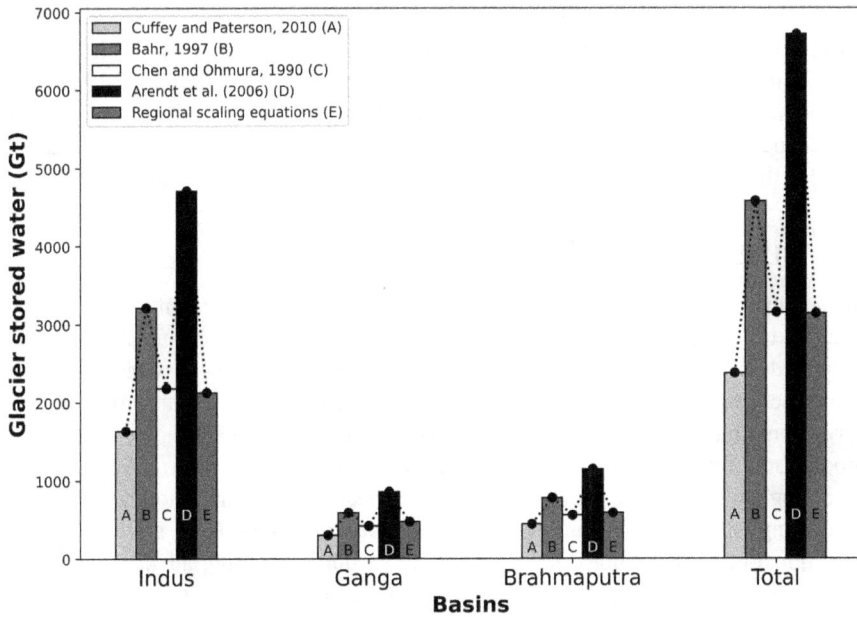

FIGURE 4.4 Glacier stored mass in the Indus, Ganga and Brahmaputra basins estimated using different scaling equations; Cuffey and Paterson (2010)[53] estimates are obtained from Kulkarni et al. (2021)[6].

respectively[8]. In addition, the results indicate that the regionally generated scaling equations aid in attaining better and more reliable ice volume estimates at a basinal extent compared to globally generated equations which may result in over- or under-estimation[6,82]. As the area-dependent scaling method is region-specific, fine-tuning in scaling parameters is required for a better volume–area relationship for the given study region[82]. Among various well-performing models, the laminar flow method integrated with the volume–area scaling method gave reliable estimates of the glacier-stored water when applied in various regions in the HKH. The laminar flow model when combined with V-A scaling (for smaller glaciers) estimated the glacier stored mass of Satluj basin (Western Himalaya) as 62.3 ± 10.8 Gt[39], Kabul basin as 107.72 ± 20 Gt[38], Bhaga basin (Western Himalaya) as 15 ± 2.8 Gt[36], Parvati basin (Western Himalaya) as 19.17 ± 3.42 Gt[83], Karakoram region as 1473 ± 17 Gt[84], Spiti basin (Western Himalaya) as 21.42 ± 3.86 Gt[37], Chenab basin (Western Himalaya) as 145.61 ± 26.2 Gt[85], Himachal Pradesh as 16 km^3 [22], Dhauliganga basin[82] (Central Himalaya) as 6.5×10^9 m^3, Satluj and Beas basin as 84 ± 15 Gt[86] and Sikkim Himalayan region as 20 ± 3 Gt[87]. The glacier-stored mass of the Alakananda basin (Central Himalaya) is estimated as 30.4 ± 9.4 Gt using the slope-dependent method[55,88]. The spatial process in hydrology model (SPHY) computed the glacier ice volume of the Bhaga basin as 13.41 km^3 in 2000 and 10.99 km^3 in 2018[89]. The Chandra basin (Western Himalaya) recorded a total glacier stored volume of 112.5 +41 km^3 using the glacier bed topography version 2 model[23].

These methods and models, along with remote sensing, help to understand HKH glaciers on a broader scale where field studies become challenging[14]. It is to be noted that the available estimates of glacier volume can vary widely from each other depending upon the parameters, methods and inventories used[8,14]. This brings out the discrepancy in the current glacier volume estimates and thus there is scope to further improve the estimates with new models by exploring the available information.

4.4   GLACIER MASS BALANCE

Glacier mass balance (GMB) is one of the most significant parameters that help in assessing glacier health, sustainability and sensitivity to climate change[14,90]. It is defined as the sum of the accumulation and ablation process in which snowfall, avalanches, snowdrifts, sublimation, calving, etc., act as the contributing factors. Various methods are available to evaluate GMB including glaciological, gravimetric, geodetic, equilibrium line altitude (ELA) and accumulation area ratio (AAR) methods. Glaciological measurements are direct in situ measurements that include traditional ice drilling, stake and pits, and field surveys. The study of GMB in the HKH region using glaciological methods is very challenging due to existing adverse conditions such as high altitude and harsh weather[6,90]. Hence, to our knowledge, only a few glaciological measurements are available over the study area and these have been mostly performed in easily accessible, smaller clean or less debris-covered glaciers[14,40]. The available glaciological measurements are listed in Table 4.4. Most of these glaciers are located in the Indus basin. Among these, the Chhota Shigri glacier in Himachal Pradesh has the most prolonged observations (2001–2019). The number of available mass balance measurements using the glaciological method and their variabilities over the years 1975–2020 are shown in Figure 4.5. A total of 214 measurements are available over the HKH region that are known to us. Among these, 24 measurements recorded a positive mass balance, while the remaining records show a negative mass balance trend. The overall trend shows an increasing mass loss rate since the 2000s, but this requires more records for a precise conclusion. The HKH region shows an average mass balance of –0.61 m.w.e. a^{-1} for the period 1975–2020, which is an average taken from all the available field measurements. These values do not deviate much from the mass balance estimates from Azam et al. (2018)[40] of 0.49 m.w.e. a^{-1}, evaluated for the period 1975–2015 and Vishwakarma et al. (2022)[14] of 0.60 m.w.e. a^{-1}, evaluated for the period 1975–2020. Further, the decadal trend in mass balance shows an overall negative trend varying from –0.36 m.w.e. a^{-1} (1980–1990) to –0.44 m.w.e. a^{-1} (1991–2000) and –0.88 m.w.e. a^{-1} after the 2000s. The increasing number of field measurements helps attain more accurate region-wide mass balance studies in HKH. However, within the available measurements, a continuous mass balance record for a prolonged period was still lacking. Hence, scientists depend on other methods to obtain comprehensive knowledge about the HKH mass budget.

The geodetic method is another widely used method, which uses glacier surface elevation (Digital Elevations Models) from different periods to estimate thickness or volume change. This enables the derivation of mass balance in a spatially distributed manner for a given time[2]. The DEMs are accessible from different remote sensing

TABLE 4.4
List of Field Measurements on Glacier Mass Balance Observed on the HKH Region

Name of glacier	Location/RGI Id	Area (km²)	Observation period	Mean mass balance (m.w.e. a⁻¹)	References
			Indus Basin		
Bara Shigri glacier	RGI60-14.15447	136.8	2017–2019	−0.33	Vishwakarma (2022)[14]
Batal glacier	RGI60-14.16042	4.9	2017–2019	−0.30	Vishwakarma (2022)[14]
Chhota Shigri glacier	32.228 N; 77.514 E	15.7	1987–1989; 2001–2019	−0.43	Dobhal (1995)[91], Azam et al. (2016)[92] Mandal et al. (2020)[93]
Gara glacier	31.463 N; 78.426 E	5.2	1974–1983	−0.27	Raina et al. (1977)[94] Sangewar and Siddiqui (2006)[95]
Gepang Gath glacier	RGI60-14.15623	13.5	2017–2019	−0.77	Vishwakarma (2022)[14]
Gor Garang Glacier	31.436 N; 78.403 E	2.02	1976–1985	−0.38	Sangewar and Siddiqui (2006)[95]
Hamtah glacier	32.238 N; 77.371 E	3.2	2000–2009; 2010–2015	−1.45	GSI (2011)[96], GSI (2016)[6,97]
Hoksar glacier	34° 10–34° 11' N 75° 17'–75° 18' E	1	2013–2018	−0.95 ± 0.39	Romshoo (2022)[90]
Kolahoi glacier	34.165 N; 75.315 E	11.9	1983–1984	−0.26	Kaul (1986)[98]
Machoi glacier	34.2767 N; 75.5274 E	5.7	2011–2014	0.14	Koul et al. (2016)[99]
Naradu glacier	31.290 N; 78.404 E	4.56	2000–2003; 2011–2015	−0.73	Koul and Ganjoo (2010)[100] GSI (2016)[6,97]
Nehnar glacier	34.147 N; 75.525 E	1.25	1975–1984	−0.51	GSI (2001)[6,101]
Patseo glacier	32.73–32.77° N 77.3–77.33° E	2.25	2010–2017	−0.34	Vishwakarma (2022)[14]

(Continued)

TABLE 4.4 (Continued)
List of Field Measurements on Glacier Mass Balance Observed on the HKH Region

Name of glacier	Location/RGI Id	Area (km²)	Observation period	Mean mass balance (m.w.e. a^{-1})	References
Pensilungpa glacier	33° 47'–33°50' N 76°15'–79°18' E	10	2016–2018	–0.36	Mehta (2021)[102]
Rulung glacier	33.137 N; 78.442 E	0.95	1979–1981	–0.11	Srivastava et al. (2001)[6,101]
Samudra Tapu glacier	RGI60-14.15613	95.1	2017–2019	–0.97	Vishwakarma (2022)[14]
Shaune Garang glacier	31.291 N; 78.339 E	4.94	1981–1991	–0.42	GSI (1992)[6,103] Sangewar and Siddiqui (2006)[95]
Shishram glacier	34.0569 N; 75.5172 E	9.9	1983–1984	–0.29	Kaul (1986)[98]
Siachen glacier	RGI60-14.07524	1220	1986–1991	–0.23	Vishwakarma (2022)[14]
Stok glacier	33.977 N 77.45 E	0.74	2014–2019	–0.39	Soheb et al. (2020)[104]
Sutri Dhaka	RGI60-14.16041	25.2	2017–2019	–0.67	Vishwakarma (2022)[14]
Ganga Basin					
AX010	27.716 N; 86.555 E	0.42	1978–1979; 1995–1999	–0.62	Fujita et al. (2001b)[105]
Chorabari	30.768 N; 79.048 E	6.6	2003–2010	–0.73	Dobhal et al. (2013)[106]
Dokriani	30.852 N;78.823 E	5.36	1992–1995; 1997–2000	–0.32	Dobhal et al. (2008)[107]
Dunagiri	30.555 N; 79.893 E	2.5	1984–1990	–1.05	Swaroop and Gautam (1990)[108]
Kangwure	28.47 N; 85.815 E	1.9	1991–1993; 2008–2010	–0.57	Liu et al. (1996)[109] Yao et al. (2012)[110]
Mera	27.723 N; 86.885 E	5.1	2007–2015	–0.03	Wagnon et al. (2013)[75] Sherpa et al. (2017)[111]
Naimona'nyi	30.45 N;81.33 E	7.8	2005–2010	–0.56	Yao et al. (2012)[110]
Pokalde	27.926 N; 86.831 E	0.1	2009–2015	–0.69	Wagnon et al. (2012)[75] Sherpa et al. (2017)[111]
Rikha Samba	28.82 N;83.49 E	5.37	1998–1999 2012–2013	–0.34	Fujita et al. (2001a)[112] Gurung et al. (2016)[113]

Satopanth	30.756 N; 79.371 E	19	2014–2015	-2	Laha et al. (2017)[114]
Tipra Bank	30.723 N;79.676 E	15.36	1981–1988	-0.14	Gautam and Mukherjee (1992)[6,115]
Trakarding-Trambau	27.9° N 86.5° E	31.7	2016–2018	-0.61	Sunako (2019)[116]
West Changri Nup	27.965 N; 86.791 E	0.9	2010–2015	-1.24	Wagnon et al. (2012)[6] Sherpa et al. (2017)[111]
Yala	28.24 N; 85.61 E	1.61	2011–2012	-0.89	Baral et al. (2014)[117]
Brahmaputra Basin					
Changme Khangpu	27.952 N; 88.686 E	5.6	1979–1986	-0.26	GSI (2001)[6,101]
Gangju La	27.94 N; 89.95 E	0.29	2003–2004; 2012–2014	-1.38	Tshering and Fujita (2016)[118]
Gurenhekou	30.195 N; 90.443 E	1.4	2005–2010	-0.31	Yao et al. (2012)[110]
Parlung Glacier No. 10	29.282 N; 96.91 E	2.1	2005–2009	-0.78	Yao et al. (2012)[110]
Parlung Glacier No. 12	29.306 N; 96.904 E	0.2	2005–2010	-1.7	Yao et al. (2012)[110]
Parlung Glacier No. 94	29.388 N; 96.975 E	2.5	2005–2010	-0.92	Yao et al. (2012)[110]
Parlung Glacier No. 390	29.357 N; 97.012 E	0.5	2006–2010	-1.02	Yao et al. (2012)[110]
Thana	RGI60-15.02578	5	2019–2020	-1.38	Vishwakarma (2022)[14]

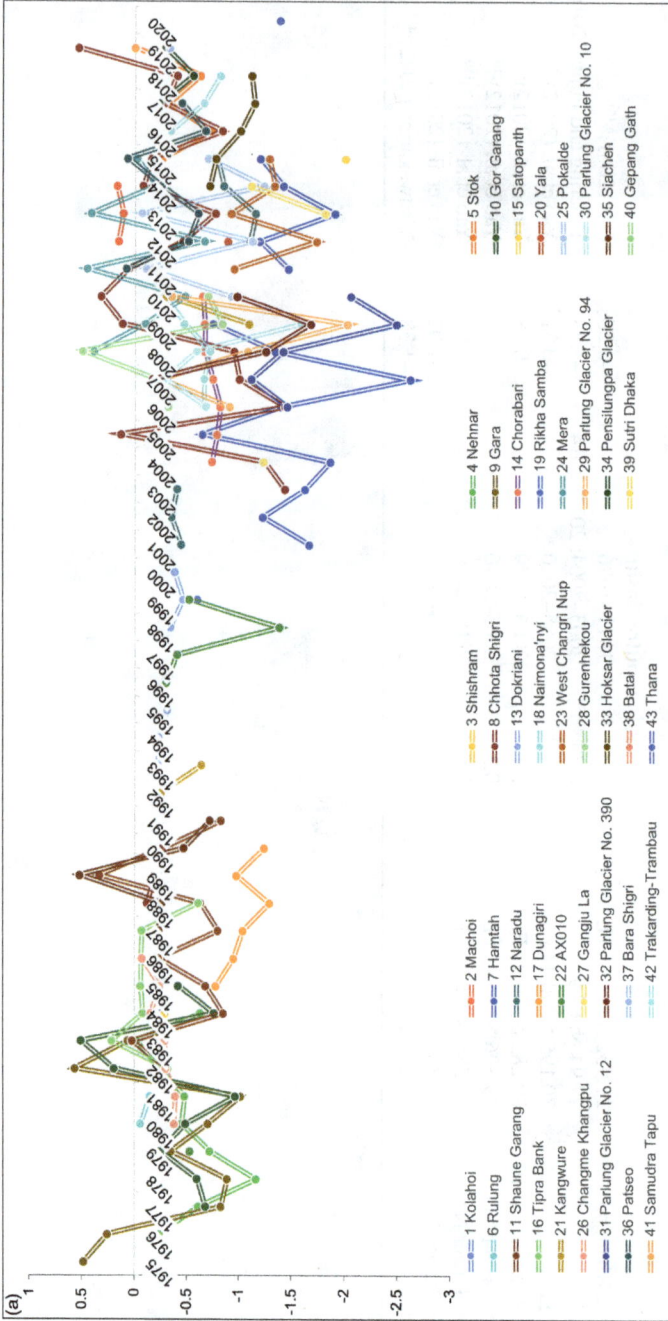

FIGURE 4.5 (A) Variability of glaciological mass balance of all the available glaciers described in Table 4.4; (B) plot of the variability of annual mass balance for the period 1975–2020; bar graph represents the available field measurements in each year.

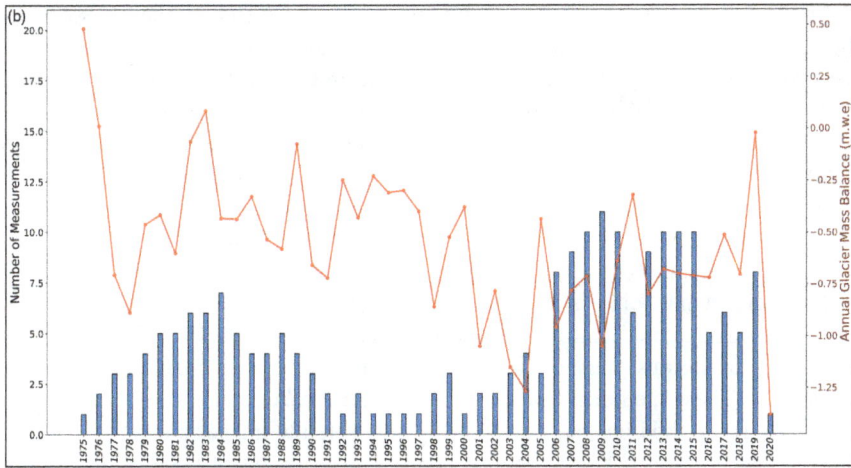

FIGURE 4.5 (Continued)

outputs such as satellite-based optical stereo images, radar interferometry, aerial photogrammetry, uncrewed aerial vehicles (SfM technique) and images obtained from laser scanning or drones[14]. These methods are beneficial in estimating decadal mass balance. The overall mass balance of High Mountain Asia has been estimated for the period 2000–2018 using the geodetic method and was observed as –0.19 ± 0.03 m.w.e. a^{-1} [6,41]. The average geodetic mass balance of the entire Himalayan region is estimated as –0.37 m.w.e. a^{-1} for the period 1975–2015[40]. The studies conducted at a regional scale report a mass balance value of −0.41 ± 0.11, −0.58 ± 0.01, −0.55 ± 0.37 and −0.10 ± 0.07 m.w.e. a^{-1} for Eastern Himalaya[119], Central Himalaya[120], Western Himalaya[121] and Karakoram[122], respectively, for the last decade. Furthermore, the glacier mass loss of the Indus, Ganga and Brahmaputra basins has been estimated and reported as −3.5 ± 1 Gt/year, −3.2 ± 0.6 Gt/year and −4.9 ± 1 Gt/year, respectively, the negative trend of which has led to an increase in annual glacier melt runoff by 18–29 percent[6,41]. At the sub-basinal scale, the studies over the Alakananda basin (Central Himalaya) reported a GMB of –0.28 ± 0.08 m.w.e. a^{-1} from 2000 to 2017[88]. Different geodetic mass balance estimates over the HKH region for the period 1975–2019 have been reported[14,123–128]. This analysis reveals an increase in mass loss since 2000. In addition, many studies provide geodetic mass balance estimates at regional, basinal or glacier-wise scales[40,129–131]. One of the limitations of this method is the vertical inaccuracy in publicly available DEMS as it limits the annual mass balance estimation[14,88]. Thus, satellite gravimetry or altimetry methods are used to estimate GMB[14,132,133]. Satellite gravimetry uses gravitational potential anomaly as the primary input, and it is sensitive only to changes in mass. This method could not be applied to a single glacier due to its poor spatial resolution (~300 km)[14]. Meanwhile the satellite altimetry method uses elevation data points as the major input for the estimation of mass balance and can also be applied to single glaciers. It provides higher accuracy in elevation measurements. However, the shorter satellite time span and uncertainty in density correction are limitations of this method[14].

Another method used for the estimation of mass balance is the albedo-based approach. This method uses the average glacier-wise albedo and long-term field-based mass balance data for estimation. It was applied in Chhota Shigri glacier (−0.68 ± 0.10 m.w.e. year⁻¹) and Mera glacier (0.13 ± 0.19 m.w.e. year⁻¹) for the period 1999–2013[134]. This method faces limitations in albedo estimation due to cloud cover and fewer field-based data. It is also not suitable for winter accumulation-type glaciers.

Other methods used for the estimation of mass balance are the accumulation area ratio (AAR) and the equilibrium line altitude (ELA) methods[135,136]. The AAR is defined as the ratio of the area covered in the accumulation zone to the total area of the glacier, while ELA is the altitude at which the accumulation zone and ablation zone are separated and the net mass balance is equal to zero. These methods are applied to various glaciers and used to estimate the mass balance of the HKH region[136-138]. These methods resulted in considerable uncertainty. Thus, an improved AAR (IAAR) method has been developed using satellite images and a temperature-index model, and this has been applied on the various basins of HKH[139]. The studies over the Satluj basin and Spiti basin (Western Himalaya) show average mass balance values of −0.40 ± 0.47 m.w.e. a⁻¹ (1984–2013)[39] and −0.56 ± 0.47 m.w.e. a⁻¹, respectively, for the period 1985–2013[37]. Further, the mass balances of the Alaknanda and Bhagirathi basins (Central Himalaya) are estimated for the period 2001–2013 and are reported as −1.1 ± 0.03 m.w.e. a⁻¹ and −1.01 ± 0.07 m.w.e. a⁻¹, respectively[140]. In addition, the mass balance estimate over the Chandra basin is reported as −0.61 ± 0.46 m.w.e. a⁻¹ for the period 1984–2012[141]. The IAAR when applied in the Chandra, Satluj, Alaknanda, Bhagirathi and Spiti basins, reported mass losses of 11.1 Gt[141], 16.4 Gt[39], 12.6 Gt[140], 8.8 Gt[140] and 8.89 Gt[37], respectively. These studies also indicated that smaller and lower-altitude glaciers are more susceptible to global warming as these glaciers exhibit higher mass loss compared to larger glaciers[6,37]. This method shows some uncertainty due to errors in ELA and AAR estimation. At the same time, it is advantageous as it can be applied to individual glaciers and also at basinal scales, and it also helps in future projections of mass balance.

The obtained glacier mass studies are used to estimate the future changes in glacier parameters by using different global climate models and representative concentration pathways (RCPs). Kraaijenbrink et al. (2017)[142] reported that about 60–70 percent of the entire glacier mass of the HKH region would disappear by the end of the 21st century if the current rate of greenhouse gases (GHGs) emission continues. Future projections of glacier area change have been reported in various basins of HKH and are described here. The future projection of glacier area using global climate models, CNRM-CM5 and GFDL-CM3[39] over the Satluj basin suggests 53% and 81% area losses by the end of the century under the RCP 8.5 scenario. Plots showing the spatial distribution of projected area loss in the Satluj basin under an RCP 8.5 climate change scenario using GFDL-CM3 are presented in Figure 4.6. Further, the studies on the Chandra basin suggest 48–50% and 55–60% area losses by 2100 under the RCP 4.5 and RCP 8.5, respectively[143]. A total glacier volume loss of 66–74% is reported under the RCP 8.5 scenario. The studies over the Spiti basin suggest an area loss of ~71.8% and volume loss of ~84.8% by 2070 under the RCP 8.5 scenario [37]. Furthermore, the studies over the Beas basin suggest a total glacier loss of ~90% at the end of the century under the RCP 8.5 scenario[144].

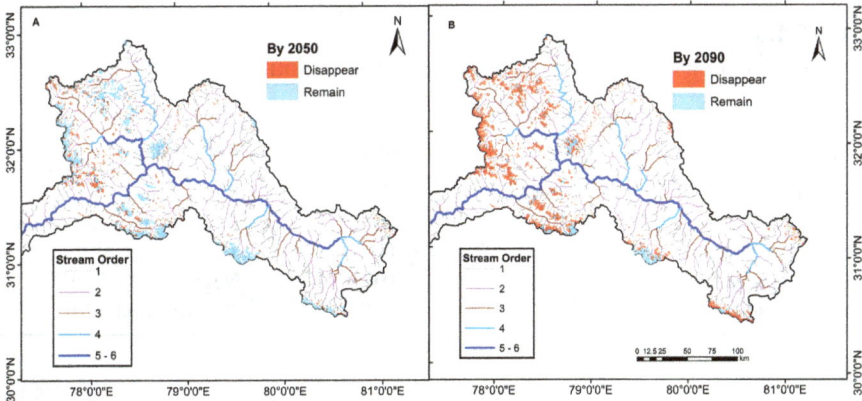

FIGURE 4.6 Spatial distribution of projected area loss of Satluj basin (Western Himalaya) under RCP 8.5 climate change scenario for (a) 2050 and (b) 2090 using the global climate model (GFDL-CM3 model); by 2050, 55% of the glaciers will disappear, marking an area loss of 475 km^2 and a mass loss of 39 Gt; by the end of the century (2090), about 97% of the glaciers will disappear, resulting in a mass loss of 56 Gt and an area loss of 1157 km^2 [39]; the region consist of 684 first-order, 313 second-order, 154 third-order, 73 fourth-order, 81 fifth-order and 52 sixth-order streams; among this approximately 310 first-order streams are likely to be affected by 2090: the datasets are taken from Prasad et al. (2019)[39].

The GMB studies over the HKH region also indicate the anomalous behaviour of the Karakoram and Pamir ranges, where glaciers are showing a near stable or positive mass balance value when compared to other regions of HKH. This phenomenon is known as the "Karakoram anomaly". Gardelle et al. (2012)[145] reported a positive mass balance in Karakoram glaciers for the period 1999–2011. Also, similar findings have been reported by different researchers[3,34,41,126,146–148]. The studies on area loss over Karakoram also show near-stable or advancing glacier conditions, and are listed in Table 4.5. This could be due to the influence of the elevation or the climate regimes of the region, as these glaciers are found at a higher altitude, and are sustained by large winter precipitation or snowfall and decreased mean summer temperature compared to the Himalayan region[6,41,122,149]. However, recent studies have reported the onset of mass loss on these glaciers since 2015, suggesting the end of the Karakoram anomaly[6,127,128].

Different factors control the spatial distribution of the mass balance of the HKH region. The spatial heterogeneity of mass balance can be strongly correlated to a set of climatic or non-climatic parameters. The glaciers in the southern part of the Tibetan plateau exhibit the highest mass loss due to the influence of the Indian summer monsoon[3,154,155]. Meanwhile, the northwestern part of High Mountain Asia experiences lower mass loss or exhibits a stable or positive nature due to the dominance of Western disturbances[3,154,155]. The strengthening of westerly winds and the weakening of monsoons are significant factors that drive the present glacier heterogeneity in HKH[154,155]. Furthermore, the changes in temperature along with precipitation strongly influence GMB, which indicates that glaciers in higher temperature

TABLE 4.5
Studies of Glacier Area Loss in the Karakoram Region

Region/glacier	Observation period	Change in area (%/year)	References
Karakoram	1969–1999	–0.14	Shangguan et al. (2006)[150]
Karakoram Region	1990–2013	0.0	Brahmbhatt et al (2015)[151]
Central Karakoram	2001–2010	0.06	Minora et al. (2013)[152]
East Karakoram	1990–2014	0.00	Negi et al. (2021)[122]
Siachen Glacier	1980-2014	0.00	Agarwal et al. (2017)[153]

zones may experience higher mass loss[90,154]. At mid or higher latitudes, GMB is more sensitive to temperature. Another critical factor that affects the GMB is supraglacial debris cover. Since not all glaciers are covered with debris, the climate is considered the dominant and critical factor affecting the heterogeneity of mass balance. In addition, topography and morphology also influence mass balance heterogeneity[154]. When assumptions of similar climatic conditions of a region are considered, the median elevation or slope are the dominant factors influencing GMB. The presence of black carbon and avalanche contribution can also influence mass balance[90,154]. The evaluation of mass balance is crucial in the HKH as it affects water availability and food security and therefore there is a great need for further investigations on the spatial heterogeneity of mass balance in the region.

Various mass balance models have been applied over the Himalayan region, including hydrological models, temperature index models, enhanced temperature index models, albedo models, energy balance models, glacier dynamics models (OGGM) and glacier evolution models (PyGEM). The aforementioned methods and models are eminently valuable for understanding the mass balance behaviour on scales varying from a single glacier to regional assessments. They are also valuable yardsticks in our comprehensive understanding of a region's dynamics and water resource assessment. Despite their usefulness, these models have limitations resulting in uncertainty. The study portrays the threat HKH glaciers are facing and also shows that most of the glaciers will disappear by 2100 if current trends of GHG emissions continue. This further raises the necessity for thorough investigations into understanding glacier mass loss, and the enactment of mitigation techniques to safeguard the HKH glaciers.

4.5 CONCLUSION

The HKH glaciers are a primary focus for the scientific community as these are the storehouses of freshwater. These glaciers are continuously changing as they exhibit significant variability in glacier mass, volume and snow cover area. These changes have a profound impact on water resources and food security. In the long run, people residing in the downstream and mountainous regions will be affected due to their dependency on these water resources for their daily needs. In order to build a comprehensive understanding, it is essential to study the current and future changes in environmental factors and assess glaciers' sensitivity to these changes. However, such studies are challenging due to the regional heterogeneity and vast span of HKH glaciers.

This study is a review that focused on the glaciers of the HKH region, particularly the glaciers in the Indus, Ganga and Brahmaputra basins. Glacier inventories from multiple sources were evaluated for the analysis. It shows that the total glaciated area of these major HKH river basins varies from 41,008 km^2 to 71,182 km^2 (ICIMOD, GAMDAM, IWMI, SAC, RGI). The region consists of a total of 39,822 glaciers based on RGI inventories. The glacier inventories exhibit discrepancies due to differences in the geographical area, methodologies and datasets. Hence, considering the dynamic nature of glacier extent and numbers, periodic modifications of the glacier outlines using high-resolution satellite imagery are suggested to obtain reliable results. The study shows that the total glacier-stored mass of the HKH region ranges between 2377 Gt to 6715.96 Gt. This variation is large, possibly due to large uncertainties introduced by the scaling methods. The estimates obtained by the flow method range from 15 to 145 Gt (basinal wise), and seem to be more realistic. The average mass balance of –0.61 m.w.e. a^{-1} is for the period 1975–2020. Due to existing adverse conditions, the field-based measurements of glacier thickness and mass balance are minimal over the HKH region. The field-based measurements on mass balance are available only for 43 glaciers, covering an area of 1696.12 km^2, accounting only for ~4% of the total glacierized area of the region. Most of these measurements are available for small-sized debris-free glaciers. Hence, more field-based, continuous and long-term measurements of mass balance, ice thickness, stream runoff, etc., are required to further understand glacier dynamics.

The geodetic and IAAR mass balance methods are some of the most extensively used methods in recent decades. The mean geodetic mass balance of the whole Himalayan region is observed as –0.37 m.w.e. a^{-1} during 1975–2015[40]. The IAAR mass balance methods are applied over various basins of HKH such as Satluj (–0.40 ± 0.47 m.w.e. a^{-1}), Spiti (–0.56 ± 0.47 m.w.e. a^{-1}), Alaknanda (−1.1 ± 0.03 m.w.e. a^{-1}), Bhagirathi (−1.01 ± 0.07 m.w.e. a^{-1}) and Chandra basin (−0.61 ± 0.46 m.w.e. a^{-1})[37,39,140,141]. Most of the modelled mass balance estimates are associated with notable uncertainties. Hence, there is a need to improve or develop new methods and models to attain more precise estimates. Other approaches based on glacier velocity, slope and ice flow equations and volume–area scaling methods are used to estimate the amounts of glacier-stored water.

Some other aspects that require more attention in the HKH region are glacier lakes and supraglacial debris cover as they significantly affect glacier dynamics in the region. The supraglacial debris cover is a less studied aspect that strongly influences surface energy balance and glacier dynamics. Furthermore, due to the wide-reaching effects of climate change, there has been the formation of new glacial lakes and the growth of existing lakes over the years[154,156]. Hence, detailed studies are needed to map the extent of glacier lakes continuously, identify the potential lake sites, alert about any risks and develop mitigation techniques. The studies report that debris-covered glaciers are more prone to lake development and retreat faster than clean glaciers[157]. Also, debris thickness significantly influences the melt processes. Hence, more studies on debris cover are required to understand the feedback of glaciers to climate change.

Climate is another critical factor that influences glacier dynamics. The changing climate conditions reveal an increase in precipitation in the form of rainfall rather than snowfall, thus significantly affecting the glacier and snow melt by reducing the stream flow[6]. The HKH glaciers marked a higher mass loss after 2000, which corresponds with the increasing temperature trends[157]. Overall, the warming climate led to an early melt of seasonal snow, increasing the contribution of rainfall-dominated runoff and a decrease in snow and ice melt[6]. Furthermore, the future projections of glacier studies indicate a significant decrease in mass and area by the end of the century under climate change scenarios. These projections build a substantial concern among planners and policymakers about future hydrological supplies and their impacts on populations and infrastructure.

Changing socio-economic dynamics or developments can also influence the water–energy–food nexus[158]. These lean on hydrological attributes, growing population, industrial demand, energy requirement and governance structure. The studies report that the Indus basin primarily relies more on snow and glacier melt compared to Ganga and Brahmaputra basins[6]. These socio-economics can affect the water withdrawal norms, so it needs to be studied further to understand these scenarios and their impact on food and water security. Studies are also required to understand future hydropower production, as the changes in glacier runoff can also affect long-term power generation.

Even though remote sensing provides valuable perspicacity into these remote areas and catchments, there exist unavoidable knowledge gaps within the HKH cryosphere. Most of the studies are focused on easily accessible areas, with the Karakoram and Eastern Himalayan regions being the least studied regions. To attenuate these gaps, high-altitude open data meteorological stations, mass balance observing sites and river flow gauges are encouraged. This review highlights the glaciers' feedback to the changing climate.

ACKNOWLEDGEMENTS

The authors are thankful to the DST-Centre for Excellence in Climate Change and Divecha Centre for Climate Change, IISc, for providing the necessary funds and requirements to carry out this research. Ashim Sattar acknowledges DST Inspire Faculty Award and DCCC, IISc Bangalore, for the necessary infrastructure and computational facilities.

REFERENCES

1. Kulkarni, A.V., Karyakarte, Y. (2014) Observed changes in Himalayan glacier. Current Science 106(2): 237–244.
2. Kulkarni, A.V., Shirsat, T. (2020) Glacier studies in India: Remote sensing applications and challenges. *Indian Journal of Geosciences* 74(3): 307–314.
3. Kumar, P., Kotlarski, S., Moseley, C., Sieck, K., Frey, H., Stoffel, M., Jacob, D. (2015) Response of Karakoram-Himalayan glaciers to climate variability and

climatic change: A regional climate model assessment. *Geophysical Research Letters* 42(6): 1818–1825.

4. Maharjan, S.B., Joya, E., Rahimi, M.M., Azizi, F., Muzafari, K.A., Bariz, M., Bromand, M.T., Shrestha, F., Shokory, A.G., Anwari, A. (2021) *Glaciers in Afghanistan: Status and Changes from 1990 to 2015*. ICIMOD Publication, 72 pp.

5. Lee, E., Carrivick, J.L., Quincey, D.J., Cook, S.J., James, W.H.M., Brown, L.E. (2021) Accelerated mass loss of Himalayan glaciers since the Little Ice Age. *Scientific Reports* 11(1): 1–8.

6. Kulkarni, A.V., Shirsat, T.S., Kulkarni, A., Negi, H.S., Bahuguna, I.M., Thamban, M. (2021) State of Himalayan cryosphere and implications for water security. *Water Security* 14(October): 100101.

7. Huss, M., Farinotti, D., Bauder, A., Funk, M. (2008) Modelling runoff from highly glacierized Alpine drainage basins in a changing climate. *Hydrological Processes* 22(19): 3888–3902.

8. Azam, M.F., Kargel, J.S., Shea, J.M., Nepal, S., Haritashya, U.K., Srivastava, S., Maussion, F., Qazi, N., Chevallier, P., Dimri, A.P., Kulkarni, A.V., Cogley, J.G., Bahuguna, I. (2021) Glaciohydrology of the Himalaya-Karakoram. *Science* 373(6557).

9. Yao, T., Bolch, T., Chen, D., Gao, J., Immerzeel, W., Piao, S., Su, F., Thompson, L., Wada, Y., Wang, L. (2022) The imbalance of the Asian water tower. *Nature Reviews Earth & Environment* 1–15.

10. Azam, M.F., Kargel, J.S., Shea, J.M., Nepal, S., Haritashya, U.K., Srivastava, S., Maussion, F., Qazi, N., Chevallier, P., Dimri, A.P. (2021) Glaciohydrology of the Himalaya-Karakoram. *Science* 373(6557).

11. Ji, Q., Yang, T., Li, M., Dong, J., Qin, Y., Liu, R. (2022) Variations in glacier coverage in the Himalayas based on optical satellite data over the past 25 years. *CATENA* 214: 106240.

12. Debnath, M., Sharma, M.C., Syiemlieh, H.J. (2019) Glacier dynamics in Changme Khangpu Basin, Sikkim Himalaya, India, between 1975 and 2016. *Geosciences* 9(6); 259.

13. Immerzeel, W.W., van Beek, L.P.H., Bierkens, M.F.P. (2010) Climate change will affect the Asian water towers. *Science* 328(5984): 1382–1385.

14. Vishwakarma, B.D., Ramsankaran, R.A.A.J., Azam, M.F., Bolch, T., Mandal, A., Srivastava, S., Kumar, P., Sahu, R., Navinkumar, P.J., Tanniru, S.R., Javed, A., Soheb, M., Dimri, A.P., Yadav, M., Devaraju, B., Chinnasamy, P., Reddy, M.J., Murugesan, G.P., Arora, M., Jain, S.K., Ojha, C.S.P., Harrison, S., Bamber, J. (2022) Challenges in understanding the variability of the cryosphere in the Himalaya and its impact on regional water resources. *Frontiers in Water* 4(July).

15. Sabin, T.P., Krishnan, R., Vellore, R., Priya, P., Borgaonkar, H.P., Singh, B.B., Sagar, A. (2020) Climate change over the Himalayas. In: *Assessment of Climate Change over the Indian Region*, pp 207–222. Springer.

16. Pepin, N., Bradley, R.S., Diaz, H.F., Baraer, M., Caceres, E.B., Forsythe, N., Fowler, H., Greenwood, G., Hashmi, M.Z., Liu, X.D., Miller, J.R., Ning, L., Ohmura, A., Palazzi, E., Rangwala, I., Schöner, W., Severskiy, I., Shahgedanova, M., Wang, M.B., Williamson, S.N., Yang, D.Q., Group, M.R.I.E.D.W.W. (2015) Elevation-dependent warming in mountain regions of the world. *Nature Climate Change* 5(5): 424–430.

17. Ambinakudige, S., Joshi, K. (2012) Remote sensing of cryosphere. *Remote Sensing-Applications, Dr. Boris Escalante* 369–380.

18. Gopika, J.S., Kulkarni, A.V., Prasad, V., Srinivasalu, P., Raman, A. (2021) Estimation of glacier stored water in the Bhaga basin using laminar flow and volume-area scaling methods. *Remote Sensing Applications: Society and Environment* 24(August): 100656.

19. Kulkarni, A.V., Bahuguna, I.M., Rathore, B.P. (2009) Application of remote sensing to monitor glaciers. *NNRMS Bulletin* 79–82.
20. Kulkarni, A.V., Bahuguna, I.M., Rathore, B.P., Singh, S.K., Randhawa, S.S., Sood, R.K., Dhar, S. (2007) Glacial retreat in Himalaya using Indian remote sensing satellite data. *Current Science* 69–74.
21. Bahuguna, I.M., Kulkarni, A.V., Nayak, S., Rathore, B.P., Negi, H.S., Mathur, P. (2007) Himalayan glacier retreat using IRS 1C PAN stereo data. *International Journal of Remote Sensing* 28(2): 437–442.
22. Kulkarni, A.V. (1991) Glacier inventory in Himachal Pradesh using satellite images. *Journal of the Indian Society of Remote Sensing* 19(3): 195–203.
23. Vatsal, S., Bhardwaj, A., Azam, M.F., Mandal, A., Bahuguna, I., Raju, N. J., Tomar, S.S. (2002) A comprehensive multidecadal glacier inventory dataset for the Chandra-Bhaga Basin, Western Himalaya, India. *Earth System Science Data*September: 1–38.
24. Sakai, A. (2019) Brief communication: Updated GAMDAM glacier inventory over high-mountain. *Cryosphere* 13(7): 2043–2049.
25. Kulkarni, A.V., Nayak, S., and Pratibha, S., (2017). Variability of glaciers and snow cover. *Observed climate variability and change over the Indian region*, pp. 193–219.
26. Kulkarni, A.V., Buch, A.M. (1991) *Glacier Atlas of Indian Himalaya, ISRO Report Number SAC/RSA/RSAG-MWRD/SN/05/91.* Space Applications Centre, Ahmedabad.
27. Sangewar, C.V., Shukla, S.P. (2009) Inventory of the Himalayan glaciers, Special Publication No. 34, Geological Survey of India. *IISN* 1: 254–436.
28. Bajracharya, S.R., Shrestha, B.R. (2011) *The Status of Glaciers in the Hindu Kush-Himalayan Region.* International Centre for Integrated Mountain Development (ICIMOD).
29. SAC Report. (2011) *Snow and Glaciers of the Himalayas, SAC Report No. SAC/RESA/MESG/SGP/TR/59/2011, 2011.*
30. Sharma, A.K., Singh, S.K., Kulkarni, A.V. (2013) Glacier inventory in Indus, Ganga and Brahmaputra Basins of the Himalaya. *National Academy Science Letters* 36(5): 497–505.
31. Pfeffer, W.T., Arendt, A.A., Bliss, A., Bolch, T., Cogley, J.G., Gardner, A.S., Hagen, J.-O., Hock, R., Kaser, G., Kienholz, C., Miles, E.S., Moholdt, G., Mölg, N., Paul, F., Radić, V., Rastner, P., Raup, B.H., Rich, J., Sharp, M.J. (2014) The Randolph Glacier Inventory: A globally complete inventory of glaciers. *Journal of Glaciology* 60(221): 537–552.
32. Arendt, A., Bolch, T., Cogley, J.G., Gardner, A., Hagen, J.-O., Hock, R., Kaser, G., Pfeffer, W.T., Moholdt, G., Paul, F. (2012) Randolph Glacier Inventory [v2. 0]: A Dataset of Global Glacier Outlines. *Global Land Ice Measurements from Space.*
33. Nuimura, T., Sakai, A., Taniguchi, K., Nagai, H., Lamsal, D., Tsutaki, S., Kozawa, A., Hoshina, Y., Takenaka, S., Omiya, S., Tsunematsu, K., Tshering, P., Fujita, K. (2015) The GAMDAM Glacier Inventory: A quality-controlled inventory of Asian glaciers. *Cryosphere* 9(3): 849–864.
34. Bolch, T., Kulkarni, A., Kääb, A., Huggel, C., Paul, F., Cogley, J.G., Frey, H., Kargel, J.S., Fujita, K., Scheel, M., Bajracharya, S., Stoffel, M. (2012) The state and fate of Himalayan glaciers. *Science* 336(6079): 310–314.
35. Savoskul, O.S., Smakhtin, V. (2013) *Glacier Systems and Seasonal Snow Cover in Six Major Asian River Basins: Hydrological Role under Changing Climate*, Vol. 150. IWMI.

36. Gopika, J.S., Kulkarni, A.V., Prasad, V., Srinivasalu, P., Raman, A. (2021) Estimation of glacier stored water in the Bhaga basin using laminar flow and volume-area scaling methods. *Remote Sensing Applications: Society and Environment* 24: 100656.

37. Kulkarni, A., Prasad, V., Shirsat, T., Chaturvedi, R.K., Bahuguna, I.M. (2021) Impact of climate change on the glaciers of Spiti River Basin, Himachal Pradesh, India. *Journal of the Indian Society of Remote Sensing* 49(8); 1951–1963.

38. Asharaf, R., Kulkarni, A.V. Estimation of glacier depth and ice volume of Kabul Basin. *Journal of Earth System Science* 132 (3): 128.

39. Prasad, V., Kulkarni, A.V., Pradeep, S., Pratibha, S., Tawde, S.A., Shirsat, T., Arya, A.R., Orr, A., Bannister, D. (2019) Large losses in glacier area and water availability by the end of twenty-first century under high emission scenario, Satluj Basin, Himalaya. *Current Science* 116(10); 1721–1730.

40. Azam, M.F., Wagnon, P., Berthier, E., Vincent, C., Fujita, K., Kargel, J.S. (2018) Review of the status and mass changes of Himalayan-Karakoram glaciers. *Journal of Glaciology* 64(243): 61–74.

41. Shean, D.E., Bhushan, S., Montesano, P., Rounce, D.R., Arendt, A., Osmanoglu, B. (2020) A systematic, regional assessment of high mountain Asia glacier mass balance. *Frontiers in Earth Science* 7(January): 1–19.

42. Cogley, J.G. (2016) Glacier shrinkage across high mountain Asia. *Annals of Glaciology* 57(71): 41–49.

43. Van Tricht, L., Huybrechts, P., Van Breedam, J., Fürst, J.J., Rybak, O., Satylkanov, R., Ermenbaiev, B., Popovnin, V., Neyns, R., Paice, C.M., Malz, P. (2021) Measuring and inferring the ice thickness distribution of four glaciers in the Tien Shan, Kyrgyzstan. *Journal of Glaciology* 67(262): 269–286.

44. Armstrong, R.L. (2010) *The Glaciers of the Hindu Kush-Himalayan Region: A Summary of the Science Regarding Glacier Melt/Retreat in the Himalayan, Hindu Kush, Karakoram, Pamir, and Tien Shan Mountain Ranges.* International Centre for Integrated Mountain Development (ICIMOD).

45. Bahr, D.B., Meier, M.F., Peckham, S.D. (1997) The physical basis of glacier volume-area scaling. *Journal of Geophysical Research Solid Earth* 102(B9): 20355–20362.

46. Farinotti, D., Huss, M., Fürst, J.J., Landmann, J., Machguth, H., Maussion, F., Pandit, A. (2019) A consensus estimate for the ice thickness distribution of all glaciers on Earth. *Nature Geoscience* 12(3): 168–173.

47. Sellevold, M.A., Kloster, K. (1964) Seismic measurements on the glacier Hardangerjøkulen, Western Norway. *Nor. Polarinst. Årb* 1964: 87–91.

48. Bahr, D.B., Pfeffer, W.T., Kaser, G. (2015) A review of volume–area scaling of glaciers. *Reviews of Geophysics* 53(1): 95–140.

49. Chen, J., Ohmura, A. (1990) Estimation of Alpine glacier water resources and their change since the 1870s. *IAHS Publ* 193: 127–135.

50. Nye, J.F. (1965) The flow of a glacier in a channel of rectangular, elliptic or parabolic cross-section. *Journal of Glaciology* 5(41): 661–690.

51. Nye, J.F. (1952) The mechanics of glacier flow. *Journal of Glaciology* 2(12); 82–93.

52. Nye, J.F. (1952) A method of calculating the thicknesses of the ice-sheets. *Nature* 169(4300): 529–530.

53. Cuffey, K.M., Paterson, W.S.B. (2010) *The Physics of Glaciers.* Academic Press.

54. Frey, H., Machguth, H., Huss, M., Huggel, C., Bajracharya, S., Bolch, T., Kulkarni, A., Linsbauer, A., Salzmann, N., Stoffel, M. (2014) Estimating the volume of glaciers in the Himalayan–Karakoram region using different methods. *Cryosphere* 8(6): 2313–2333.

55. Haeberli, W., Hoelzle, M. (1995) Application of inventory data for estimating characteristics of and regional climate-change effects on mountain glaciers: A pilot study with the European Alps. *Annals of Glaciolpgy* 21: 206–212.
56. Li, H., Ng, F., Li, Z., Qin, D., Cheng, G. (2012) An extended "perfect-plasticity" method for estimating ice thickness along the flow line of mountain glaciers. *Journal of Geophysical Research: Earth Surface* 117 (F1).
57. Gantayat, P., Kulkarni, A.V., Srinivasan, J. (2014) Estimation of ice thickness using surface velocities and slope: Case study at Gangotri Glacier, India. *Journal of Glaciology* 60(220): 277–282.
58. Maanya, U.S., Kulkarni, A.V., Tiwari, A., Bhar, E.D., Srinivasan, J. (2016) Identification of potential glacial lake sites and mapping maximum extent of existing glacier lakes in Drang Drung and Samudra Tapu Glaciers, Indian Himalaya. *Current Science* 553–560.
59. Budd, W.F., Allison, I.F. (1975) An empirical scheme for estimating the dynamics of unmeasured glaciers. In: *Proceedings of the Moscow Symposium "Snow and Ice*, pp. 246–256. IAHS-AISH Publ. No. 104
60. Farinotti, D., Huss, M., Bauder, A., Funk, M., Truffer, M. (2009) A method to estimate the ice volume and ice-thickness distribution of Alpine glaciers. *Journal of Glaciology* 55(191): 422–430.
61. Rasmussen, L.A. (1988) Bed topography and mass-balance distribution of Columbia Glacier, Alaska, U.S.A., determined from sequential aerial photography. *Journal of Glaciology* 34(117): 208–216.
62. Kulkarni, A., Ajanta, G., Raaj, R., Gulab, S., Ajai, D., Pradeep, S., Ramya, N., Nagashri, K., Arya, R., Anisha G. (2019) *Himalayan Glacier Thickness Mapper (Higthim) User Manual*.
63. Linsbauer, A., Paul, F., Hoelzle, M., Frey, H., Haeberli, W., Purves, R.S. (2009) *The Swiss Alps without Glaciers–a GIS-Based Modelling Approach for Reconstruction of Glacier Beds*. Zurich, CH: Department of Geography, University of Zurich, pp. 243–247.
64. Pandit, A., Ramsankaran, R. (2020) Modeling ice thickness distribution and storage volume of glaciers in Chandra Basin, Western Himalayas. *Journal of Mountain Science* 17(8): 2011–2022.
65. Huss, M., Farinotti, D. (2012) Distributed ice thickness and volume of all glaciers around the globe. *Journal of Geophysical Research: Earth Surface* 117(F4).
66. Zou, X., Gao, H., Zhang, Y., Ma, N., Wu, J., Farhan, S. Bin. (2021) Quantifying ice storage in Upper Indus River Basin using ground-penetrating radar measurements and glacier bed topography model version 2. *Hydrological Processes* 35(4): e14145.
67. Vincent, C., Wagnon, P., Shea, J.M., Immerzeel, W.W., Kraaijenbrink, P., Shrestha, D., Soruco, A., Arnaud, Y., Brun, F., Berthier, E., Sherpa, S.F. (2016) Reduced melt on debris-covered glaciers: Investigations from Changri Nup Glacier, Nepal. *Cryosphere* 10(4): 1845–1858.
68. Azam, M.F., Wagnon, P., Ramanathan, A., Vincent, C., Sharma, P., Arnaud, Y., Linda, A., Pottakkal, J.G., Chevallier, P., Singh, V.B., Berthier, E. (2012) From balance to imbalance: A shift in the dynamic behaviour of Chhota Shigri Glacier, Western Himalaya, India. *Journal of Glaciology* 58(208): 315–324.
69. Singh, S.K., Rathore, B.P., Bahuguna, I.M., Ramnathan, A.L. and Ajai (2022) Estimation of glacier ice thickness using ground penetrating radar in the Himalayan region *Current Science* 103(1): 68–73.
70. Gergan, J.T., Dobhal, D.P., Kaushik, R. (1999) Ground penetrating radar ice thickness measurements of Dokriani bamak (glacier), Garhwal Himalaya. *Current Science* 77(1): 169–173

71. Swain, A.K., Mukhtar, M.A., Majeed, Z., Shukla, S.P. (2018) Depth profiling and recessional history of the Hamtah and Parang Glaciers in Lahaul and Spiti, Himachal Pradesh, Indian Himalaya. *Geological Society of London, Special Publication* 462(1): 35–49.

72. Ma, L.L., Tian, L. De, Pu, J.C., Wang, P.L. (2010) Recent area and ice volume change of Kangwure Glacier in the middle of Himalayas. *Chinese Science Bulletin* 55(20): 2088–2096.

73. Pritchard, H.D., King, E.C., Goodger, D.J., McCarthy, M., Mayer, C., Kayastha, R. (2020) Towards Bedmap Himalayas: Development of an airborne ice-sounding radar for glacier thickness surveys in high-mountain Asia. *Annals of Glaciology* 61(81): 35–45.

74. Prakash, S., Sharma, M.C., Sreekesh, S., Chand, P., Pandey, V.K., Latief, S.U., Deswal, S., Manna, I., Das, S., Mandal, S.T., Bahuguna, I.M. (2022) Decadal terminus position changes and ice thickness measurement of Menthosa Glacier in Lahaul Region of North-Western Himalaya. *Geocarto International* 37(22): 6422–6441.

75. Wagnon, P., Vincent, C., Arnaud, Y., Berthier, E., Vuillermoz, E., Gruber, S., Ménégoz, M., Gilbert, A., Dumont, M., Shea, J.M., Stumm, D., Pokhrel, B.K. (2013) Seasonal and annual mass balances of Mera and Pokalde glaciers (Nepal Himalaya) since 2007. *Cryosphere* 7(6): 1769–1786.

76. Singh, K.K., Kulkarni, A.V., Mishra, V.D. (2010) Estimation of glacier depth and moraine cover study using ground penetrating radar (GPR) in the Himalayan region. *Journal of the Indian Society of Remote Sensing* 38(1): 1–9.

77. Singh, K.K., Singh, D.K., Negi, H.S., Kulkarni, A.V., Gusain, H.S., Ganju, A., Babu Govindha Raj, K. (2018) Temporal change and flow velocity estimation of Patseo glacier, Western Himalaya, India. *Current Science* 776–784.

78. Mishra, A., Negi, B.D.S., Banerjee, A., Nainwal, H.C., Shankar, R. (2018) Estimation of ice thickness of the Satopanth glacier, Central Himalaya using ground penetrating radar. *Current Science* 114(4): 785–791.

79. Sugiyama, S., Fukui, K., Fujita, K., Tone, K., Yamaguchi, S. (2013) Changes in ice thickness and flow velocity of Yala glacier, Langtang Himal, Nepal, from 1982 to 2009. *Annals of Glaciology* 54(64): 157–162.

80. Millan, R., Mouginot, J., Rabatel, A., Morlighem, M. (2022) Ice velocity and thickness of the world's glaciers. *Nature Geoscience* 15(2): 124–129.

81. Arendt, A., Echelmeyer, K., Harrison, W., Lingle, C., Zirnheld, S., Valentine, V., Ritchie, B., Druckenmiller, M. (2006) Updated estimates of glacier volume changes in the Western Chugach Mountains, Alaska, and a comparison of regional extrapolation methods. *Journal of Geophysical Research: Earth Surface* 111(F3).

82. Sattar, A., Goswami, A., Kulkarni, A.V., Das, P. (2019) Glacier-surface velocity derived ice volume and retreat assessment in the Dhauliganga basin, Central Himalaya–A remote sensing and modeling based approach. *Frontiers in Earth Science* 7: 105.

83. Srinivasalu, P., Kulkarni, A., Srinivas, V.V., Satheesh, S.K. (2023) *An assessment of the water availability for mountain communities in the Parvati basin, Western Himalaya using a distributed hydrological model.* EGU General Assembly 2023, Vienna, Austria, 24–28 Apr 2023, EGU23-523.

84. Singh, K.K., Negi, H.S., Singh, D.K. (2019) Assessment of glacier stored water in Karakoram Himalaya using satellite remote sensing and field investigation. *Journal of Mountain Science* 16(4): 836–849.

85. Gopika, J.S., Kulkarni, A.V., Prasad, V. Glacier volume estimation using laminar-flow and volume-area scaling techniques in the Chenab basin. *Journal of the Indian Society of Remote Sensing*

86. Srinivasalu, P., Kulkarni, A., Godha, A., Goswami, A., Namboodiri, R., Shirsat, T., Krishnamurthy, N. (2018) An estimate of glacier stored ice in the Satluj and Beas basins of the Himalaya. In: *EGU General Assembly Conference Abstracts*, p. 434.

87. Srinivasalu., P., Kulkarni., A., Namboodiri., R. (2018) *An Estimate of Glacier stored ice in the Sikkim Himalaya. Poster presentation at State of the Cryosphere in the Himalaya: With a focus on Sikkim and eastern Himalayas – gaps, challenges, and opportunities.*

88. Remya, S.N., Kulkarni, A.V., Hassan Syed, T., Nainwal, H.C. (2022) Glacier mass loss in the Alaknanda basin, Garhwal Himalaya on a decadal scale. *Geocarto International* 37(10): 3014–3032.

89. Singh, V., Jain, S.K., Shukla, S. (2021) Glacier change and glacier runoff variation in the Himalayan Baspa river basin. *Journal of Hydrology* 593: 125918.

90. Romshoo, S.A., Murtaza, K.O., Abdullah, T. (2022) Towards understanding various influences on mass balance of the Hoksar Glacier in the Upper Indus Basin using observations. *Scientific Reports* 12(1): 1–14.

91. Dobhal, D.P., Kumar, S., Mundepi, A.K., Azam, M.F., Ramanathan, A.L., Wagnon, P., Vincent, C., Linda, A., Berthier, E., Sharma, P., Mandal, A., Angchuk, T., Singh, V.B., Ramanathan, A.L., Azam, M.F., Angchuk, T., Soheb, M., Kumar, N., Pottakkal, J.G., Vatsal, S., Mishra, S., Singh, V.B. (1995) Meteorological conditions, seasonal and annual mass balances of Chhota Shigri Glacier, Western Himalaya, India. *Current Science* 57(259): 727–741.

92. Azam, M.F., Ramanathan, A.L., Wagnon, P., Vincent, C., Linda, A., Berthier, E., Sharma, P., Mandal, A., Angchuk, T., Singh, V.B. (2016) Meteorological conditions, seasonal and annual mass balances of Chhota Shigri Glacier, Western Himalaya, India. *Annals of Glaciology* 57(71): 328–338.

93. Mandal, A., Ramanathan, A., Azam, M.F., Angchuk, T., Soheb, M., Kumar, N., Pottakkal, J.G., Vatsal, S., Mishra, S., Singh, V.B. (2020) Understanding the interrelationships among mass balance, meteorology, discharge and surface velocity on Chhota Shigri Glacier over 2002–2019 using in situ measurements. *Journal of Glaciology* 66(259); 727–741.

94. Raina, V.K., Kaul, M.K., Singh, S. (1977) Mass-balance studies of Gara Glacier. *Journal of Glaciology* 18(80): 415–423.

95. Sangewar, C.V., Siddique, N.S. (2006) Thematic compilation of mass balance data on glaciers of Satluj catchment in Himachal Himalaya. *Records of the Geological Survey of India* 141(pt 8): 159–161.

96. GSI. (2014) *GSI (Geological Survey of India). (2011). Annual General Report, Part 8. Volume 144..* GSI.

97. Mishra, R., Kumar, A., Singh, D. (2014) Long term monitoring of mass balance of Hamtah glacier, Lahaul and Spiti district, Himachal Pradesh. *Geological Survey of India* 147(Pt 8): 230–231.

98. Kaul, M.N. (1986) Mass balance of Liddar glaciers. *Transatcions of the Institute of Indian Geographers* 8: 95–111.

99. Koul, M.N., Bahuguna, I.M., Rajawat, A.S., Ali, S., Koul, S. (2016) Glacier area change over past 50 years to stable phase in Drass Valley, Ladakh Himalaya (India). *American Journal of Climate Change* 5(01): 88.

100. Koul, M.N., Ganjoo, R.K. (2010) Impact of inter- and intra-annual variation in weather parameters on mass balance and equilibrium line altitude of Naradu Glacier (Himachal Pradesh), NW Himalaya, India. *Climate Change* 99(1): 119–139.

101. Srivastava, D. (2001) *Glaciology of Indian Himalaya: A Bilingual Contribution in 150th Year of Geological Survey of India.* Kolkata: Geological Survey of India.

102. Mehta, M., Kumar, V., Garg, S., Shukla, A. (2021) Little Ice Age glacier extent and temporal changes in annual mass balance (2016–2019) of Pensilungpa Glacier, Zanskar Himalaya. *Regional Environmental Change* 21(2).

103. GSI. (1992) *GSI (Geological Survey of India). (1992). Annual General Report. Part 8. Volume 125.* GSI.

104. Soheb, M., Ramanathan, A., Angchuk, T., Mandal, A., Kumar, N., Lotus, S. (2020) Mass-balance observation, reconstruction and sensitivity of Stok Glacier, Ladakh Region, India, between 1978 and 2019. *Journal of Glaciology* 66(258): 627–642.

105. Fujita, K., Kadota, T., Rana, B., Kayastha, R.B., Ageta, Y. (2001) Shrinkages of Glacier AX010 in Shorong Region, Nepal Himalaya in the 1990s. *Bulletin of Glaciological Research* 51–54.

106. Dobhal, D.P., Mehta, M., Srivastava, D. (2013) Influence of debris cover on terminus retreat and mass changes of Chorabari Glacier, Garhwal Region, Central Himalaya, India. *Journal of Glaciology* 59(217): 961–971.

107. Dobhal, D.P., Gergan, J.T., Thayyen, R.J. (2008) Mass balance studies of the Dokriani Glacier from 1992 to 2000, Garhwal Himalaya, India. *Bulletin of Glaciological Research* 25: 9–17.

108. Swaroop, S., and Gautam, C.K. (1990) Glaciological studies on Dunagiri Glacier, Chamoli District, Uttar Pradesh. *Geological Survey of India,* 24(8).

109. Liu, S., Xie, Z., Song, G., Ma, L., Ageta, Y. (1996) Mass balance of Kangwure (flat-top) Glacier on the north side of Mt. Xixiabangma, China. *Bulletin of Glacier Research* 14: 37–43.

110. Yao, T., Thompson, L., Yang, W., Yu, W., Gao, Y., Guo, X., Yang, X., Duan, K., Zhao, H., Xu, B. (2012) Different glacier status with atmospheric circulations in Tibetan plateau and surroundings. *Nature Climate Change* 2(9): 663–667.

111. Sherpa, S.F., Wagnon, P., Brun, F., Berthier, E., Vincent, C., Lejeune, Y., Arnaud, Y., Kayastha, R.B., Sinisalo, A. (2017) Contrasted surface mass balances of debris-free glaciers observed between the southern and the inner parts of the Everest region (2007–15). *Journal of Glaciology* 63(240): 637–651.

112. Fujita, K., Nakazawa, F., Rana, B. (2001) Glaciological observations on Rikha Samba Glacier in Hidden Valley, Nepal Himalayas, 1998 and 1999. *Bulletin ig Glaciological Research* 18(5).

113. Gurung, S., Bhattrai, B.C., Kayastha, R.B., Stumm, D., Joshi, S., Mool, P.K. Study of annual mass balance (2011–2013) of Rikha Samba Glacier, Hidden Valley, Mustang, Nepal. *Sciences in Cold and Arid Regions* 8(4): 311–318.

114. Laha, S., Kumari, R., Singh, S., Mishra, A., Sharma, T., Banerjee, A., Nainwal, H .C., Shankar, R. (2017) Evaluating the contribution of avalanching to the mass balance of Himalayan Glaciers. *Annals of Glaciology* 58(75): 110–118.

115. Gautam, C. K., and Mukherjee, B.K. (1992) *Synthesis of Glaciological Studies on Tipra Bank Glacier Bhyundar Ganga Basin, District Chamoli.* Uttar Pradesh (FS 1980–1988) Geological Survey of India, Northern Region, Lucknow. Geological Survey of India.

116. Sunako, S., Fujita, K., Sakai, A., Kayastha, R.B. (2019) Mass balance of Trambau Glacier, Rolwaling region, Nepal Himalaya: In-situ observations, long-term reconstruction and mass-balance sensitivity. *Journal of Glaciology* 65(252): 605–616.

117. Baral, P., Kayastha, R.B., Immerzeel, W.W., Pradhananga, N.S., Bhattarai, B.C., Shahi, S., Galos, S., Springer, C., Joshi, S.P., Mool, P.K. (2014) Preliminary results of mass-balance observations of Yala glacier and analysis of temperature and precipitation gradients in Langtang Valley, Nepal. *Annals of Glaciology* 55(66): 9–14.

118. Tshering, P., Fujita, K. (2016) First in situ record of decadal glacier mass balance (2003–2014) from the Bhutan Himalaya. *Annals of Glaciology* 57(71): 289–294.

119. Maurer, J.M., Rupper, S.B., Schaefer, J.M. (2016) Quantifying ice loss in the Eastern Himalayas since 1974 using declassified spy satellite imagery. *Cryosphere* 10(5); 2203–2215.

120. Bandyopadhyay, D., Singh, G., Kulkarni, A.V. (2019) Spatial distribution of decadal ice-thickness change and glacier stored water loss in the Upper Ganga Basin, India during 2000–2014. *Scientific Reports* 9(1).

121. Vijay, S., Braun, M. (2016) Elevation change rates of glaciers in the Lahaul-Spiti (Western Himalaya, India) during 2000–2012 and 2012–2013. *Remote Sensing* 8(12): 1038.

122. Negi, H.S., Kumar, A., Kanda, N., Thakur, N.K., Singh, K.K. (2021) Status of glaciers and climate change of East Karakoram in early twenty-first century. *Science of the Total Environment* 753: 141914.

123. Maurer, J.M., Schaefer, J.M., Rupper, S., Corley, A. (2019) Acceleration of ice loss across the Himalayas over the past 40 years. *Science Advances* 5(6): eaav7266.

124. King, O., Bhattacharya, A., Bhambri, R., Bolch, T. (2019) Glacial lakes exacerbate Himalayan glacier mass loss. *Scientific Reports* 9(1): 1–9.

125. King, O., Bhattacharya, A., Ghuffar, S., Tait, A., Guilford, S., Elmore, A.C., Bolch, T. (2020) Six decades of glacier mass changes around Mt. Everest are revealed by historical and contemporary images. *One Earth* 3(5): 608–620.

126. Brun, F., Berthier, E., Wagnon, P., Kääb, A., Treichler, D. (2017) A spatially resolved estimate of high mountain Asia glacier mass balances from 2000 to 2016. *Nature Geoscience* 10(9): 668–673.

127. Hugonnet, R., McNabb, R., Berthier, E., Menounos, B., Nuth, C., Girod, L., Farinotti, D., Huss, M., Dussaillant, I., Brun, F. (2021) Accelerated global glacier mass loss in the early twenty-first century. *Nature* 592(7856): 726–731.

128. Bhattacharya, A., Bolch, T., Mukherjee, K., King, O., Menounos, B., Kapitsa, V., Neckel, N., Yang, W., Yao, T. (2021) High mountain Asian glacier response to climate revealed by multi-temporal satellite observations since the 1960s. *Nature Communications* 12(1): 1–13.

129. Bolch, T., Shea, J.M., Liu, S., Azam, F.M., Gao, Y., Gruber, S., Immerzeel, W.W., Kulkarni, A., Li, H., Tahir, A.A. (2019) Status and change of the cryosphere in the extended Hindu Kush Himalaya region. In: *The Hindu Kush Himalaya Assessment*, pp. 209–255. Springer;

130. Bhushan, S., Syed, T.H., Kulkarni, A.V., Gantayat, P., Agarwal, V. (2017) Quantifying changes in the Gangotri Glacier of Central Himalaya: Evidence for increasing mass loss and decreasing velocity. *IEEE Journal of Selected Topics in Applied Earth Obsservations and Remote Sensing* 10(12): 5295–5306.

131. Gaddam, V.K., Kulkarni, A.V., Gupta, A.K. (2020) Assessment of the Baspa basin glaciers mass budget using different remote sensing methods and modeling techniques. *Geocarto International* 35(3): 296–316.

132. Kääb, A., Berthier, E., Nuth, C., Gardelle, J., Arnaud, Y. (2012) Contrasting patterns of early twenty-first-century glacier mass change in the Himalayas. *Nature* 488(7412): 495–498.

133. Jakob, L., Gourmelen, N., Ewart, M., Plummer, S. (2021) Spatially and temporally resolved ice loss in high mountain Asia and the Gulf of Alaska observed by CryoSat-2 Swath altimetry between 2010 and 2019. *Cryosphere* 15(4): 1845–1862.

134. Brun, F., Dumont, M., Wagnon, P., Berthier, E., Azam, M.F., Shea, J.M., Sirguey, P., Rabatel, A., Ramanathan, A. (2015) Seasonal changes in surface albedo of

Himalayan glaciers from MODIS data and links with the annual mass balance. *Cryosphere* 9(1): 341–355.

135. Kulkarni, A.V. (1992) Mass balance of Himalayan glaciers using AAR and ELA methods. *Journal of Glaciology* 38(128): 101–104.

136. Kulkarni, A.V., Rathore, B.P., Alex, S. (2004) Monitoring of glacial mass balance in the Baspa basin using accumulation area ratio method. *Current Science* 185–190.

137. Mir, R.A., Jain, S.K., Saraf, A.K., Goswami, A. (2014) Detection of changes in glacier mass balance using satellite and meteorological data in Tirungkhad basin located in Western Himalaya. *Journal of the Indian Society of Remote Sensing* 42(1): 91–105.

138. Tak, S., Keshari, A.K. (2020) Investigating mass balance of Parvati glacier in Himalaya using satellite imagery based model. *Scientific Reports* 10(1): 1–16.

139. Tawde, S.A., Kulkarni, A.V., Bala, G. (2016) Estimation of glacier mass balance on a basin scale: An approach based on satellite-derived snowlines and a temperature index model. *Current Science* 1977–1989.

140. Raman, A., Kulkarni, A.V., Prasad, V. (2022) Glacier mass balance estimation in Garhwal Himalaya using improved accumulation area ratio method. *Environmental Monitoring and Assessment* 194(8): 583.

141. Tawde, S.A., Kulkarni, A.V., Bala, G. (2017) An estimate of glacier mass balance for the Chandra basin, Western Himalaya, for the period 1984–2012. *Annals of Glaciology* 58(75pt2): 99–109.

142. Kraaijenbrink, P.D.A., Bierkens, M.F.P., Lutz, A.F., Immerzeel, W.W. (2017) Impact of a global temperature rise of 1.5 degrees Celsius on Asia's glaciers. *Nature* 549(7671): 257–260.

143. Tawde, S.A., Kulkarni, A.V., Bala, G. (2019) An assessment of climate change impacts on glacier mass balance and geometry in the Chandra basin, Western Himalaya for the 21st century. *Environmental Research Communications* 1(4): 41003.

144. Dixit, A., Sahany, S., Kulkarni, A.V. (2021) Glacial changes over the Himalayan Beas basin under global warming. *Journal of Environmental Management* 295: 113101.

145. Gardelle, J., Berthier, E., Arnaud, Y., Kääb, A. (2013) Region-wide glacier mass balances over the Pamir-Karakoram-Himalaya during 1999–2011. *Cryosphere* 7(4): 1263–1286.

146. Gardelle, J., Berthier, E., Arnaud, Y. (2012) Slight mass gain of Karakoram glaciers in the early twenty-first century. *Nature Geoscience* 5(5): 322–325.

147. Hewitt, K. (2005) The Karakoram anomaly? Glacier expansion and the 'elevation effect,' Karakoram Himalaya. *Mountain Research and Development* 25(4); 332–340.

148. Neckel, N., Kropáček, J., Bolch, T., Hochschild, V. (2014) Glacier mass changes on the Tibetan plateau 2003–2009 derived from ICESat laser altimetry measurements. *Environmental Research Letters* 9(1); 14009.

149. Shrestha, A.B., Agrawal, N.K., Alfthan, B., Bajracharya, S.R., Maréchal, J., Oort, B. van. (2015) *The Himalayan Climate and Water Atlas: Impact of Climate Change on Water Resources in Five of Asia's Major River Basins.* ICIMOD, GRID-Arendal and CICERO, pp. 1–96.

150. Shangguan, D., Liu, S., Ding, Y., Ding, L., Xiong, L., Cai, D., Li, G., Lu, A., Zhang, S., Zhang, Y. (2006) Monitoring the glacier changes in the Muztag Ata and Konggur Mountains, East Pamirs, based on Chinese glacier inventory and recent satellite imagery. *Annals of Glaciology* 43: 79–85.

151. Brahmbhatt, R.M., Bahuguna, I.M., Rathore, B.P., Singh, S.K., Rajawat, A.S., Shah, R.D., Kargel, J.S. (2015) Satellite monitoring of glaciers in the Karakoram from 1977 to 2013: An overall almost stable population of dynamic glaciers. *Cryosphere Discussions* 9(2): 1555–1592.

152. Minora, U., Bocchiola, D., D'Agata, C., Maragno, D., Mayer, C., Lambrecht, A., Mosconi, B., Vuillermoz, E., Senese, A., Compostella, C. (2013) 2001–2010 glacier changes in the Central Karakoram National Park: A contribution to evaluate the magnitude and rate of the "Karakoram anomaly". *Cryosphere Discussions* 7(3): 2891–2941.

153. Agarwal, V., Bolch, T., Syed, T.H., Pieczonka, T., Strozzi, T., Nagaich, R. (2017) Area and mass changes of Siachen glacier (East Karakoram). *Journal of Glaciology* 63(237): 148–163.

154. Zhang, Z., Gu, Z., Hu, K., Xu, Y., Zhao, J. (2022) Spatial variability between glacier mass balance and environmental factors in the High Mountain Asia. *Journal of Arid Land* 1–14.

155. Shaw, T.E., Miles, E.S., Chen, D., Jouberton, A., Kneib, M., Fugger, S., Ou, T., Lai, H.W., Fujita, K., Yang, W. (2022) Multi-decadal monsoon characteristics and glacier response in High Mountain Asia. *Environmental Research Letters* 17(10): 104001.

156. Gardelle, J., Arnaud, Y., Berthier, E. (2011) Contrasted evolution of glacial lakes along the Hindu Kush Himalaya Mountain Range between 1990 and 2009. *Global and Planetary Change* 75(1–2): 47–55.

157. Basnett, S., Kulkarni, A.V., Bolch, T. (2013) The influence of debris cover and glacial lakes on the recession of glaciers in Sikkim Himalaya, India. *Journal of Glaciology* 59(218): 1035–1046.

158. Nie, Y., Pritchard, H.D., Liu, Q., Hennig, T., Wang, W., Wang, X., Liu, S., Nepal, S., Samyn, D., Hewitt, K. (2021) Glacial change and hydrological implications in the Himalaya and Karakoram. *Nature Reviews Earth & Environment* 2(2): 91–106.

5 Response of Global Warming on Glaciers of Garhwal Himalaya, India
A Remote Sensing and GIS Approach

Dhirendra Kumar, Anoop Kumar Singh, Pawan Kumar Gautam, Balkrishan Vishawakarma, Chetan Anand Dubey, and Dhruv Sen Singh

5.1 INTRODUCTION

Climate change and global temperature rise affect the health of glaciers worldwide (the disappearance of the Okjokull glacier in western Iceland and the rapid retreat of glaciers confirm this), whereas the glaciers of Karakoram offer some contradictions to this effect (Hewitt, 2005; Cogley ,2011; Scherler et al., 2011; Gardelle et al., 2013). The record suggests that the world's average surface temperature has increased between 0.3 and 0.6°C over the past 100 years (IPCC, 1990, 1992, 2007). The scientists of the International Union for Conservation of Nature (IUCN) have reported that almost half of the World Heritage glacial sites (21 out of 46) will disappear by 2100 if the emissions of greenhouse gases continues as usual. These sites involve some of the world's most iconic glaciers, such as Grosser Aletschgletscher of the Swiss Alps, Khumbu Glacier of the Himalayas, and Jakobshavn Isbrae of Greenland. Their study also suggests that about 33–60% of the total ice volume present in 2017 will be lost by 2100. Recently, Maurer et al. (2019) explained that about 650 glaciers of the Himalayas have been experiencing significant ice loss for the last 40 years and the rate of ice loss has increased drastically since 2000.

Himalayan glaciers are the best-studied regions analysed for climate change and recessional patterns (Singh and Mishra, 2002a,b; Singh, 2004; Singh, 2013; Singh and Awasthi 2011a,b; Singh et al., 2017; Singh et al., 2019; Singh et al., 2020; Kumar et al., 2020 a,b, 2021, etc.). Over the last two decades, some groups have also worked on the palaeoclimate reconstruction and melting of glaciers (Eyles et al., 1983; Kick, 1986; Singh and Mishra, 2001; Singh et al., 2015, 2017; Saxena and Singh, 2017,

Singh et al., 2018, 2019; Richards et al., 2000). In other areas, satellite data have been used as effective tools for interpreting the depositional processes and reconstructing climate change and monsoon variability (Singh and Singh, 2005; Singh et al., 2009; Singh et al., 2010; Singh and Awasthi, 2011a,b; Singh et al., 2015; Saxena and Singh, 2017; Singh et al., 2018, 2019). Similar monitoring and surface processes have been identified for the glacial deposits in the Arctic (Singh and Ravindra, 2011a,b). However, the remote sensing technique in identifying the pattern of retreat for a group of glaciers has not attracted much attention. Therefore, this chapter evaluates the effects of global warming and climate change on the Gangotri, Meru, Thelu, Swetvarn and Chorabari glaciers of the Garhwal Himalayan region using RS and GIS techniques.

5.2   STUDY AREA

Gangotri glacier is the longest valley glacier of Garhwal Himalaya. It originates from the Chaukhamba group of peaks and flows towards the northwest. The melt-water of the Gangotri glacier at Gaumukh is the origin of the Bhagirathi River, which forms the River Ganga after the confluence with the Alkahnanda River at Devprayag. Geographically it lies between 30°44'–30°56' N latitude and 79°04'–79°15'E longitude (Figure 5.1) 4000–7000 m above mean sea level (msl). The Maiandi glacier, Swachhand glacier, Chaturangi glacier, and Raktvarn glacier feed from the right, while the Ghanohim glacier, Kirti glacier, and Meru glacier feed from the left to the Gangotri glacier. Singh and Mishra (2002a,b) have classified tributaries of the Gangotri glaciers into two categories: active and inactive. Active tributary glaciers are those which are still connected to the main glacier, contributing ice budget to it and also forming new landforms. Inactive tributary glaciers are those which are now detached from the main glacier but were connected to it in the past. These are neither contributing ice budget to the main glacier nor forming any new landform.

Meru glacier is a small tributary glacier, which meets with the Gangotri glacier from the left side. The length of the Meru glacier is ~ 7.50 km and it lies between 30° 51' 00"–30° 55' 30" N and 79° 01' 00"–79° 04' 30" E having an NE sloping trend (Figure 5.1). It was considered that the Meru glacier actively contributed to the ice budget of the Gangotri glacier but now it has detached from the main glacier and has become an inactive tributary (Singh and Mishra, 2002a,b).

Thelu and Swetvarn are the tributary glaciers of the Raktvarn glacier, situated within a Bhagirathi catchment of the Garhwal Himalayan region. Raktvarn glacier is one of the major tributaries of the Gangotri glacier, which meets it from the right side. The length of the Thelu glacier from the accumulation zone to its confluence with the trunk glacier is ~ 4.17 km which covers ~ 2.30 km^2 total area. The Swetvarn glacier is ~ 5.70 km in length and covers more than 6 km^2 of the total area. Both these glaciers flow from north to south at an elevation range between 6000 m and 4600 m.

Chorabari glacier is a part of the Mandakini River basin in the Alaknanda catchment. It is a ~7.50 km long valley glacier located between 79° 02' 0" E–79° 04' 0" E longitude and 30° 44' 0" N–30° 48' 0" N latitude and flows from north to south (Figure 5.1). The slope map of Chorabari glacier ranges between 0.34° and 59.75°, but

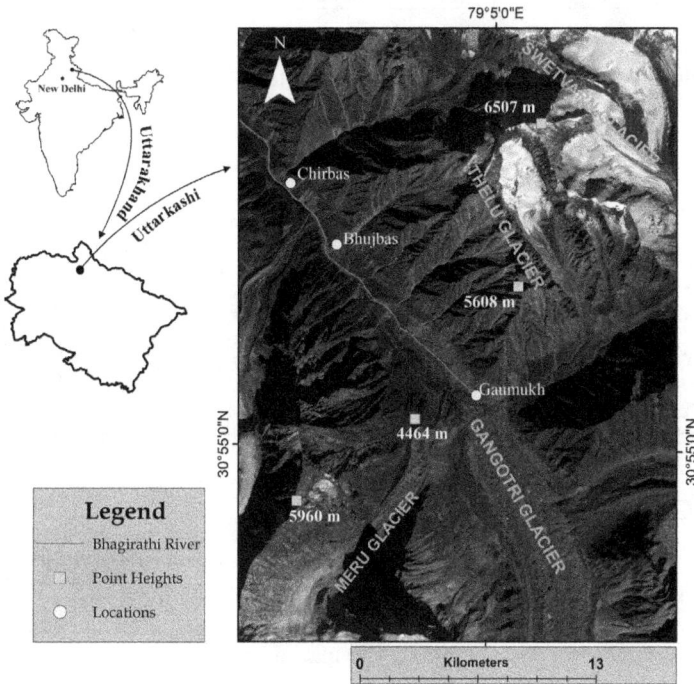

FIGURE 5.1   Satellite data (LISS IV) showing the location map of the study area.

most of the area comes under a slope range of between 0.34° and 23.87° (Figure 5.2). The general climate of the study area is humid-temperate in summer and dry-cold in winter (Dobhal et al., 2013).

5.3   METHODOLOGY

The declassified images of CORONA KHA, orthorectified images of LANDSAT (MSS, TM, ETM+, OLI), and LISS IV (Table 5.1) between 1960 and 2019 were utilized to carry out this work. The cloud-free data of the peak ablation period (September to November) were used to carry out the research work. The Shuttle Radar Topographic Mission (SRTM) Global Digital Elevation Model version-3 (GDEM-v3) was also utilized for the slope analysis and elevation changes.

5.3.1   Pre-processing of Data

Data pre-processing involves co-registration of images and conversion of digital numbers (DN) to physical units. Here the medium-resolution image of Thematic Mapper (TM) 1990 is treated as a principal image and rectified into the UTM projection system and WGS-1984 datum. The other downloaded images were rectified with respect to the principal image (TM 1990) through the image-to-image co-registration method in ERDAS Imagine software.

TABLE 5.1
The Satellite Data Set Used for the Present Study

Images	Path/row or scene ID	Resolution (m)	Date of acquisition
Corona KHA	DS1048-1134D F108_108_d	4	27/09/1968
Landsat MSS	156/39	57	23/11/1976
Landsat TM	145/39	28.5	15/11/1990
Landsat ETM+	145/39	30	20/10/2001
Landsat TM	146/39	28.5	12/11/2011
Resources at II LISS IV	097/49	5.8	17/10/2017
Landsat OLI	146/39	30m (15m for PAN)	13/11/2016

5.3.2 GLACIAL MAPPING

An onscreen manual digitization process for extracting the desirable information from the downloaded images was adopted to get the best results (Paul et al., 2004; Shukla et al., 2019), maximum information (Kääb, 2005), and was a more accurate method than an automated one (Raup et al., 2007). The glacial boundaries of the selected glaciers were delineated using different false colour composites (FCC), such as SWIR–NIR–Green, NIR–SWIR–RED, and NIR–Red–Green. The accumulation and ablation zone of the selected glaciers were demarcated based on high surface reflectance (ice of the accumulation zone appears white in FCC; ice of the ablation zone appears greyish white in FCC). It is very difficult to identify the correct snout position on raster images (satellite images), therefore various permutations and combinations adopted by the earlier works (the presence of stream water, ice-wall shadow, supraglacial lakes, and proglacial morphological features) were applied to locate the correct snout position. In the case of selected glaciers, the emerging stream helped to monitor the position of the snout.

5.3.3 UNCERTAINTY ESTIMATION

The uncertainty that occurred during length measurement was resolved by the following formula suggested by Hall et al. (2003), Bhambri et al. (2012), and Garg et al. (2017):

$$U_{retreat} = \sqrt{a^2 + b^2} + \sigma$$

where $U_{retreat}$ is uncertainty in retreat estimation, "a" and "b" are the resolutions of temporal images, and σ is the error in image registration.

5.4 RESULTS AND DISCUSSION

Figures 5.2 A and B clearly show how the retreating pattern and retreating rate of the Gangotri glacier have changed over time. It lost an overall 1460 m total length

(A)

(B)

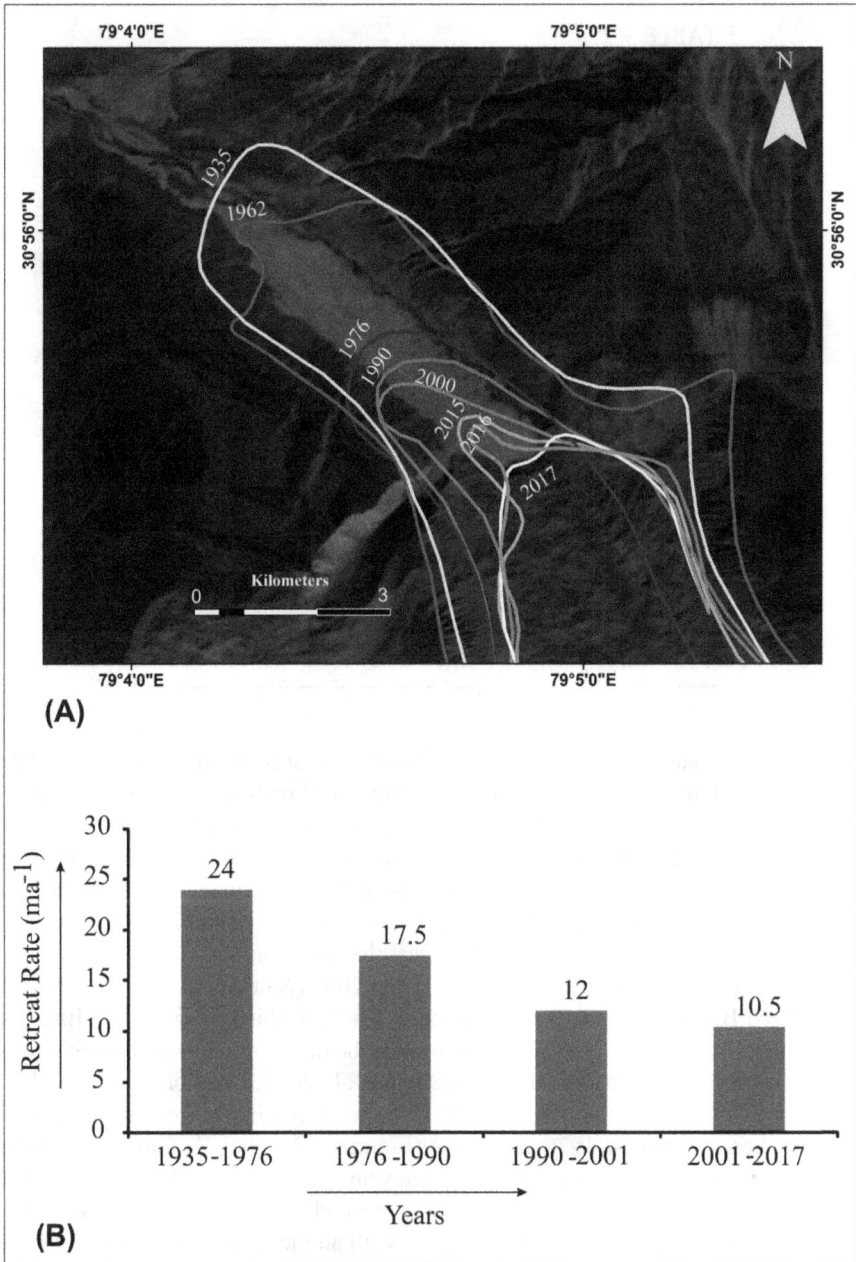

FIGURE 5.2 (A and B) The retreating scenario of the Gangotri glacier at different time intervals.

TABLE 5.2
Annual Retreat Rate of the Gangotri Glacier
during Different Time Intervals

S. no.	Year	Retreat rate (m year^{-1})
1	1935–1976	24
2	1976–1990	17.5
3	1990–2001	12
4	2001–2017	10.5

TABLE 5.3
Annual Retreat Rate of the Meru Glacier during
Different Time Intervals

S. no.	Year	Shift (m)	Change/year (m year^{-1})
1	1976–1990	174	12.42
2	1990–2001	90	7.72
3	2001–2011	60	6.00
4	2011–2019	92	11.50
5	1976–2019	416	9.67

between 1935 and 2017 with a variable rate of melt. It receded 24 m year^{-1}, 17.50 m year^{-1}, 13 m year^{-1}, 12 m year^{-1}, and 10.50 m year^{-1} between the time intervals of 1935–1976, 1976–1990, 1990–2001, and 2001–2017, respectively (Table 5.2).

The Gangotri glacier is one of the most studied and well-documented glaciers in the vicinity of Garhwal Himalaya and several published articles and journals have revealed that the Gangotri glacier has been in a state of retreat since 1850 (Mayewski and Jeschke, 1979). Available records suggest that the terminus of the Gangotri glacier was retreating at a higher rate from 1968 to 2000 (Naithani et al., 2001; Tangari et al., 2004; Bahuguna et al., 2007; Cruz et al., 2007; Ambinakudige, 2010; Bhambri et al., 2012; Bhattacharya et al., 2016). However, the rate of retreat was lessening from 2000 to 2015 (Bhattacharya et al., 2016; Singh et al., 2017; Bhambri et al., 2012).

The data analysis of the Meru glacier also shows continuous retreat, it retreated 12.42 m year^{-1} between 1976 and 1990 (Figures 5.3A and B; Table 5.3). Meru receded at a diminished retreat rate pf 8.18 m year^{-1} from 1990 to 2001. The trend of a diminished retreat rate lasted up to 2011 and receded to 6.00 m year^{-1} during 2001 to 2011. Recently the Meru glacier has receded with an increased rate of retreat (11.50 m year^{-1}).

Chitranshi et al. (2004) monitored the recession of the Meru glacier using SPOT, IRS–1C and IRS–1D satellites, and their observation suggested that the snout of the Meru glacier shifted 275 m (27.50 m year^{-1}) during 1977–1987, 120 m (9.23 m year^{-1}) during 1987–2000 and overall lost 395 m (17.17 m a^{-1}) length between 1977 and 2000. However, current research work suggests that the terminus of the Meru glacier

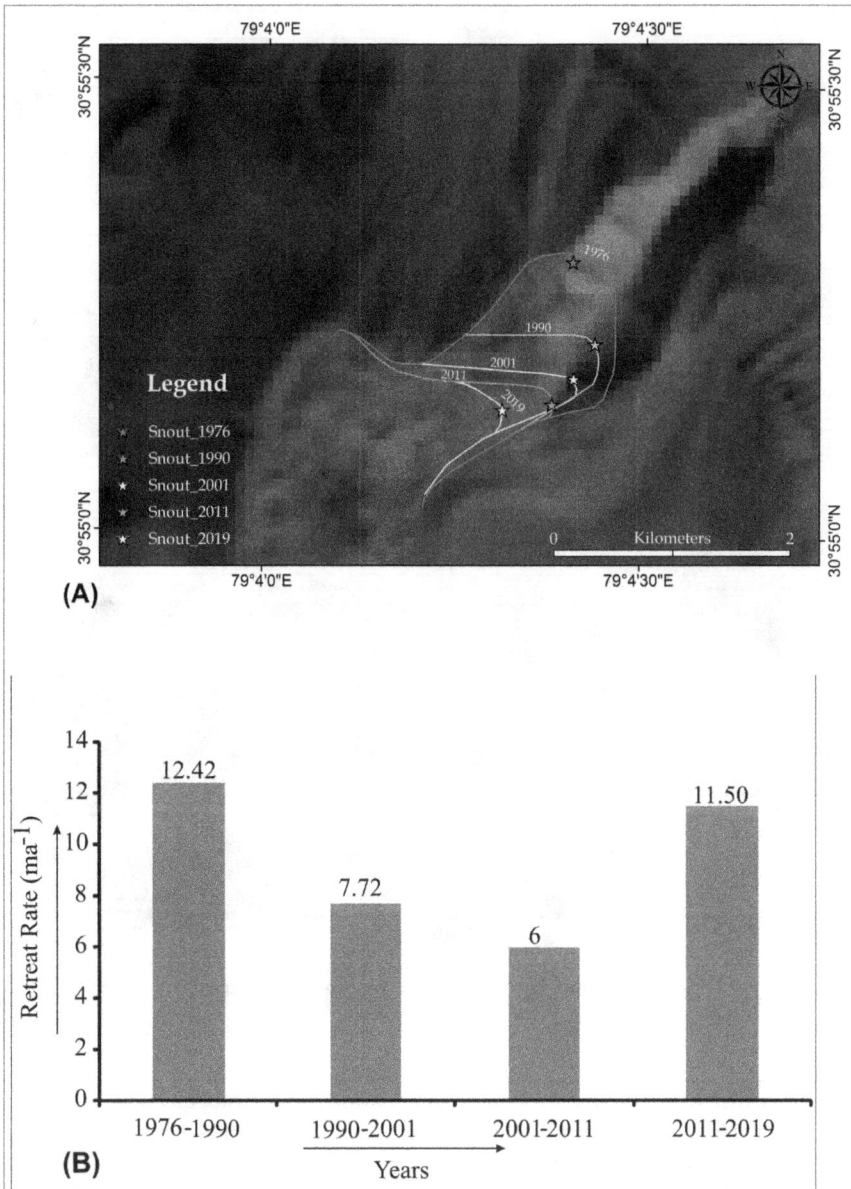

FIGURE 5.3   (A and B) The retreating pattern of the Meru glacier at different time intervals.

receded 174 m (12.42 m year⁻¹) in length during 1976–1990 and 90 m (8.18 m year⁻¹) length during 1990–2001, which contradicts the results of Chitranshi et al. (2004). This difference may be due to the difference in the time period, data resolution, and methodology used for deducing the changes. The observation of Bhattacharya et al. (2016) suggests that the Meru glacier overall lost 378 m length between 1965 and

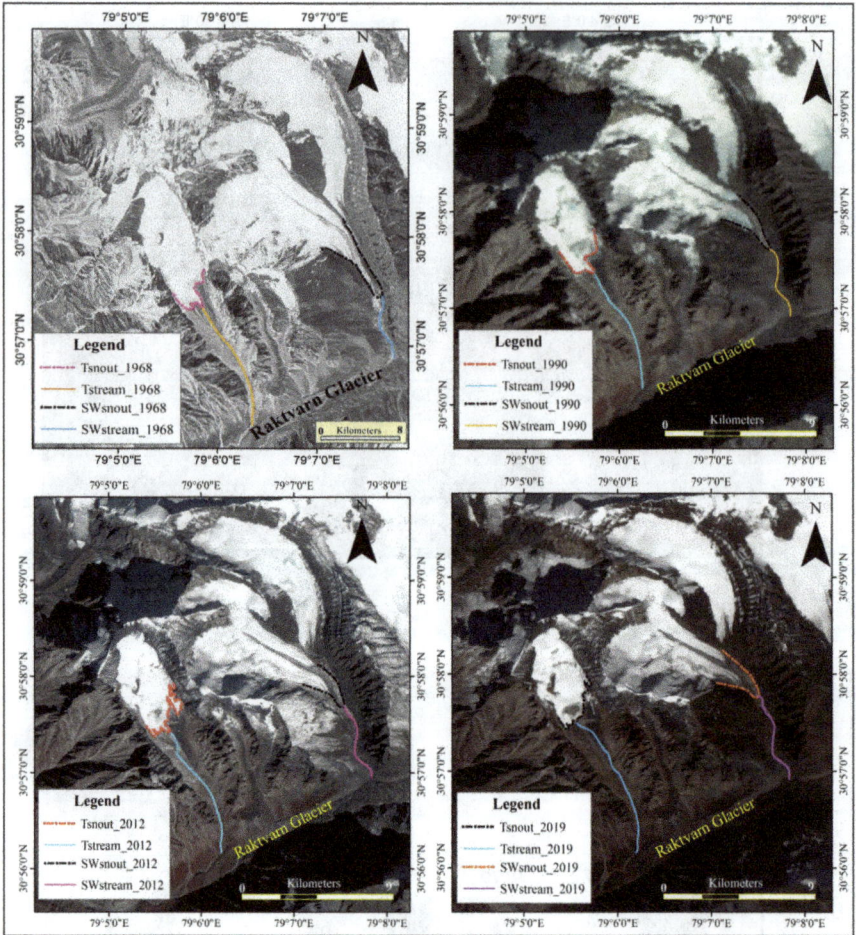

FIGURE 5.4 The retreating scenario of the Thelu and Swetvarn glaciers (during 1968, 1990, 2012, and 2019, respectively).

2015 and melted at 7.56 m year^{-1}. However, their analysis did not calculate the retreat rate of the Meru glacier temporally; our result for the first time calculates the recession rate of the Meru glacier for the time intervals of 1976–1990, 1990–2001, 2001–2011, and 2011–2019.

Thelu and Swetvarn glaciers retreated with retreat rates of 9.5 m year^{-1} and 12.40 m year^{-1}, respectively, during the time period of 1968–1990. It has been observed that the retreat rate of the selected glaciers reduced from 9.5 m year^{-1} to 7.18 m year^{-1} for Thelu glacier and 12.40 m year^{-1} to 8.63 m year^{-1} for the Swetvarn glacier during 1990–2012. However, during 2012–2019, the retreat rates of both glaciers increased very significantly (31.71 m year^{-1} for the Thelu glacier and 22.85 m year^{-1} for the Swetvarn glacier) (Figures 5.4 and 5.5A and B; Tables 5.4 and 5.5).

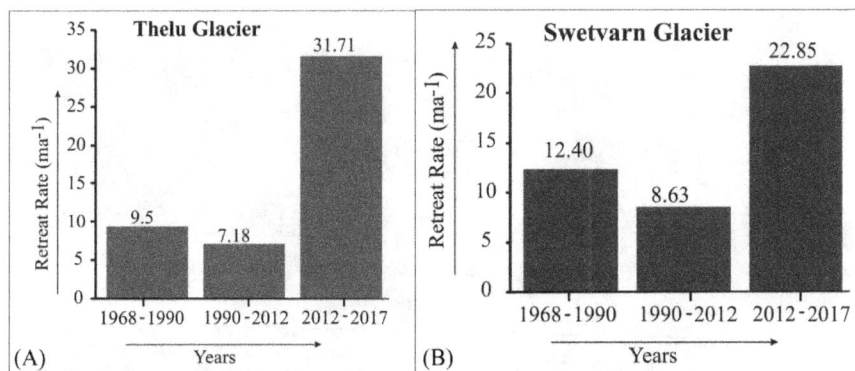

FIGURE 5.5   (A and B) The retreating rates of the Thelu and Swetvarn glaciers.

TABLE 5.4
Annual Retreat Rate of the Thelu Glacier during Different Time Intervals

S. no.	Year	Shift (m)	Change/year (m year⁻¹)
1	1968–1990	210 ± 36	9.5 ± 1.63
2	1990–2012	158 ± 37	7.18 ± 1.68
3	2012–2019	222 ± 23	31.71 ± 3.28
4	1968–2019	590	11.56

TABLE 5.5
Annual Retreat Rate of the Swetvarn Glacier during Different Time Intervals

S. no.	Year	Shift (m)	Change/year (m year⁻¹)
1	1968–1990	273 ± 36	12.40 ± 1.63
2	1990–2012	190 ± 37	8.63 ± 1.68
3	2012–2019	160 ± 23	22.85 ± 3.28
4	1968–2019	623	12.21

The recession rate of Thelu glacier was earlier monitored by Kotlia et al. (2009) and their observations suggested that the Thelu glacier receded 30.66 m year⁻¹ during 1962–2004 and lost 1288 m in length. Our results completely contradict the results of Kotlia et al. (2009) and suggest that the Thelu glacier melted with slow retreat rates of 7.40 m year⁻¹ and 7.18 m year⁻¹ between 1968–1990 and 1990–2012, respectively. This difference may be due to the use of the Survey of India (SOI) toposheet by Kotlia et al. (2009), because of its high uncertainty value (Bhambri et al., 2011; Chand and Sharma, 2015), while the current work excluded use of the SOI toposheet.

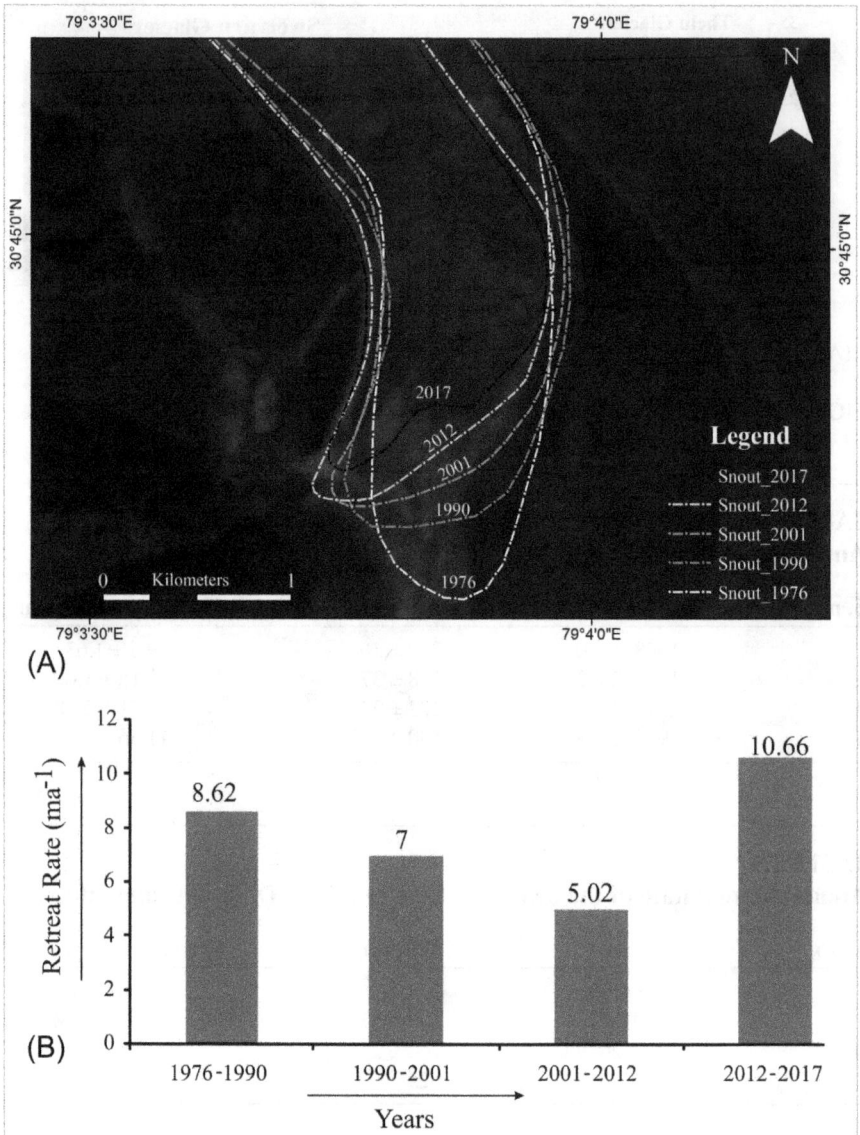

FIGURE 5.6   (A and B) Pattern of retreat of the Chorabari glacier between 1976 and 2017.

   The retreat of the Chorabari glacier has been monitored for the years 1976, 1990, 2001, 2012, and 2017, and a study revealed that it has receded continuously since 1976 (Figure 5.6A) at an inconsistent rate (Figure 5.6B). Chorabari retreated rapidly during 1976–1990 (8.62 m year[-1]) and the retreat rate decreased between 1990–2001 (7.00 m year[-1]) and 2001–2012 (5.02 m year[-1]). However, during 2012–2017, the glacier further retreated rapidly with a 10.66 m year[-1] retreat rate (Table 5.6).

TABLE 5.6
Annual Retreat Rate of the Chorabari Glacier during Different Time Intervals

S. no.	Year	Shift (m)	Change/year (m year^{-1})
1	1976–1990	120.81 ± 78	8.62 ± 5.57
2	1990–2001	77.06 ± 45	7.00 ± 4.09
3	2001–2012	55.22 ± 21	5.02 ± 1.90
4	2012–2017	53.34 ± 12	10.66 ± 2.40

The Chorabari glacier was monitored by Dobhal et al. (2013) between 1962 and 2010. Their observations suggested that the snout of the Chorabari glacier receded variably at different time intervals; it receded at 6.43 m year^{-1} (1962–1990), 6.31 m year^{-1} (1990–2003), and 9.30 m year^{-1} (2003–2010), with a total loss of length of 327 m. Mehta et al. (2014) also reported a retreat rate of 6 m year^{-1} from 1962 to 2003 and 9 m year^{-1} from 2003 to 2012 for this glacier. Our estimates are slightly different to the previous work. This difference may be due to the difference in the time span, resolution of satellite imageries, and methodology adopted for quantifying the changes. Dobhal et al. (2013) and Mehta et al. (2014) both utilized the Survey of India (SOI) toposheet (1962) and medium-resolution satellite images of ETM^{+} and ASTER to monitor the snout position of the Chorabari glacier. Since the SOI toposheet gives a high uncertainty value (Bhambri et al., 2011; Chand and Sharma, 2015), the current work completely excluded it and preferred the use of satellite images only (Table 5.1) for analysis.

In addition, the Geological Survey of India (GSI) also monitored the secular movement of the Chorabari glacier from 1992 to 1997 (Swaroop et al., 1999). Their records suggest that about 55 m length (11 m year^{-1}) and ~ 0.006 km^2 area have been lost during the study tenure (1992–1997). The current work completely contradicts these results.

The analysis of annual rainfall and annual mean temperature data between 1911 and 2011 in the Garhwal Himalaya region shows frequent catastrophic events for temperature and precipitation from 1960 to 1990 (Figure 5.7). These events directly influenced the melting process of the Garhwal Himalayan region to a great level. Bhutiyani et al. (2009), Singhvi and Krishnan (2014), and Dyurgerov and Meier (2000) have also mentioned that most parts of the Indian subcontinent experienced catastrophic variations in rainfall and temperature during 1960–1990.

The rainfall data also reflect a positive correlation with tree ring width chronology (Shekhar et al., 2018) of Garhwal Himalaya (Figure 5.8). While comparing both these proxies (rainfall data and tree ring data), we found many wet (1965–66, 1973, 1984, 2000, 2004, 2010) and dry (1972, 1977, 2006) events during 1960–2010.

The statistical analysis (Figure 5.7) further suggests that the reduced trend in precipitation and increased trend in temperature became more prominent after 1990 in the Uttarakhand region. Mishra (2017) has also mentioned ~ 13.05 cm annual rainfall decrease and ~ 0.46°C mean annual temperature increase during the last 100 years in the Uttarakhand region. The studies of Shekhar et al. (2010) and Negi et al. (2017)

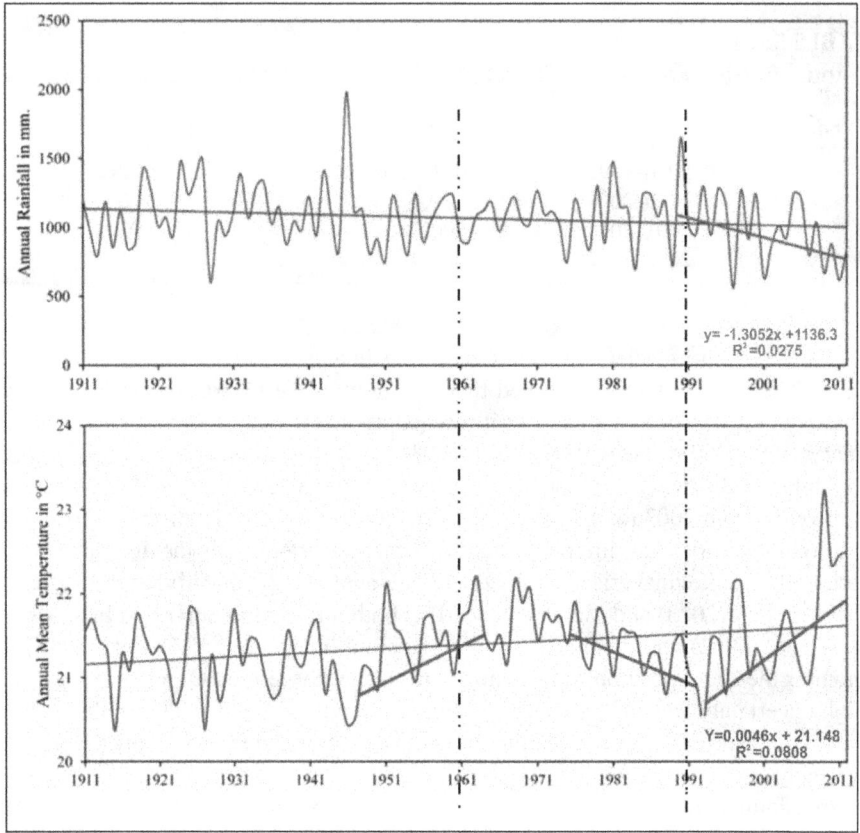

FIGURE 5.7   Trend of average annual rainfall and temperature of Uttarakhand region from 1901 to 2011.

also suggests that the maximum, minimum, and mean temperatures of Garhwal Himalaya followed rising trends, with total increases of 0.9°C, 0.19°C, and 0.65°C, respectively, during the last 25 years.

It was found that rising temperature resulted in increased liquid precipitation and decreased snowfall over the Himalayas during the winter season (Dimri and Mohanty, 2007; Thayyen et al., 2005). The instances of increased liquid precipitation and reduced solid precipitation are mainly attributed to rising temperatures, and such climatic conditions are seen as an impact of El Niño (Francou et al., 2004). It has been suggested that rising trends in liquid precipitation have a negative influence on the Himalayan glaciers and the frequency of hazards, such as avalanches and landslides, is expected to increase during the late winter season (Shekhar et al., 2010; Kulkarni and Karyakarte, 2014). Increased temperature has enhanced the deglaciation process and shifted the snowline upward (Kulkarni et al., 2005; Kumar et al., 2007; Wagnon et al., 2013).

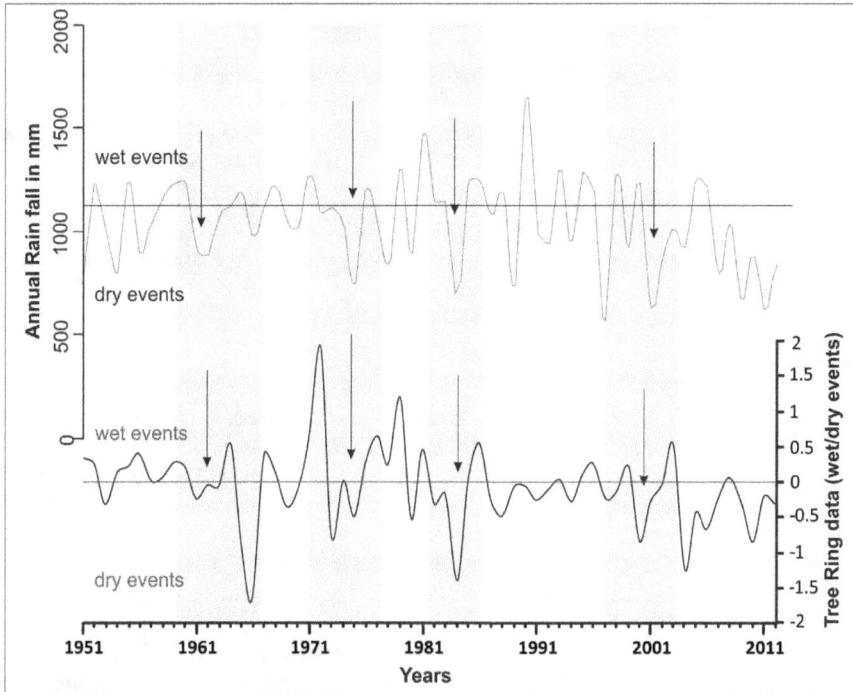

FIGURE 5.8   Tree ring width chronology positively correlates with the rainfall data.

5.5   CONCLUSIONS

Based on the above work it can be concluded that the health of the selected glaciers of Garhwal Himalaya is deteriorating continuously due to the impacts of climate change and global warming. The Gangotri Glacier lost 1460 m total length between 1935 and 2017 (17.80 m year^{-1}) and the rate of retreat has been variable in different time intervals. It receded 24 m year^{-1}, 17.50 m year^{-1}, 13 m year^{-1}, 12 m year^{-1}, and 10.50 m year^{-1} between the intervals of 1935–1976, 1976–1990, 1990–2001, and 2001–2017, respectively. Meru glacier lost 416 m of its total length during 1976–2019 at a 9.67 m year^{-1} retreat rate. Meanwhile, the Swetvarn glacier lost ~ 623 m total length with a 12.21 m year^{-1} retreat rate and the Thelu glacier lost ~ 590 m total length (11.56 m year^{-1}) from 1968 to 2019. The Chorabari glacier also lost ~ 306.43 ± 61 m total length and retreated at 8.62 m year^{-1} during 1976–1990, 7.00 m year^{-1} during 1990–2001, 5.02 m year^{-1} during 2001–2012, 10.66 m year^{-1} from 2012–2017, and considerable growth has been seen after 2012.

The melting of the glaciers in this region is mainly controlled by three factors: (1) catastrophic variation between rainfall and temperature between 1960 and 1990; (2) gradual increase in temperature and decrease in precipitation; and (3) increase in rainfall and decrease in snowfall.

ACKNOWLEDGEMENTS

We are thankful to the Head, Department of Geology, University of Lucknow, Lucknow, for providing the working facilities. Dhirendra Kumar gratefully acknowledges the UGC for financial assistance (PDFSS-2015-17-UTT-10685) and Anoop Kumar Singh is thankful for SERB N-PDF (PDF/2020/000251). Pawan Kumar Gautam gratefully acknowledged the UGC RGNF Fellowship SRF (RGNF-2017-18-SC-UTT-34263). We are also thankful to Dr N. Khare, Scientist G, Ministry of Earth Sciences Prithvi Bhawan, New Delhi, for allowing us to submit this chapter to this prestigious book.

REFERENCES

Ambinakudge, S. (2010) A study of the Gangotri glaciers retreat in the Himalayas using Landsat satellite images. *International Journal of Geoinformatics* 6(3): 7–12.

Bahuguna, I.M., Kulkarni, A.V., Nayak, S., Rathore, B.P., Negi, H.S., Mathur, P. (2007) Himalayan glacier retreat using IRS 1c PAN stereo data. *International Journal of Remote Sensing* 28: 437–442.

Bhambri, R., Bolch, T., Chaujar, R.K., Kulshreshtha, S.C. (2011) Glacier changes in the Garhwal Himalayas, India, from 1968 to 2006 based on remote sensing. *Journal of Glaciology* 57: 543–556.

Bhambri, R., Bolch, T., Chaujar, R.K. (2012) Frontal recession of Gangotri Glacier, Garhwal Himalayas, from 1965 to 2006 measured through high-resolution remote sensing data. *Current Science* 102: 489–494.

Bhattacharya, A., Bolch, T., Mukherjee, K., Pieczonka, T., KropáčEk, J., Buchroithner, M.F. (2016) Overall recession and mass budget of Gangotri Glacier, Garhwal Himalayas, from 1965 to 2015 using remote sensing data. *Journal of Glaciology* 62(236): 1115–1133.

Bhutiyani, M.R., Kale, V.S., Pawar, N.J. (2009) Climate change and the precipitation variations in the north-western Himalaya: 1866–2006. *International Journal of Climatology* 30: 535–548.

Burns, P., Nolin, A. (2014) Using atmospherically-corrected Landsat imagery to measure glacier area change in the Cordillera Blanca, Peru from 1987 to 2010. *Remote Sensing of Environment* 140: 165–178.

Chand, P., Sharma, M.C. (2015) Glacier changes in the Ravi basin, north-western Himalaya (India) during the last four decades (1971–2010/13). *Global and Planetary Change* 135: 133–147.

Chitranshi, A., Sangewar, C.V., Srivastava, D., Puri, V.M.K., Dutta, S.S. (2004) Recession pattern of Meru Bamak Glacier, Bhagirathi Basin, Uttaranchal. In: Srivastava D., Gupta KR., Mukerji S. (Eds.) *Geological Survey of India Special Publication* 80: 33–38.

Cogley, J.G. (2011) Present and future states of Himalaya and Karakoram glaciers. *Annals of Glaciology* 52(59): 69–73.

Cruz, R.V., Harasawa, H., Lal, M., Wu, S., Anokhin, Y., Punsalmaa, B., Honda, Y., Jafari, M., Li, C., Huu, N.N. (2007) Asia, in Climate Change 2007: Impacts, Adaptation and Vulnerability. In: *Contribution of Working Group II to the Fourth Assessment Report of the intergovernmental panel on Climate Change*, Parry, M., Canziani, O., Palutikof, J., van der Linden, P., Hansen, C. (Eds.), pp. 469–506.Cambridge: Cambridge University Press.

Dimri, A.P., Mohanty, U.C. (2007) Location-specific prediction of maximum and minimum temperature over the Western Himalayas. *Meteorological Applications* 14: 79–93.

Dobhal, D.P., Mehta M., Srivastava, D. (2013) Influence of debris cover on terminus retreat and mass changes of Chorabari Glacier, Garhwal region, central Himalaya, India. *Journal of Glaciology* 59(217): 961–971.

Dyurgerov, M.B., Meier, M.F. (2000) Twentieth-century climate change: evidence from small glaciers. *Proceedings of the National Academy of Science* 97(4): 1406–1411.

Eyles, N., Eyles, C.H., Miall, A.D. (1983) Lithofacies types and vertical profile models; an alternative approach to the description and environmental interpretation of glacial diamict and diamictite sequences. *Sedimentology* 30: 393–410.

Francou, B., Vuille, M., Favier, V., Ca´ceres, B. (2004) New evidence for an ENSO impact on low-latitude glaciers: Antizana 15, Andes of Ecuador, 0°28'S. *Journal of Geophysical Research* 109: 1–17.

Gardelle, J., Berthier, E., Arnaud, Y., Kaab, A. (2013) Region-wide glacier mass balances over the Pamir-Karakoram-Himalaya during 1999–2011. *Cryosphere* 7(6): 1263–1286.

Garg, P.K., Shukla, A., Tiwari, R.K., Jasrotia, A.V. (2017) Assessing the status of glaciers in part of the Chandra basin, Himachal Himalaya: A multiparametric approach. *Geomorphology* 284: 99–114.

Hall, D.K., Bayr, K.J., Schoner, W., Bindschadler, R.A., Chien, J.Y.L. (2003) Consideration of the errors inherent in mapping historical glacier positions in Austria from the ground and space (1893–2001). *Remote Sensing of Environment* 86: 566–577.

Hewitt, K. (2005) The Karakoram anomaly? Glacier expansion and the 'elevation effect', Karakoram Himalaya. *Mountain Research and Development* 25: 332–340.

IPCC (1990) Climate change. In: Houghton J.T., Jenkins G.J., Ephraums J.J. (Eds.) The *IPCC Scientific Assessment*, 362 pp. Cambridge: Cambridge University Press.

IPCC (1992) Climate change 1992. In: Houghton J.T., Callander B.A., Varney S.K. (Eds) *The Supplementary Report to the IPCC Scientific Assessment*, 200 pp. Cambridge: Cambridge University Press.

Kääb, A. (2005) Combination of SRTM3 and repeat ASTER data for deriving alpine glacier flow velocities in the Bhutan Himalaya. *Remote Sensing of Environment* 94: 463–474.

Kick, W. (1986) Glacier mapping for an inventory of the Indus drainage basin: current state and future possibilities. *Annals of Glaciology* 8: 102–105.

Kotlia, B.S., Dumka, R.K., Miral, M.S., Joshi, L.M., Kumar, K. (2009) Recession rate of a tributary of Gangotri glacier system, Garhwal Himalaya-India. In: Srivastava, A., Roy, I. (Eds.) *Bio-Nano-Geo Science the future challenge*, pp.141–146. Ane Books Pvt. Ltd: New Delhi.

Kulkarni, A.V., Rathore, B.P., Mahajan, S., Mathur, P. (2005) Beas basin, Himachal Pradesh. *Current Science* 88(11): 1844–1850.

Kulkarni, A.V., Karyakarte, Y. (2014) Observed changes in Himalayan glaciers. *Current Science* 106(2): 237–244.

Kumar, D., Singh, A.K., Singh, D.S. (2020a) Spatio-temporal fluctuations over Chorabari glacier, Garhwal Himalaya, India between 1976 and 2017. *Quaternary International* 575–576: 178–189.

Kumar, D., Singh, A.K., Taloor, A., Singh, D.S. (2020b) Recessional pattern of Thelu and Swetvarn glaciers between 1968 and 2019, Bhagirathi basin, Garhwal Himalaya, India. *Quaternary International* 575-576: 227–235.

Kumar, R., Hasnain, S., Wagnon, P., Arnaud, Y., Chevallier, P., Linda, A., Sharma, P. (2007) *Climate change signal detected through mass balance measurement on benchmark glacier, Himachal Pradesh, India. Climate and anthropogenic impacts on the variability of water resources*, pp. 65–74. Technical Document in Hydrology 80. Paris: UNESCO/UMR 5569, Hydro Sciences Montpellier.

MaMayewski, P.A., Jeschke, P.A. (1979) Himalayan and trans-Himalayan glacier fluctuations since A.D. 1812. *Arctic, Antarctic, and Alpine Research* 11(3): 267–287.

Maurer, J.M., Schaefer, J.M., Rupper, S., Corley, A. (2019) Acceleration of ice loss across the Himalayas over the past 40 years. *Science Advances* 5: 1–12.

Mehta, M., Dobhal, D.P., Kesarwani, K., Pratap, B., Kumar, A., Verma, A. (2014) Monitoring of glacier changes and response time in Chorabari Glacier, Central Himalaya, Garhwal, India. *Current Science* 107(2): 281–289.

Mishra, A. (2017) Changing temperature and rainfall patterns of Uttarakhand. *International Journal of Environmental Sciences & Natural Resources* 7(4): 1–5.

Naithani, A.K., Nainwal, H.C., Sati, K. K., Prasad, C. (2001) Geomorphological evidence of retreat of the Gangotri glacier and its characteristics. *Current Science* 80: 87–94.

Negi, H.S., Datt, P., Thakur, N.K., Ganju, A., Bhatia, V.K., Vinay, K.G. (2017) Observed spatio-temporal changes of winter snow albedo over the northwest Himalayas. *International Journal of Climatology* 37(5): 2304–2317.

Paul, F., Huggel, C., Kääb, A. (2004) Combining satellite multispectral image data and a digital elevation model for mapping debris-covered glaciers. *Remote Sensing of Environment* 89(4): 510–518.

Raup, B., Kääb, A., Kargel, J.S., Bishop, M.P., Hamilton, G., Lee, E., Paul, F., Rau, F., Soltesz, D,, Khalsa, S.J., Beedle, M. (2007) Remote sensing and GIS technology in the Global Land Ice Measurements from Space (GLIMS) project. *Computational Geosciences* 33: 104–125.

Richards, B.W.M., Benn, D.I., Owen, L.A., Rhodes, E.J., Spencer, J.Q. (2000) Timing of Late Quaternary glaciations south of Mount Everest in the Khumbu Himal, Nepal. *Geological Society of America Bulletin* 112: 1621–1632.

Saxena, A., Singh D.S. (2017) Multiproxy records of vegetation and monsoon variability from the lacustrine sediments of eastern Ganga Plain since 1350 A.D. *Quaternary International* 444(A): 24–34.

Scherler, D., Bookhagen, B., Strecker, M.R. (2011) Spatially variable response of Himalayan glaciers to climate change affected by debris cover. *Nature Geoscience* 4(3): 156–159.

Shekhar, M.S., Chand, H., Kumar, S., Srinivasan, K., Ganju, A. (2010) Climate change studies in the Western Himalayas. *Annals of Glaciology* 51(54): 105–112.

Shekhar, M., Pal, A.K., Bhattacharyya, A., Ranhotra, P.S., Roy, I. (2018) Tree-ring-based reconstruction of winter drought since 1767 CE from Uttarkashi, Western Himalaya. *Quaternary International* 479: 58–69.

Shukla, A., Garg, P. (2019) Evolution of a debris-covered glacier in the western Himalaya during the last four decades (1971–2016): A multiparametric assessment using remote sensing and field observations. *Geomorphology* 341: 1–14.

Singh, A.K., Kumar, D., Kumar, V., Singh, D.S. (2020) Study of temporal response (1976–2019) and associated mass movement event (during 2017) of Meru glacier, Bhagirathi valley, Garhwal Himalaya, India. *Quaternary International* 565: 12–21.

Singh, D. S., Dubey, C.A., Kumar, D., Vishawakarma, B., Singh, A.K., Tripathi, A., Gautam, P.K., Bali, R., Agarwal, K.K., Sharma, R. (2019) Monsoon variability and major climatic events between 25 and 0.05 Ka BP using sedimentary parameters in the Gangotri glacier region, Garhwal Himalaya, India. *Quaternary International* 507: 148–155.

Singh, D.S. (2004) Late Quaternary morpho-sedimentary processes in the Gangotri glacier area Garhwal Himalaya, India. *Geological Survey of India* 80: 97–103.

Singh, D.S. (20130 Snow melts ephemeral streams in the Gangotri glacier area, Garhwal Himalaya, India. In: Kotlia, B.S. (Ed.) *Holocene: Perspectives, Environmental Dynamics and Impact Events*, pp. 157–164. Nova Science Publishers: USA.

Singh, D.S., Awasthi, A. (2011a) Implication of drainage basin parameters of Chhoti Gandak river, Ganga Plain, India. *Journal of the Geological Society of India* 78(2): 370–378.

Singh, D.S., Awasthi, A. (2011b) Natural hazards in the Ghaghara river area, Ganga Plain, India. *Nature Hazards* 57: 213–225.

Singh, D.S., Awasthi, A., Bhardwaj, V. (2009) Control of tectonics and climate on Chhoti Gandak River Basin, East Ganga Plain, India. *Himalayan Geology* 30(2): 147–154.

Singh, D.S., Awasthi, A., Nishat, R. (2010) Impact of climate change on the rivers of Ganga plain. *International Journal of Rural Development and Management Standards* 4(1): 1–8.

Singh, D.S., Dubey, C.A., Kumar, D., Kumar, P., Ravindra, R. (2018) Climate events between 47.5 and 1 ka BP in glaciated terrain of the Ny-Alesund region, Arctic, using geomorphology and sedimentology of diversified morphological zones. *Polar Science* 18:, 123–134.

Singh, D.S., Gupta, A.K., Sangode, S.J., Clemens, S.C., Srivastava, P., Prajapati, S.K., Prakasam, M. (2015) Multiproxy record of monsoon variability from the Ganga Plain during 400-1200 A.D. *Quaternary International* 371: 157–163.

Singh, D.S., Mishra, A. (2001) Gangotri glacier characteristics, retreat and processes of sedimentation in the Bhagirathi valley. *Geological Survey of India* 65(3): 17–20.

Singh, D.S., Mishra, A. (2002a) Role of tributary glaciers on landscape modification in the Gangotri glacier area, Garhwal Himalaya, India. *Current Science* 82(5): 101–105.

Singh, D.S., Mishra, A. (2002b) Gangotri glacier system: An analysis using GIS technique. In: Pant, C.C., Sharma, A.K. (Eds.) *Aspects of Geology and Environment of the Himalaya*, pp. 349–358. Gyano: Prakashan, Nainital.

Singh, D.S., Ravindra, R. (2011a) Control of glacial and fluvial environments in the Ny-Alesund region. *The Arctic. Mausam* 62(4): 641–646.

Singh, D.S., Ravindra, R. (2011b) Geomorphology of the Midre loven glacier, ny-Alesund, Svalbard, Arctic. In: Singh, D.S., Chhabra, N.L. (Eds.) *Geological Processes and Climate Change*, pp. 269–281. Macmillan Publishers India Ltd.

Singh, D.S., Singh, I.B. (2005) Facies architecture of the Gandak Megafan, Ganga Plain, India. *Palaeontological Society of India* 12: 125–140.

Singh, D.S., Tangri, A.K., Kumar, D., Dubey, C.A., Bali, R. (2017) The pattern of retreat and related morphological zones of Gangotri glacier, Garhwal Himalaya, India. *Quaternary International* 444: 172–181.

Singhvi, A.K., Krishnan, R. (2014) *Past and the Present Climate of India. Landscapes and Landforms of India*, pp. 15–23. Springer: Netherlands.

Swaroop, S., Oberoi, L.K., Srivastava, D., Gautam, C.K. (1999) Recent fluctuations in snout front of Dunagiri and Chaurabari glaciers, Dhauliganga and Mandakini-Alaknanda basins, Chamoli district, Uttar Pradesh. *Geological Survey of India, Special Publication* 53: 77–81

Tangri, A.K., Chandra, R., Yadav, S.K.S. (2004) Temporal monitoring of the snout, equilibrium line and ablation zone of Gangotri glacier through remote sensing and GIS techniques– an attempt at deciphering the climatic variability. *Geological Survey of India, Special Publication* 80: 145–153.

Thayyen, R.J., Gergan J.T., Dobhal D.P. (2005) Monsoonal control on glacier discharge and hydrographic characteristics, a case study of Dokriani glacier, Garhwal Himalaya, India. *Journal of Hydrology* 36(1–4): 37–49.

Wagnon, P., Vincent, C., Arnaud, Y., Berthier, E., Vuillermoz, E., Gruber, S., Dumont, M., Shea, J.M., Stumm, D. (2013) Seasonal and annual mass balances of Mera and Pokalde glaciers (Nepal Himalaya) since 2007. *Cryosphere* 7(6): 1769–1786.

6 Understanding Global Warming through Polar Ice

Shabnam Choudhary, Syed Mohammad Saalim, and Neloy Khare

6.1 INTRODUCTION

Global warming is the long-term heating of the Earth's climate system that has been observed since the pre-industrial period (between 1850 and 1900) due to human activities, primarily fossil fuel burning, which increases heat-trapping greenhouse gas levels in the Earth's atmosphere. It is most commonly measured as the average increase in the Earth's global surface temperature. Changes observed in the Earth's climate since the early 20th century are primarily driven by human activities, particularly fossil fuel burning, which increases heat-trapping greenhouse gas levels in the Earth's atmosphere, thereby raising the Earth's average surface temperature. These human-produced temperature increases are commonly referred to as global warming. Since the pre-industrial period, anthropogenic activities are estimated to have increased the Earth's global average temperature by about 1 degree Celsius, a figure that is currently increasing by 0.2 degrees Celsius per decade. Most of the current warming trend is primarily (greater than 95 per cent probability) the result of human activity since the 1950s and it is proceeding at an unprecedented rate over decades to millennia.

Natural processes can also contribute to climate change, including internal variability (e.g., cyclical ocean patterns like El Niño, La Niña and the Pacific Decadal Oscillation) and external forcings (e.g., volcanic activity, changes in the Sun's energy output, variations in the Earth's orbit), however, this change in climate does not take place at an unprecedented rate. Climate data records provide evidence of climate change key indicators, such as global land and ocean temperature increases; rising sea levels; ice loss at the Earth's poles and in glaciers; increased frequency and severity of changes in extreme weather such as hurricanes, heatwaves, wildfires, droughts, floods and precipitation; and cloud and vegetation cover changes.

When solar radiation hits snow and ice, approximately 90% of it is reflected into space. As global warming causes more snow and ice to melt each summer, the ocean and land that were underneath the ice are exposed at the Earth's surface. Because they are darker in colour, the ocean and land absorb more incoming solar radiation

DOI: 10.1201/9781003364115-6

and then release heat to the atmosphere. This causes more global warming. In this way, melting ice causes more warming and more melting of ice. This is known as ice–albedo feedback.

Global warming is causing soils in the polar regions that have been frozen for as long as 40,000 years to thaw. As they thaw, carbon trapped within the soils is released into the atmosphere as carbon dioxide and methane. These gases, released into the atmosphere, cause more warming, which then thaws the frozen soil.

Sea levels have been rising by about 1–2 millimetres each year as the Earth has become warmer. Some of this rise in sea levels is due to melting glaciers and ice sheets which add water to the oceans that was once trapped on land. Some glaciers and ice sheets are particularly vulnerable. Global warming has caused them to be less stable, to move faster toward the ocean, and add more ice into the water. These areas with less stable ice include the Greenland Ice Sheet and the West Antarctic Ice Sheet. If the Greenland Ice Sheet melted or moved into the ocean, the global sea level would rise approximately 6.5 metres. If the West Antarctic Ice Sheet were to melt or move into the ocean, the global sea level would rise by approximately 8 metres. Therefore, polar ice plays an important role in the climate of the Arctic and Antarctic regions.

6.2   SEA ICE AND ITS ROLE IN MODIFYING THE CLIMATE

Sea ice is frozen seawater that floats on the ocean surface. It forms in both the Arctic and the Antarctic in each hemisphere's winter; it retreats in the summer but does not completely disappear. This floating ice has a profound influence on the polar environment, influencing ocean circulation, weather, and regional climate. About 15% of the world's oceans are covered by sea ice for part of the year.

As ice crystals form at the ocean surface, they expel salt, which increases the salinity of the underlying waters. This cold, salty water is dense and can sink to the ocean floor, where it flows back toward the equator. The sea ice layer also restricts wind and wave action near coastlines, lessening coastal erosion and protecting ice shelves. Sea ice also creates an insulating cap across the ocean surface, which reduces evaporation and heat loss to the atmosphere. As a result, the weather over ice-covered areas tends to be colder and drier than it would be without ice.

Icebergs and frozen seawater melt in warm temperatures but are not significant contributors to sea level rise. This is because the volume of water they displace as ice is about the same as the volume of water they add to the ocean when they melt. As a result, the sea level does not rise when sea ice melts. Sea ice does not influence sea level, because it is already floating in the ocean and already displacing its weight. Melting sea ice won't raise ocean levels more than melting glaciers and ice sheets.

The influence of sea ice on the Earth is not regional, rather it's global. The white surface reflects far more sunlight to space than ocean water does, i.e., ice has a high albedo. As more ice melts and exposes more dark water, the water absorbs more sunlight. The sun-warmed water then melts more ice. Over several years, this positive feedback cycle (the ice–albedo feedback) can influence the global climate.

6.3   ICE LOSS EFFECTS ON THE ARCTIC

The Arctic is regarded as one of the fastest-warming regions worldwide and is heating at twice the global average. The extension of summer sea ice in the Arctic is shrinking (Stroeve et al., 2012), most likely caused by the anthropogenic emission of greenhouse gases inducing global warming (Notz and Marotzke, 2012; Notz and Stroeve, 2016). In modelled scenarios of the future, this global warming is accelerating the melting of Arctic ice sheets (especially, the Greenland Ice Sheet), which is assumed to contribute largely to global sea level rise (Alley et al., 2005; Gregory and Huybrechts, 2006). These examples illustrate that the Arctic plays an important role in the global climate system. Due to its exceptionally sensitive responses to climate variability, it can be regarded as an "early warning system", especially in the context of the ongoing anthropogenic climate change. Temperatures in the Arctic are rising twice as rapidly as the global mean owing to positive feedback mechanisms – this effect is known as "Arctic amplification" (Serreze and Barry, 2011). Arctic sea ice decline can lead to a slow-down of the Atlantic meridional overturning circulation (AMOC). The current Arctic Sea ice decline could contribute about 40% to the AMOC weakening over the next 60 years. This effect is related to the warming and freshening of the upper ocean in the Arctic, and the subsequent spread of generated buoyancy anomalies downstream where they affect the North Atlantic deep convection sites and hence the AMOC on multi-decadal timescales. In the North Atlantic, water heated near the equator travels north at the surface of the ocean into cold, high latitudes where it becomes cooler. As it cools, it becomes denser and, because cold water is denser than warm water, it sinks to the deep ocean where it travels south again. More warm surface water flows in to take its place, cools, sinks, and the pattern continues. However, melting Arctic Sea ice and melting Greenland glaciers could change this pattern of ocean currents, or stop it altogether. Recent research shows that Arctic Sea ice is melting due to climate warming. The melting ice causes freshwater to be added to the seawater in the Arctic Ocean which flows into the North Atlantic. The added freshwater makes the seawater less dense. This has caused the North Atlantic to become fresher over the past several decades and has caused the currents to slow. Less dense water will not be able to sink and flow through the deep ocean, which may disrupt or stop the pattern of ocean currents in the region. Scientists estimate that, given the current rate of change, these currents could stop within the next few decades. Even though warming is disrupting ocean currents, stopped or slowed currents in the North Atlantic would cause regional cooling in Western Europe and North America. The ocean currents carry warmth from the tropics up to these places, which would no longer happen. If the currents were to stop completely, the average temperature of Europe would cool 5–10 degrees Celsius. There would also be impacts on fisheries and hurricanes in the region. The currents in the North Atlantic are part of a global pattern called thermohaline circulation, or the global ocean conveyor. If the thermohaline circulation was to stop, this would not be the first time that the global ocean conveyor belt has halted. There is evidence from sedimentary rocks and ice cores that it has shut down several times in the past which caused climate changes. One of the most well-known, called the Younger Dryas Event, happened about 12,700 years ago and caused temperatures to cool about 5°C in the region.

6.3.1 MELTING OF GREENLAND ICE SHEET

The Greenland ice sheet is a vast body of ice covering 1,710,000 square kilometres, roughly 79% of the surface of Greenland. It is the second largest ice body in the world, after the Antarctic ice sheet (Figure 6.1). The ice sheet is almost 2,900 kilometres long, oriented in a north–south direction, and its greatest width is 1,100 kilometres at a latitude of 77°N, near its northern margin. The mean altitude of the ice is 2,135 metres. The thickness is generally more than 2 km, and it is over 3 km at its thickest point. In addition to the large ice sheet, smaller ice caps, as well as glaciers, cover between 76,000 and 100,000 square kilometres around the periphery. If the entire 2,850,000 cubic kilometres of ice were to melt, it would lead to a global sea level rise of 7.2 m (Houghton et al., 2001).

Greenland's melting glaciers plunge into Arctic waters via steep-sided inlets, or fjords, which are among the main contributors to global sea level rise in response to climate change. Gaining a better understanding of how warming ocean water affects these glaciers will help improve predictions of their fate. Such predictions could in turn be used by communities around the world to better prepare for flooding and mitigate coastal ecosystem damage. At the edges of Greenland, the vast glaciers extending from the ice sheet travel slowly down valleys like icy conveyor belts, which pour into the fjords and then melt or break off (or calve) as icebergs. The ice is replenished by snowfall that is compressed over time into the ice pack. If the ice sheet were in

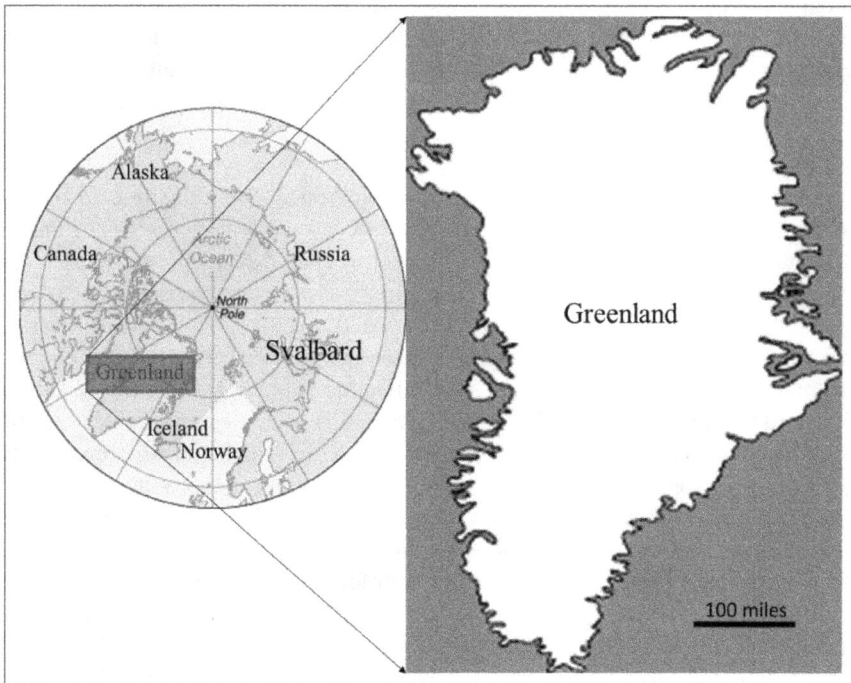

FIGURE 6.1   Map showing Greenland Ice Sheet.

balance, the amount of snow accumulating on the top would roughly equal the ice lost from melt, evaporation, and calving. However, previous observations have shown that the ice sheet has been out of balance since the 1990s. Melting has accelerated and calving has increased. The rate of ice being lost to the ocean is exceeding the supply from the ice sheet. This is causing the ice sheet to shrink and the glaciers to retreat toward land. The main cause of such glacier retreat is the process of under-cutting, which is driven by factors such as the amount of meltwater flowing from the glacier and the warm layer of salty water at the base of the fjord. During the summer months, increasing air temperatures heat the glacier's surface, creating pools of melt-water. These pools leak through the ice and flow from the glacier into rivers below the surface. As the meltwater flows into the sea, it encounters the warmer salty water at the bottom of the fjord. Glacial meltwater doesn't contain salt, so it is less dense than saltwater and thus rises as a plume. The plume drags the warmer ocean water into contact with the glacier's base. The amount of undercutting depends on the depth of the fjord, the warmth of the ocean water, and the amount of meltwater flowing out from beneath the glacier. As the climate warms, the amount of meltwater will increase and the ocean temperature will rise, two factors that boost the undercutting process.

The melt water contributes to rising sea level as: (1) water from the ice surface finds its way to the ocean, contributing directly to sea level rise, and (2) the water's ability to speed flowing glacial ice towards the ocean also contributes to sea level rise. If all the ice that is on Greenland were to melt or calve into the ocean, the global sea level would rise 7.2 meters (21 feet). Currently, melting Greenland ice increases the global sea level by about 0.5 millimetres each year. If the ice melts faster, then the sea level will rise faster. A better understanding of how Greenland's ice is changing will help us understand more about how sea levels will change in the future.

6.3.2   ICE LOSS EFFECTS ON ANTARCTICA

Similarly to the Arctic, the surface of the ocean around Antarctica freezes over in the winter and melts back each summer. Antarctic sea ice usually reaches its annual maximum extent in mid-to-late September and reaches its annual minimum in late February or early March. The 2020 maximum extent (on September 28, 2020) was 7.32 million square miles, which was above the 1981–2010 climatological average but not a record high. The 2021 minimum extent, on or near February 21, 2021, was 1.01 million square miles, below the 1981–2010 climatological average but well above the record low recorded in 2017 (NOAA). The timing of the seasonal cycles isn't the only way that Antarctic sea ice differs from the Arctic. The major difference is the larger range between austral winter maximum extent and summer minimum extent. Antarctic sea ice extends to about 7.2 million square miles in winter, versus 6 million square miles in the Arctic; the Antarctic summer minimum is about 1.1 million square miles versus 2.5 million square miles for the Arctic. The difference between the sea-sonal extremes of the Arctic and Antarctic is attributed to the basic geography. The Arctic is an ocean basin largely surrounded by land. Sea ice forms over the North Pole itself, however, its expansion is checked by Eurasia, North America, and Greenland. The Antarctic is a continent surrounded by a vast ocean. Sea ice can expand freely across the Southern Ocean in winter, but it cannot reach closer to the South Pole as the Antarctic coastline will not allow it. Because it forms at lower, warmer latitudes, less Antarctic sea ice survives the summer. On average, about 40 per cent of the Arctic Ocean's winter ice cover remains at the summer minimum, whereas in the Southern Ocean only about 15% does. Because so little Antarctic ice persists through the summer, the majority of Antarctica's sea ice is only one winter old at most. As a result, Antarctic sea ice is relatively thin, often 1 metre (about 3 feet) or less. (In the Arctic, multiyear ice that survives at least one summer is generally 3–4 metres thick, and even seasonal ice that formed since the previous summer can often reach about 2 metres in thickness.) So overall, the average Antarctic ice thickness is much lower than Arctic sea ice. However, snowfall often thickens Antarctic sea ice. The heavy snow burden can depress ice floes, and seawater can subsequently flood those floes.

Sea ice waxes and wanes with the seasons, but minimum and maximum extents rarely match from year to year; over years and decades, summer and winter extents vary. Compared to the Arctic, Antarctic sea ice shows less variability in summer and more variability in winter. These changes largely result from the geographic differences mentioned above, namely Antarctic sea ice's distance from the pole (sea ice can melt back to the coast in summer, making for less summer-to-summer variability) and unconstrained growth potential in winter. Weather events often drive variability, but have different effects in the Northern and Southern Hemispheres. Weather exerts a greater influence on the Arctic minimum and the Antarctic maximum.

Land–sea configurations affect sea ice extents not only by limiting where ice can form but also by introducing their effects. In the Arctic, landmasses surround and influence the sea ice in the Arctic Ocean. Ice and (especially) snow are highly reflective, bouncing much of the Sun's energy back into space. As the Northern Hemisphere spring and summer snow cover declines, the underlying land surface absorbs more energy and warms. Warmer conditions on land affect the nearby ocean, and more sea ice melts as a result. The melt–warmth–melt feedback cycle means that the Arctic is warming faster than the rest of the globe. Such a *polar amplification* effect has not occurred on a large scale in the Southern Hemisphere, however, Antarctica is surrounded by ocean, not a land surface that is losing its reflective snow and ice cover in the spring and summer. Generally, summertime sea ice melts back nearly to the Antarctic coastline, leaving large expanses of the Southern Ocean exposed to heating from the summer sun. By contrast, the loss of reflective snow and ice in high northern latitudes surrounding the Arctic Basin represents a profound change from the normal scenario. Natural cycles in the Southern Ocean can have pronounced effects on Antarctic sea ice. Atmospheric patterns are partly influenced by greenhouse gas emissions.

6.4   SOUTHERN ANNULAR MODE (SAM)

The Southern Annular Mode is usually defined as the difference in the zonal mean sea level pressure at 40°S (mid-latitudes) and 65°S (Antarctica) (Abram et al., 2014; Lee et al., 2019). Changes in air pressure distribution cause changes in the strength and position of the westerly winds. The SAM index is effectively a measure of the strength of the Southern Westerly Winds, and as it increases, the westerlies have been moving south and increasing (Holland et al., 2020). This is influenced by El Niño-Southern

Oscillation conditions, so it is partly driven by natural oscillations. At the same time, anthropogenic global warming shifts SAM into a more frequent positive mode, and the resulting wind effects generally increase the Antarctic sea ice extent.

A positive phase of the Southern Annular Mode will continue to drive changes in the Southern Westerly Winds, causing warming and drying over Patagonia, and increased upwelling of warm Circumpolar Deep Water and glacier recession in western Antarctica and the Antarctic Peninsula.

The Southern Annular Mode describes the north–south movement of this wind belt over a timescale of decades to centuries. It is a key climatic component that will strongly affect how glaciers in the Southern Hemisphere respond to climate change. It explains the key drivers for glaciation in the Southern Hemisphere, and why glacier advances are asynchronous with those in the Northern Hemisphere (Darvill et al., 2016; Davis et al., 2020).

6.4.1 Positive Southern Annular Mode

In a positive phase of the Southern Annular Mode, there is lower anomalous air pressure over Antarctica and higher anomalous air pressure over the mid-latitudes.

In a positive Southern Annular Mode (the situation today), the belt of strong westerly winds strengthens and contracts towards Antarctica. It weakens at the northern boundary in the mid-latitudes (40–50°S). It is drier over Patagonia, driving glacier recession. In Antarctica, increased Circumpolar Deep Water upwells onto the continental shelf, driving glacier and ice sheet recession (Rignot et al., 2018).

6.4.2 Negative Southern Annular Mode

In a negative Southern Annular Mode, the belt of strong Southern Westerly Winds expands northwards towards the equator, bringing cold, wet weather to Patagonia and glacier advance, and decreased Circumpolar Deep Water upwelling on the Antarctic Continental Shelf. The winds are weaker in this phase. This was the situation during Holocene neoglaciations in Patagonia and during the Last Glacial Maximum.

6.4.3 Effect of Greenhouse Gases (GHG) on the Southern Annular Mode (SAM)

The Southern Annular Mode is currently in a positive phase (Marshall, 2003), and this is projected to continue due to increased greenhouse gas emissions. The leading mode of Southern Hemisphere (SH) climatic variability, the Southern Annular Mode (SAM), has recently seen a shift towards its positive phase due to stratospheric ozone depletion and increasing greenhouse gas (GHG) concentrations. The stratospheric ozone layer is Earth's primary protection from harmful ultraviolet radiations. Ozone normally heats the stratosphere as solar radiation is absorbed, due to ozone depletion, the stratosphere over the South Pole has become 6°C cooler. This cooling of the stratosphere lifts the polar tropopause (Son et al., 2008, Thompson et al., 2011). As a result, over the past four decades, atmospheric pressure has declined over the South Pole and increased over the mid-latitudes, causing the SH polar jet stream to shift poleward, especially during the austral summer (Sexton, 2001; Gillett and Thompson, 2003; Jones, 2012; Kang et al., 2011; Marshall, 2003,). While the loss of stratospheric ozone happens in the austral spring, the greatest impact at the surface is felt during the summer and autumn due to the lag associated with the downward propagation of the signal (Orr et al., 2012). The mode of climate variability that captures differences in atmospheric pressure between the mid- and high latitudes of the SH is known as the Southern Annular Mode (SAM or Antarctic Oscillation; defined as the zonal mean sea-level pressure difference between the latitudes of 40°S and 60°S). The Antarctic ozone hole shifts the polar jet south by 1–2° of latitude, consistent with a more positive SAM, in austral summer (Orr et al., 2012; Thompson et al., 2011; Turner et al., 2014). Since the late 1970s, the strength of the Southern Ocean polar jet has increased by 15–20% (Korhonen, 2010; Turner and Marshall, 2011). Climate models, which can separate, via different model configurations, the contribution due to ozone depletion from GHG forcing, suggest that the bulk of these SAM-related changes, in the austral summer, have occurred because of the development of the Antarctic ozone hole with only a small contribution to date from increases in GHG (Gonzalez et al., 2014; Lee and Feldstein, 2013; McLandress et al., 2011; Polvani et al., 2011; Son et al., 2010; Son et al., 2008; Son et al., 2009). By cooling the polar stratosphere, the Antarctic ozone hole thus increases the thermal gradient between the Pole and the mid-latitudes and helps to seal off the Antarctic continent from lower latitudes by strengthening and tightening the vortex of westerlies that flow around the polar cap. The strong vortex locks very cold air on the high Antarctic land mass, sheltering the coldest region on Earth from the effects of greenhouse warming and explaining the cooling trend observed over much of East Antarctica in the past 30 years (Convey et al., 2009; Turner et al., 2014; Wu et al., 2013). Thus the loss of stratospheric ozone and the resultant reduced poleward heat flux is manifest as a slight cooling at stations around the coast of East Antarctica. Effectively, ozone depletion has helped to shield most of Antarctica from the bulk of SH warming with consequences that radiate across the globe. Maintaining frigid temperatures over Antarctica has implications for both the ecosystems of that continent and the rest of the planet in terms of ice melt and consequent sea level rise (Turner et al., 2014). Ozone effects on SAM should become weaker as ozone concentrations recover over the next century, reducing Antarctica's buffer against SH warming, but this may well be countered by the effect of increasing GHG on the SAM index (Abram et al., 2014; Dixon et al., 2012; Perlwitz, 2011). Whilst East Antarctica has cooled slightly, on the west of the continent this tightening of the vortex draws milder maritime air over the Antarctic Peninsula and onto the Larsen Ice Shelf, resulting in parts of this region becoming some of the fastest warming regions on the planet (Orr et al., 2008). Such rapid warming contributes to enhanced melting and break up of ice sheets, e.g. the collapse of the Larsen B ice shelf in 2002.

6.4.4 MELTING OF THE WEST ANTARCTIC ICE SHEET

The West Antarctic Ice Sheet (WAIS) is capable of rapid change as it is a marine ice sheet and therefore could be unstable. It has the potential to raise the global sea level by 3.3 m over a matter of centuries (Bamber et al., 2009). The Transantarctic Mountains

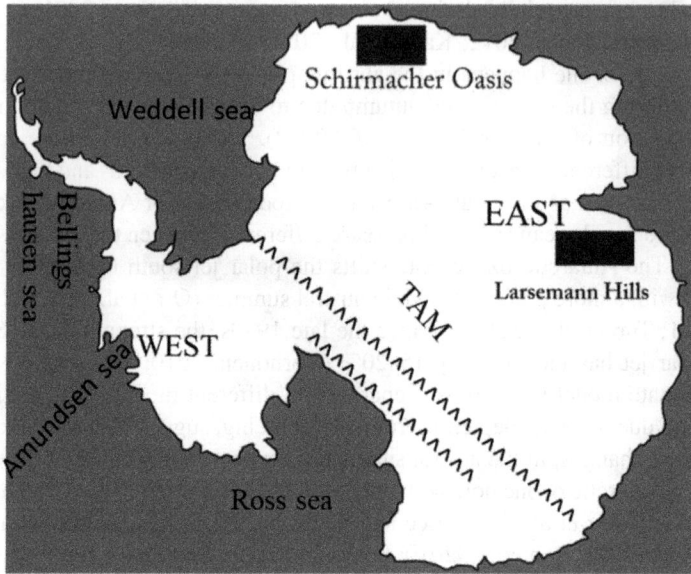

FIGURE 6.2 East Antarctica and West Antarctica separated by the Trans-Antarctic Mountains (TAM).

divide the West Antarctic Ice Sheet from the East Antarctic Ice Sheet (Bindschadler, 2006) (Figure 6.2). West Antarctica is approximately 97% ice-covered, and is 1.97 × 106 km² in area. The West Antarctic Ice Sheet flows into the Bellingshausen, Weddell, Amundsen and Ross Seas.

There are principally three sectors of the ice sheet, which flow northeast-ward into the Weddell Sea, westward into the Ross Ice Shelf and northward into the Amundsen/Bellingshausen seas. The highest elevations reached are 3000 m above sea level (Bindschadler, 2006) occurring at the divides between these sectors. The size of the West Antarctic Ice Sheet is limited, despite its high average snow falls, by the faster speeds of its ice streams. The West Antarctic Ice Sheet is drained by several large ice streams. The basal sediments of West Antarctica comprise soft marine sediments. Combined with geothermal heating at the base, this is sufficient to allow glaciers to slide rapidly. This ice flow is partly constrained by buttressing ice shelves. The ice streams flow from an inland reservoir of ice towards the ocean, passing over a grounding line and, in places, into an ice shelf. Nearly all the precipitation received in West Antarctica eventually passes through these ice streams.

Largely, West Antarctica drains through the Pine Island Glacier and Thwaites ice streams into Pine Island Bay. These ice shelves are warmed from below by Circumpolar Deep Water (Jacobs et al., 2011), which has resulted in system imbalances, more intense melting, glacier acceleration and drainage basin drawdown (Shepherd et al., 2001, 2004; Wingham et al., 2009). This is the "Weak Underbelly" of the West Antarctic Ice Sheet (Hughes, 1981), which may be prone to collapse. Pine Island Glacier is currently thinning (Pritchard et al., 2009, 2012), and, combined with rapid basal melting of the Amundsen Sea ice shelves (Pritchard et al., 2012), means that there is a concern for the future viability of its fringing ice shelves.

6.5 CONCLUSION

Sea ice exists primarily in the polar regions and influences the global climate. The bright surface of sea ice reflects a lot of sunlight into the atmosphere and, importantly, back into space. Because this solar energy "bounces back" and is not absorbed into the ocean, temperatures nearer the poles remain cool relative to the equator. Changes in the amount of sea ice can disrupt normal ocean circulation, thereby leading to changes in global climate. Even a small increase in temperature can lead to greater warming over time, making the polar regions the most sensitive areas to climate change on Earth. Thus, polar ice plays a crucial role in changing climate.

REFERENCES

Abram, N.J., Mulvaney, R., Vimeux, F., Phipps, S.J., Turner, J., England, M.H. (2014) Evolution of the Southern Annular Mode during the past millennium. *Nature Climate Change* 4: 564–569.

Alley, R.B., Clark, P.U., Huybrechts, P., Joughin, I. (2005) Ice-sheet and sea-level changes. *Science* 310(5747): 456–460.

Bamber, J.L., Riva, R.E.M., Vermeersen, B.L.A., Le Brocq, A.M. (2009) Reassessment of the potential sea-level rise from a collapse of the West Antarctic Ice Sheet. *Science* 324(5929): 901–903.

Bindschadler, R. (2006) The environment and evolution of the West Antarctic ice sheet: setting the stage. *Philosophical Transactions of the Royal Society A: Mathematical, Physical and Engineering Sciences* 364(1844): 1583–1605.

Center, W.R.C. and Center, H.P., 2003. National Oceanic and Atmospheric Administration (NOAA). *Reno, NV.* www.Wrcc.sage.Dri.Edu.

Convey, P., Chown, S.L., Clarke, A., Barnes, D.K., Bokhorst, S., Cummings, V., Ducklow, H.W., Frati, F., Green, T.A., Gordon, S., Griffiths, H.J. (2014) The spatial structure of Antarctic biodiversity. *Ecological Monographs* 84(2): 203–244.

Darvill, C.M., Bentley, M.J., Stokes, C.R., Shulmeister, J. (2016) The timing and cause of glacial advances in the southern mid-latitudes during the last glacial cycle based on a synthesis of exposure ages from Patagonia and New Zealand. *Quaternary Science Reviews* 149: 200–214.

Davies, B.J., Darvill, C.M., Lovell, H., Bendle, J.M., Dowdeswell, J.A., Fabel, D., García, J.L., Geiger, A., Glasser, N.F., Gheorghiu, D.M., Harrison, S. (2020) The evolution of the Patagonian Ice Sheet from 35 ka to the present day (PATICE). *Earth-Science Reviews* 204: 103152.

Dixon, D.A., Mayewski, P.A., Goodwin, I.D., Marshall, G.J., Freeman, R., Maasch, K.A., Sneed, S.B. (2012) An ice-core proxy for northerly air mass incursions into West Antarctica. *International Journal of Climatology* 32: 1455–1465.

Gillett, N.P., Thompson, D.W.J. (2003) Simulation of recent Southern Hemisphere climate change. *Science* 302: 273–275.

Gonzalez, P.L.M., Polvani, L.M., Seager, R., Correa, G.J.P. (2014) Stratospheric ozone depletion: a key driver of recent precipitation trends in South Eastern South America. *Climate Dynamics* 42: 1775–1792.

Gregory, J.M., Huybrechts, P. (2006) Ice-sheet contributions to future sea-level change. *Philosophical Transactions of the Royal Society A: Mathematical, Physical and Engineering Sciences* 364(1844): 1709–1732.

Holland, D.M., Nicholls, K.W., Basinski, A. (2020) The Southern Ocean and its interaction with the Antarctic Ice Sheet. *Science* 367(6484): 1326–1330.

Houghton, J.T., Ding, Y.D.J.G., Griggs, D.J., Noguer, M., van der Linden, P.J., Dai, X., Maskell, K., Johnson, C.A. (2001) *Climate Change 2001: The Scientific Basis*. The Press Syndicate of the University of Cambridge.

Hughes, T.J. (1981) The weak underbelly of the West Antarctic Ice Sheet. *Journal of Glaciology* 27: 518–525.

Jacobs, S.S., Jenkins, A., Giulivi, C.F., Dutrieux, P. (2011) Stronger ocean circulation and increased melting under Pine Island Glacier ice shelf. *Nature Geoscience* 4(8): 519–523.

Jones, J. (2012) Tree rings and storm tracks. *Nature Geoscience* 5: 764–765.

Kang, S.M., Polvani, L.M., Fyfe, J.C, Sigmond, M. (2011). Impact of polar ozone

Korhonen H., Carslaw, K.S., Forster, P.M., Mikkonen, S., Gordon, N.D., Kokkola, H. (2010) Aerosol climate feedback due to decadal increases in Southern Hemisphere wind speeds. *Geophysical Research Letters* 37.

Lee, S., Feldstein, S.B. (2013) Detecting ozone- and greenhouse gas-driven wind trends with observational data. *Nature Geoscience* 339: 563–567.

Lee, D.Y., Petersen, M.R., Lin, W. (2019) The Southern Annular Mode and Southern Ocean Surface Westerly Winds in E3SM. *Earth and Space Science* 6: 2624–2643.

Marshall, G.J. (2003) Trends in the Southern Annular Mode from observations and reanalyses. *Journal of Climatology* 16: 4134–4143.

McLandress, C., Shepherd, T.G., Scinocca, J.F., Plummer, D.A., Sigmond, M., Jonsson, A.I., Reader, M.C. (2011) Separating the dynamical effects of climate change and ozone depletion. Part II: Southern hemisphere troposphere. *Journal of Climate* 24: 1850–1868.

Notz, D., Marotzke, J. (2012) Observations reveal external drivers for Arctic sea-ice retreat. *Geophysical Research Letters* 39(8).

Notz, D., Stroeve, J. (2016) Observed Arctic sea-ice loss directly follows anthropogenic CO_2 emission. *Science* 354(6313): 747–750.

Orr, A., Bracegirdle, T.J., Hosking, J.S. (2012) Possible dynamical mechanisms for Southern Hemisphere climate change due to the ozone hole. *Journal of the Atmospheric Sciences* 69: 2917–2932.

Perlwitz, J. (2011) Tug of war on the jet stream. *Nature Climate Change* 1: 29–31.

Polvani, L.M., Waugh, D.W., Correa, G.J.P., Son, S.-W. (2011) Stratospheric ozone depletion: The main driver of twentieth-century atmospheric circulation changes in the Southern Hemisphere. *Journal of Climate* 24: 795–812.

Pritchard, H.D., Arthern, R.J., Vaughan, D.G., Edwards, L.A. (2009) Extensive dynamic thinning on the margins of the Greenland and Antarctic ice sheets. *Nature* 461(7266): 971–975.

Pritchard, H.D., Ligtenberg, S.R.M., Fricker, H.A., Vaughan, D.G., van den Broeke, M.R., Padman, L. (2012) Antarctic ice-sheet loss is driven by basal melting of ice shelves. *Nature* 484(7395): 502–505.

Rignot, E., Mouginot, J., Scheuchl, B., Van Den Broeke, M., Van Wessem, M.J., Morlighem, M. (2018) Four decades of Antarctic Ice Sheet mass balance from 1979–2017. *Proceedings of the National Academy of Sciences* 116(4): 1095–1103.

Serreze, M.C., Barry, R.G. (2011) Processes and impacts of Arctic amplification: A research synthesis. *Global and Planetary Change* 77(1–2): 85–96.

Sexton, D.M.H. (2001) The effect of stratospheric ozone depletion on the phase of the Antarctic Oscillation. *Geophysical Research Letters* 28.

Shepherd, A., Wingham, D., Rignot, E. (2004) Warm ocean is eroding West Antarctic Ice Sheet. *Geophysical Research Letters* 31(23): L23402.

Shepherd, A., Wingham, D.J., Mansley, J.A.D., Corr, H.F.J. (2001) Inland thinning of Pine Island Glacier, West Antarctica. *Science* 291: 862–864.

Simpkins, G.R., Karpechko, A.Y. (2012) Sensitivity of the southern annular mode to greenhouse gas emission scenarios. *Climate Dynamics* 38(3–4), 563–572.

Son, S.W., Gerber, E.P., Perlwitz, J., Polvani, L.M., Gillett, N.P., Seo, K.H., Eyring, V., Shepherd, T.G., Waugh, D., Akiyoshi, H., Austin, J. (2010) Impact of stratospheric ozone on Southern Hemisphere circulation change: A multimodel assessment. *Journal of Geophysical Research: Atmospheres* 115(D3).

Son, S.W., Tandon, N.F., Polvani, L.M., Waugh, D.W. (2009) Ozone hole and Southern Hemisphere climate change. *Geophysical Research Letters* 36.

Son, S.W., Polvani, L.M., Waugh, D.W., Akiyoshi, H., Garcia, R., Kinnison, D., Pawson, S., Rozanov, E., Shepherd, T.G., Shibata, K. (2008) The impact of stratospheric ozone recovery on the Southern Hemisphere westerly jet. *Science* 320(5882): 1486–1489.

Stroeve, J.C., Serreze, M.C., Holland, M.M., Kay, J.E., Malanik, J., Barrett, A.P. (2012) The Arctic's rapidly shrinking sea ice cover: a research synthesis. *Climatic Change* 110(3): 1005–1027.

Thompson, D.W.J, Solomon, S., Kushner, P.J., England, M.H., Grise, K.M., Karoly, D.J. (2011) Signatures of the Antarctic ozone hole in Southern Hemisphere surface climate change. *Nature Geoscience* 726(4): 741–749.

Turner, J., Barrand, N.E., Bracegirdle, T.J., Convey, P., Hodgson, D.A., Jarvis, M., Jenkins, A., Marshall, G., Meredith, M.P., Roscoe, H., Shanklin, J. (2014) Antarctic climate change and the environment: an update. *Polar Record* 50(3): 237–259.

Turner, J., Marshall, G.J. (2011) *Climate Change in the Polar Regions*. Cambridge University Press.

Wingham, D.J., Wallis, D.W., Shepherd, A. (2009) Spatial and temporal evolution of Pine Island Glacier thinning, 1995–2006. *Geophysical Research Letters* 35: L17501.

Wu, Y., Polvani, L.M., Seager, R. (2013) The importance of the Montreal Protocol in protecting Earth's hydroclimate. *Journal of Climate* 26: 4049–4068.

7 Antarctic Sea Ice
A Climate Change Perspective

Mihir Kumar Dash

7.1 INTRODUCTION

The cryosphere collectively designates different components of the Earth system where water exists in its frozen state. It is comprised of snow, glaciers and ice caps, ice sheets, ice shelves, river and lake ice, sea ice, and frozen ground that exists on the land and beneath the sea, i.e. permafrost. The cryosphere covers 52.0–55.0% of the global land surface and 5.3–7.3% of the total ocean area. The lifespan of different components of the cryosphere varies from a season to millions of years. For example: the lake, river and sea ice can survive for one winter, whereas the East Antarctic ice sheet is believed to have been stable for the last 14 million years (Barrett, 2013). The cryosphere in general and sea ice in particular has an integrated response to climate change and can often be referred as a "natural climate meter", responsive not only to temperature but also to precipitation. Nevertheless, each component of the cryosphere is intrinsically sensitive to variations in air temperature and precipitation, and hence to climate change. Hence, they form an integral part of the global climate system and perform a vital and extremely dynamic role in its shaping.

Among the oceanic cryosphere, sea ice is more dynamic. Sea ice forms an insulative, high-albedo surface cover on the polar ocean. Sea ice along with the seasonal snow cover over it, modulates the heat, mass (gases) and momentum exchange between the ocean and atmosphere in high latitudes. Additionally, it also impacts the radiative, dynamic and thermodynamic properties of the surface. Moreover, brine rejection to the underneath sea water during the formation of sea ice and freshwater input during ice melt are significant determinants of the upper ocean freshwater budget in polar seas. Additionally, the latent heat released to the sea surface during the formation of an ice layer of, for example, one meter thick, is an order of magnitude smaller than the mean annual total of either short-wave or long-wave radiation received at the surface. Even the interannual variation of radiation is of the order of 10% of mean annual value. Thus, it is not unexpected that small changes in climatic forcing are accompanied by large changes in the extent of the ice-covered area.

Especially, sea ice advances in winter and retreats in summer (Figure 7.1). However, sea ice in some areas of the Antarctic survives more than a melt season to become multi-year ice and lasts for several years (Figure 7.1). The sea ice cover in the Antarctic is highly dynamic and extends over ~19 million km² in September (austral

DOI: 10.1201/9781003364115-7

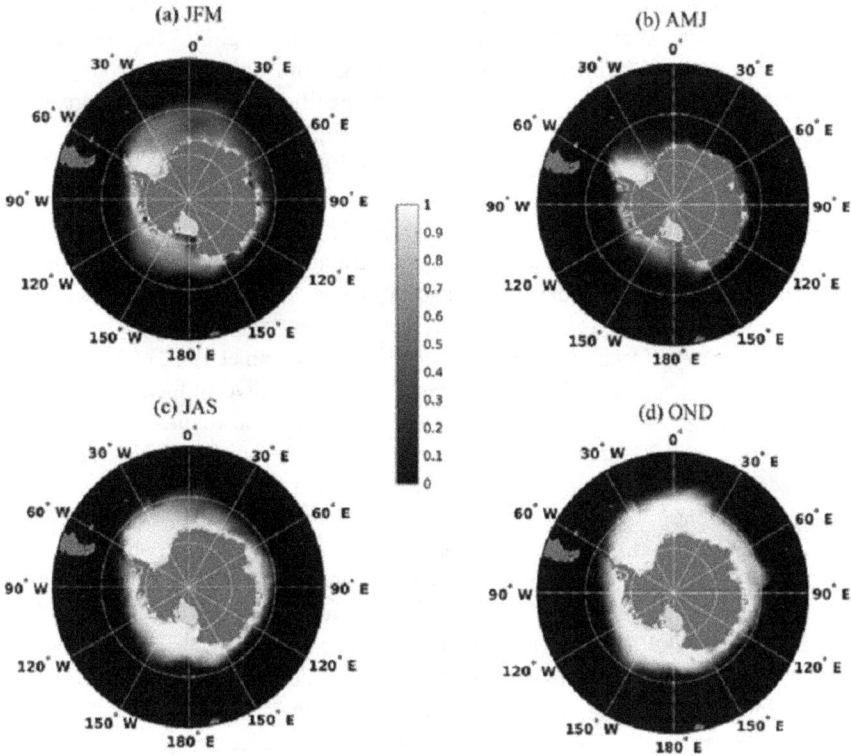

FIGURE 7.1   Seasonal climatology of sea ice cover in the Antarctic. (a) Austral summer (January, February and March [JFM]). (b) Austral autumn (April, May and June [AMJ]). (c) Austral winter (July, August and September [JAS]). (d) Austral spring (October, November and December [OND]).

winter) to ~3–4 million km² in February (austral summer) (Comiso and Nishio, 2008; Gloersen et al., 1992). Further, certain key coastal areas in the Antarctic are the sites for the formation of cold, saline, dense and oxygen-rich Antarctic bottom water (AABW) which is a crucial driver of the global thermohaline circulation, the "conveyor belt" (Lubin and Massom, 2006; and references therein). Seasonality along with the interannual variation of sea ice presents one of the largest physical and albedo variations in the world. The magnitude of the impact of sea ice on these processes (i.e. the formation of AABW, energy balance over the region, etc.) depends not only on the areal extent, concentration, ice type, thickness of ice in different sectors of the Antarctic, but also its dynamics, degree of deformation and the characteristics of snow cover and their complex interrelationships with the environment and with itself, and spatio-temporal variation.

Further, sea ice properties and the seasonal cycle play a pivotal role in controlling the regional biogeochemical cycle in the Antarctic (Thomas et al., 2009; Moline et al., 2008; Tynan et al., 2009). Sea ice significantly contributes to the variability in primary productivity of the Southern Ocean by assisting the growth of algal blooms through nutrient dynamics, ocean stratification and the availability of photosynthetic radiation upon its melt in spring/summer (Arrigo et al., 2008; Smith and Comiso, 2008; Morley et al., 2020). In fact, the habitat, growth and sustainability of Antarctic krill are highly dependent on sea ice cover at various stages of its life cycle (Quetin and Ross, 2009). However, the role of sea ice in regulating outgassing and ocean acidification (Fabry et al., 2009), the atmospheric CO_2 sequestration and above all the biological pump in the Southern Ocean (Tre´guer and Pondaven, 2002), are poorly understood; although emerging research advocates that sea ice is not just an impermeable, non-conducting barrier between the two media, ocean and atmosphere, that inhibits the exchanges mass, momentum and energy, but also plays a significant role in biogeochemical cyc-ling of the Southern Ocean (Delille et al., 2007; Dieckmann et al., 2008).

The large variations in sea ice cover can participate in diverse feedback (positive/negative) processes between the ocean and atmosphere, and regulate its key cause. For instance, a positive anomaly of sea ice extent reduces the absorption of net short-wave radiation at the ocean surface due to the albedo feedback, cooling the ocean surface and pitching for more freezing and further expansion of ice cover (positive feedback). Simultaneously, the colder surface diminishes the outgoing long-wave radiation, while not affecting the incoming long-wave radiation; thus resulting in a net increase in long-wave radiation (or radiation loading) nearer to the surface (nega-tive feedback). Additionally, the cold surface builds a temperature inversion layer close to it, directing more sensible heat flux toward the surface (negative feedback). The interplay between the positive and negative feedback decides the extent of ice cover. From the above description it is evident that sea ice is engaged in a com-plex interaction with its environment which regulates its growth. Moreover, climate-change feedbacks such as an increase in surface air temperature, warming of the ocean, enhancement or reduction in sea ice coverage, extension in melt season, and variation in the pattern of deepwater ventilation can impact the sea ice cover in the Antarctic. Still, the noticeable and prevalent nature of changes in the Antarctic sea ice cover remains hazy, however they are recurrently attributed to the impact of climate change. It is thus imperative that we understand and evaluate the current changes within the framework of past changes and natural variability.

7.2   FORMATION AND GROWTH OF SEA ICE

Sea ice is a complex viscous-plastic solid, fragile and dynamic, formed in the polar seas under suitable thermodynamic conditions. Moreover, sea ice formation has different stages of growth and decay dependent upon the meteorological condition, radiation of energy to the atmosphere at the snow–ice surface and advection of heat by ocean at the bottom of sea ice (Figure 7.2). A simple way to estimate the vertical growth of sea ice is to associate its growth and decay with negative daily air tempera-ture in average snow conditions (Lebedev, 1938) which can be expressed as:

$$h = 1.33\left(\sum \theta\right)^{0.58} \tag{7.1}$$

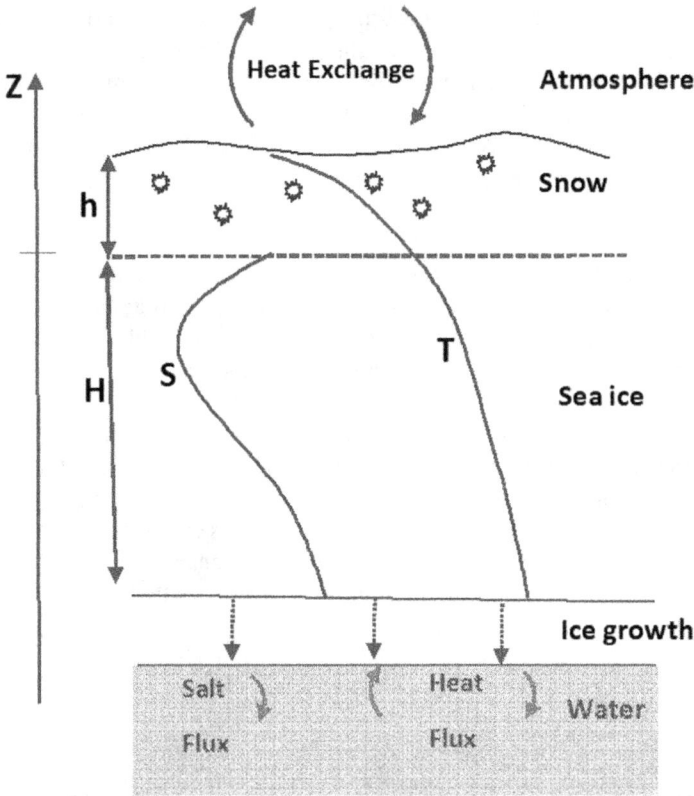

FIGURE 7.2 Schematic illustration of thermodynamic growth of sea ice. S and T represent the salinity and temperature, respectively. H and h represent the thickness of sea ice and snow cover, respectively. Z axis is taken to be vertically upward and growth is considered along the Z axis. (Adapted from Leppäranta [1993].)

where h is the thickness of sea ice in cm, and θ represents the negative daily air temperature. Negative daily air temperature essentially measures how cold a day has been and for how long. For example: the typical value of freezing point of sea water is $-1.8°C$. Assume that the average daily temperature was $-6.8°C$, then "θ" would be 5 [$(-1.8) - (-6.8) = 5$] degrees Celsius below freezing point for day 1. Let the daily mean temperature the next day be 8 degrees below the freezing point. Then, $\Sigma\theta$ for day 2 is 13 (i.e., $5 + 8 = 13$).

As the temperature of the polar ocean surface reaches below the freezing point of sea water ($-1.8°C$ or $28.8°F$), ice formation starts. The sea ice thus formed acts as an insulating layer between the atmosphere and the ocean. Moreover, the growth/melt rate of sea ice depends on the way heat is exchanged within the sea ice and the environment (i.e. snow and air at the top, and ocean at the bottom [Figure 7.2]). However, radiation loss from the ocean surface and the cold air overlying initiate sea ice formation, but the rate of true growth of sea ice crystals is probably restricted by the dissipation of heat and solute (i.e. brine [concentrated salt solution], chlorophyll, etc.) through the liquid. In fact, freezing of the ocean below the sea ice requires the heat from the ocean to be conducted through the sea ice and emitted to the atmosphere or to be transported/advected to another region by ocean currents. The vertical growth of ice slows down as the ice thickens because it takes longer for the water below the ice to reach the freezing point. Further, the mass balance of sea ice in an area is a key diagnostic for climate change. The sea ice mass balance in a given region is determined by the sea ice dynamics and thermodynamics. On one hand, the thermodynamics deals with sea ice growth and melt and dynamics, and on the other hand it provides information on ice importing into or exporting out of the region.

Sea ice thermodynamics is a coupled temperature and salinity system. It corresponds to all the processes, heat conduction in the ice, radiation through snow cover (Figure 7.2), heat and salt exchange between the ocean present below the sea ice (Figure 7.2) etc., and involves energy transfer through and storage into the ice. Excesses or negative heat available may lead to net growth or melt of ice. The rates of sea ice formation and melting ("F") are determined by the balance between external heat fluxes and internal conduction fluxes. This can be modelled as a change in the internal energy of the sea ice. Internal energy change can be converted into mass within the sea ice components. If the sea ice reaches the melting temperature, it melts without any additional energy (Bitz and Lipscomb, 1999). Therefore, the accretion or ablation depends on the internal energy expressed as specific energy of melting (*q*).

$$q(S,T) = \rho_i c_0 (-\mu S - T) + \rho_i L_0 \left(1 + \frac{\mu S}{T} \right) + c_w \mu S \qquad (7.2)$$

where temperature (*T*) and salinity (*S*) are in °C and ppt (parts per thousand) ,respectively. ρ_i is the density of fresh ice in kg/m³, c_0 and c_w are the specific heat (in J kg⁻¹ K⁻¹) of fresh ice and sea-water, respectively, L_0 is the latent heat of fusion (in J kg⁻¹), and μ is an empirical constant that relates the freezing point of seawater to its salinity ($T_m = -\mu S$).

Net upward/downward flux ($F(T)$) towards the top/bottom surface of sea-ice of thickness "*H*" at temperature *T* is expressed as

$$F(T) = -q(S,T) \frac{dH}{dt} \qquad (7.3)$$

The growth or melt rate of sea ice (*F*) is a balance between external fluxes and the conduction of heat in a vertical direction and heat storage in the sea ice–snow system and is given by

$$\rho c \frac{\partial T}{\partial t} = \frac{\partial}{\partial z} k \frac{\partial T}{\partial z} + \kappa I_0 e^{-\kappa z} \qquad (7.4)$$

where c represents the heat capacity of sea ice $c(T,S)$, k represents the thermal conductivity ice, I_0 is the solar radiation that penetrates the upper surface, and κ is the extinction coefficient of radiation from Beer's law ($\kappa = 1.5\ m^{-1}$) (Untersteiner, 1961). However, the heat capacity of sea ice is a function of temperature, hence heat capacity in Equation (7.4) can be expressed as

$$\int_{T_0}^{T} \rho c dT = \int_{T_0}^{T} (\frac{\partial}{\partial z} k \frac{\partial T}{\partial z} + \kappa I_0 e^{-\kappa z}) dT \tag{7.5}$$

where T_0 and T are the initial and final temperatures. The second term in the right-hand side represents the internal source term and presents the solar radiation penetrating into the ice. The thermal conductivity of sea ice is a function of temperature and salinity and can be expressed as

$$k(T,S) = k_0 + \frac{\beta S}{T} \tag{7.6}$$

where k_0 is 2.034 Wm^{-1} deg^{-1} is the conductivity of fresh ice and $\beta = 0.117$ wm^{-1}‰ following Untersteiner (1964).

The boundary conditions for Equation (7.5) are determined by flux balance at the top of the ice surface and the temperature of the ocean layer present at the bottom of the sea ice. The energy balance at the top of the sea ice, i.e., air-snow/ice interface, is given by:

$$F^{net}(T_{su}) = F^{sw}(1-\alpha)(1-i_0) + F^{\downarrow lw} - \epsilon \sigma T_{su}^4 - F^{sh} - F^{lh} + k \frac{\partial T}{\partial z} \tag{7.7}$$

where F^{sw} is the short-wave downwelling radiation reaching the ice surface (positive downwards), α is the surface albedo, i_0 is the fraction of the solar radiation penetrating within the ice (i.e., which does not contribute to the surface energy balance), $F^{\downarrow lw}$ is the downwelling long-wave radiation flux (positive downwards), ϵ is the emissivity of the surface, σ is the Stefan–Boltzmann constant, $\epsilon \sigma T_{su}^4$ is the outgoing long-wave radiation (for T_{su} in Kelvin), F^{sh} and F^{lh} are the turbulent sensible and latent heat fluxes, respectively, taken positive upward, and the term $k \frac{\partial T}{\partial z}$ is the conductive flux from the sea ice/snow interior towards the top surface. Any imbalance in the heat fluxes present in Equation (7.5) is converted into ice growth or melt.

The bottom surface temperature of sea is assumed to be at its freezing point and the flux balance at the surface is

$$F_w - k \frac{\partial T}{\partial Z} = -q(S,T) \frac{dh}{dt} \tag{7.8}$$

where F_w is the heat flux from the ocean and $k\dfrac{\partial T}{\partial Z}$ is the conductive flux from the bottom of the sea ice towards the interior.

The dynamics of sea ice is caused by the wind and the ocean current. Under the influence of stress from the wind and ocean current the sea ice deforms. The ice velocity can be expressed as

$$m\frac{\partial \boldsymbol{u}}{\partial t} = \nabla.\sigma + C\left(T_a + T_w\right) - mf\boldsymbol{k} \times \boldsymbol{u} - mg\nabla\eta \qquad (7.9)$$

where m is the mass of sea ice per unit area, $\boldsymbol{u}$ is the velocity of sea ice, σ is the internal stress tensor, C is the ice concentration, $T_a + T_w$ are the stresses from air and ocean, respectively, f is the Coriolis parameter, $\boldsymbol{k}$ is a unit vector pointing upwards, g is the acceleration due to gravity, and η is the ocean surface elevation with respect to zero sea level. More general information on ice dynamics can be found in Hunke and Dukowicz (1997), Leppäranta (2005), and Bouillon et al. (2009).

Before discussing the sea ice model results we discuss the observation of sea ice in the Antarctic and its trend.

7.3 OBSERVATION OF SEA ICE IN THE ANTARCTIC

A number of metrics, such as sea ice extent (SIE), sea ice concentration (SIC), sea ice thickness, sea ice volume, etc., have been used to study the variability of sea ice cover locally and on a large scale. Unique dielectric properties of sea water and sea-ice in the microwave spectral region enable remote sensing of a number of characteristics of the ice in the Antarctic region. Robust and profound sea ice information of the Southern Ocean started with the launch of scanning multi-channel microwave radiometer (SMMR) onboard Nimbus-7 satellite on 24 October 1978. Following this, Defence Meteorological Satellite Program (DMSP) satellite missions carrying special sensor microwave imager (SSMI) instruments were used to observe the sea ice cover in the polar regions. The first such series that carried the SSMI instrument was DMSP F8 satellite launched on 18 June 1987. Later, the SSMI instrument was replaced by SSMI sounder (SSMIS) instruments and was launched for the first time on board DMSP F16 satellite on 18 October 2003. To generate consistent long-term sea ice data, the successive instruments were intercalibrated (Cavalieri et al., 1999, 2012).

With the lunch of Oceansat-1: Multi-frequency Scanning Microwave Radiometer (MSMR), India's first oceanographic satellite on 26 May 1999, additional independent radiometric observations of the polar regions were available to the global scientific community. Spectral signatures of MSMR are not only used to demarcate open water and sea ice regions in the Antarctic but also to identify several large and prominent continental features such as Gamburtsev sub-glacial mountain, the Trans-Antarctic Mountain ranges, and Wilkes and Aurora sub-glacial basins (Dash et al., 2001; Vyas and Dash, 2000; Vyas et al., 2001). Further, Sharma et al. (2011) used

MSMR brightness temperature to measure ice melt over Greenland. Details of sea ice studies from an Indian perspective are provided by Oza et al. (2017).

In the pre-passive microwave remote sensing era, sea ice variability in the Antarctic was based on human judgment of the ice edge from poor-resolution visible images. These images were constrained by cloud cover and had biases compared to satellite passive microwave observations. Further, these images are available for short time periods, thus internal variability is difficult to ascertain.

To obtain the information in the pre-satellite era, proxies are used to reconstruct sea ice change and its variability. One such proxy is the historical whaling records (1930s to the 1980s) with the premise that the whaling fleet favoured ice edge for the catch and hence can be used as a proxy for ice edge location (de la Mare, 1997; Cotte and Guinet, 2007; de la Mare, 2009). However, this estimate has been challenged on the basis of summer-time biases between ship and satellite estimates of ice edge location (Ackley et al., 2003). Further, chemical proxies like methane sulfonic acid (MSA) and excess deuterium in different layers of the ice core are also used as proxies for sea ice extent. The MSA proxy records indicate that maximum SIEs have declined since the mid-20th century in the East Antarctic sector (Curran et al., 2003) and the Bellingshausen Sea (Abram et al., 2010). Moreover, excess deuterium records from a Whitehall Glacier ice core indicate that current sea ice expansion started in the mid-1960s (Sinclair et al., 2014) and was confirmed by recent independent ice cores (Thomas and Abram, 2016). The obvious limitation of MSA reconstructions is that the most intense recent trends have occurred in the warmer seasons.

On the other hand, satellite passive-microwave observations over the Antarctic provide many advantages for regular and continuous monitoring of sea ice in this region.

- First, frequent coverage of the whole region: Satellites provide full images of Antarctic ice cover in 1 or 2 days.
- Second, potentiality to discriminate and distinguish sea ice in the background of the open ocean.
- Third, sea ice can be monitored throughout the year as the satellite sensor senses emitted radiation from the Earth–atmosphere system.
- Fourth, selection of appropriate microwave frequency(ies), emitted radiation from the Earth's surface can penetrate most of the cloud cover, enabling sea ice measurements.
- Fifth, availability of long-term, calibrated and uninterrupted observations of sea ice cover since October 1978.

Therefore, researchers predominantly use passive microwave observations to study the long-term changes in sea ice cover in the Antarctic (Parkinson, 2019; Parkinson, 2014; Parkinson and Cavalieri, 2012; Meehl et al., 2016). The availability of consistent long-term records of sea ice concentration (SIC), sea ice extent (SIE), and sea ice area (SIA) from series of passive microwave sensors helps to explore the impact of climate change on Antarctic sea ice and vice versa.

7.3.1 SEA ICE CONCENTRATION (SIC)

Sea ice concentration (SIC) is defined as the percentage of the area of an ocean pixel/grid-scale covered by sea ice. It is the principal sea ice parameter retrieved using the passive microwave observations obtained at different frequencies from radiometers/imagers flown on the satellites. SIC is the preliminary data from which most other metrics, such as sea ice extent (SIE) and sea ice area (SIA) are derived. The retrieval of SIC using microwave radiometry works on the following two principles:

1. Different surface types have different microwave signatures, which are strongly clustered.
2. The radiometric signature observed by the instrument is a linear combination of energies that are emitted from different surface types, with the weights depending on their relative concentrations.

The radiometric signature at the instrument head is normally measured in brightness temperature. Operationally, SIC is available from two different algorithms: (1) the NASA team algorithm and (2) Bootstrap algorithm. The NASA team algorithm was originally developed to compute SIC using Nimbus-7 SMMIl radiances (Cavalieri et al., 1984; Gloersen and Cavalieri, 1986). This algorithm uses the polarization (PR) and the spectral gradient ratio (GR) to compute the SIC.

$$PR = \frac{T_B(19V) - T_B(19H)}{T_B(19V) + T_B(19H)} \tag{7.10}$$

$$GR = \frac{T_B(37V) - T_B(19V)}{T_B(37V) + T_B(19V)} \tag{7.11}$$

where V and H represent vertical and horizontal polarizations, respectively.

The ratio of sum and difference of radiances makes the NASA team algorithm slightly nonlinear, but the influence of temperature variation is mitigated, because the brightness temperature has a linear relation with physical temperature when all other properties are equal. However, it was modified subsequently for DMSP SSMI and SSMIS radiances also. Bhandari et al. (2005) used the spectral signatures (i.e. brightness temperatures) of MSMR at 10 and 18 GHz in both (horizontal and vertical) polarizations to derive the SIC. Also, they used polarization (PR) and gradient (GR) ratios, computed using the 10 GHz and 18 GHz dual-polarized observations to derive SIC. They found that SIC derived using both methods matched well with that of SSM/I.

On the other hand, the Bootstrap technique (Comiso, 1995) is built on the distributions of clusters of brightness temperature of sea ice formed using multichannel observations. An inherent assumption of the algorithm is that there are large regions in the central Arctic in winter in which the ice concentration is 100%. Similar assumptions are also considered for the Antarctic region. The Bootstrap algorithm takes advantage of the fact that, in the scatter plot of 37 GHz(H) versus 37 GHz(V), the data points in the Arctic are highly correlated.

The basic differences between the two techniques are:

(i)   The choice and utilization of observations at different frequencies and polarizations, i.e. use of 19 GHz(V), 19 GHz(H) and 37 GHz(V) in the NASA team algorithm vs 37 GHz (V) and 37 GHz(H) in Bootstrap algorithm.
(ii)   Different reference brightness temperatures (or emissivity).
(iii)   Sensitivities to changes in physical temperature of the Earth surfaces.
(iv)   Use of different weather filters.
(v)   Seasonal adjustment of Bootstrap tie points.

The Bootstrap technique takes advantage of the higher spatial resolution available in 37-GHz channels to obtain better results in the perennial ice region. Meanwhile, the NASA team algorithm used polarization and gradient ratios, to minimize the effect of temporal and spatial changes in the ice surface temperature. Both algorithms use the 22-GHz channel in masking out the open ocean for the SSMI data. A detailed comparison of two techniques is described by Comiso et al. (1997). Comparisons with high-resolution AVHRR observations and other data sets show that the accuracies of SIC derived from passive microwave sensors have an error rate of 5% (absolute) (Comiso et al., 1997; Andersen et al., 2006; Heygste et al., 2009). A number of factors are responsible for reducing the accuracy of the retrieval of SIC, such as snow layer present over the sea ice, variations in brine and moisture content, the presence/absence of melt ponds, and also surface temperature variation. Further, adverse weather conditions (such as cloud and humidity) and wave activities in the ocean affect the radiometric signature over open water. However, the availability of long-term records of SIC from the NOAA/NSIDC Climate Data Record (CDR) product (Meier et al., 2014) enable researchers to study the SIC trend from a climate change perspective.

In order to study the regional trend, the Antarctic region is divided into five sectors (Figure 7.3). The year is divided into four seasons. Austral summer season is considered as January, February, and March (JFM). The months April, May, and June are considered as autumn season, and winter season spans from July to September (JAS). Finally, October, November, and December (OND) are considered as austral spring season. The seasonal variation is clearly seen in the ice cover area, with the minimum area in the summer season and the maximum area in the winter season (Figure 7.1). Broadly, a spatial periodic pattern is seen in the concentration trend (Figure 7.4). Roughly, two diagonally opposite regions of increased/decreased SIC trend are observed with the Ross Sea and Western/Eastern Weddell Sea (depending on seasons) showing an increasing trend and decrease or no change around East Antarctica and the Amundsen-Bellingshausen Seas. A similar pattern was also reported by Zwally et al. (2002). Except austral summer, mostly high positive/negative SIC trends are observed to the outer edge of the ice-covered regions (Figure 7.4). Irrespective of season, a decreasing (increasing) trend in concentration is noticed in

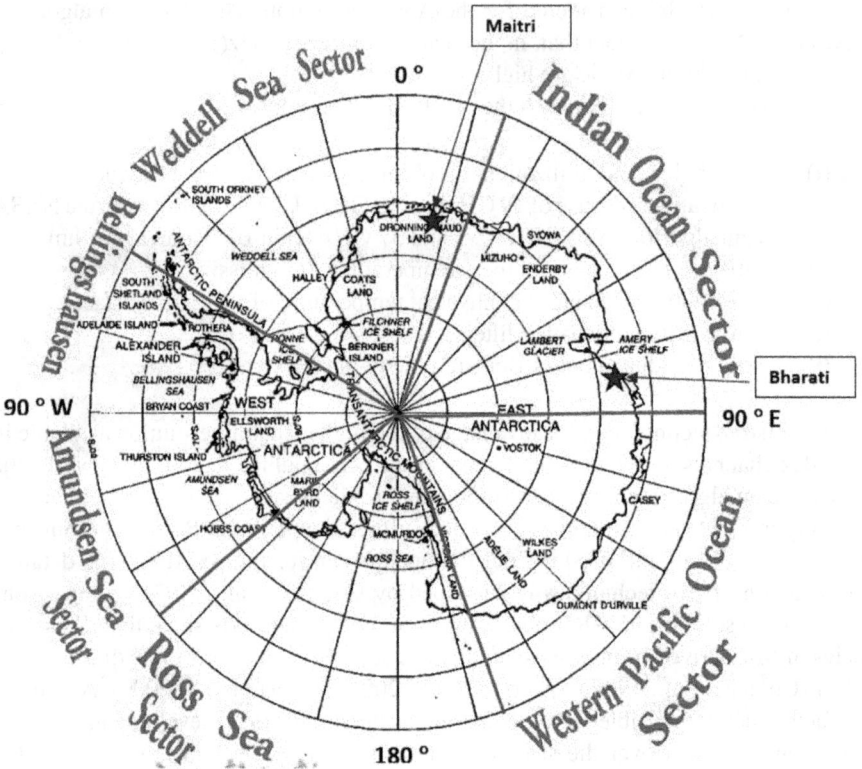

FIGURE 7.3 Definition of various sectors of the Southern Polar Ocean (after Gloersen et al., 1992). The stars represent the Indian Antarctic stations.

the western side of the Antarctic Peninsula (Ross Sea sector). A diffused positive trend is noticed in the Weddell sector. The positive SIC trend in the Weddell Sea during summer is attributed to the positive trend in intensification of concentration in the spring season (Holland, 2014). However, a small or no concentration trend is observed in autumn and winter. The Amundsen Sea experiences a decreasing trend in summer concentration that disappears in winter (Figure 7.4). This negative trend in summer is attributed to enhanced spring ice loss (Holland, 2014). The Western Pacific sector shows a decreasing trend in SIC except for the Austral spring. There is a see-saw trend between the Indian Ocean Sector and the Western Pacific sector except for the spring season. Summarizing the SIC trend, during austral summer and autumn the largest trends are positive in the Weddell and western Ross Sea and negative in the Amundsen and Bellingshausen Seas (ABS) (Holland, 2014). In winter and spring, the statistically significant positive trends are more or less confined to the ice edge of the Ross Sea, with a reduction along the western coast of the Antarctic Peninsula. The Indian Ocean sector experiences an increased trend in SIC in all seasons.

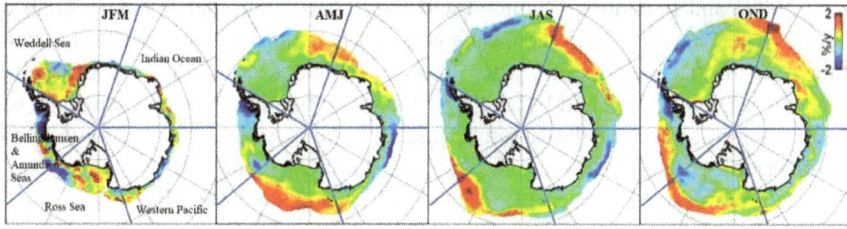

FIGURE 7.4 Linear trends in ice concentration in different sectors of the Antarctic observed over the period of 1992–2010. (Adapted and modified from Holland et al. [2014].) The sea ice concentrations used are calculated using bootstrap algorithm (Comiso, 2000).

7.3.2 SEA ICE EXTENT (SIE) AND SEA ICE AREA (SIA)

Often, the sea ice variability is described in terms of sea ice extent (SIE). SIE is defined as the integrated area of circumpolar ocean region covered within the outer sea ice edge in a specific sector (such as those presented in Figure 7.3) or as a continuous function of longitude. The sea ice edge pixels are typically defined as 15% of SIC.

Region	Longitude Range
Wedell Sea Sector	60° W to 20° E
Indian Ocean Sector	20° E to 90° E
Western Pacific Ocean Sector	90° E to 160° E
Ross Sea Sector	160° E to 130° W through 180°
Bellingshausen & Amundsen Sea	60° W to 130° W

The other metric, less used to describe the sea ice variability, is sea ice area (SIA). SIA is calculated as the total area of ice coverage for a given domain i.e.,

$$SIA = \sum SIC(i,j) \times A(i,j) \tag{7.12}$$

where $SIC(i,j)$ and $A(i,j)$ are the concentration and area of i-th and j-th pixel/grid cells, respectively.

These two metrics are slightly different, SIA presents the area excluding regions of open water within the ice pack (i.e. area of polynyas). Moreover, SIE is widely cited because uncertainties in the radiometric retrievals have less of an effect on it (Hobbs et al., 2016). However, SIA can be used to derive ice volume and ice mass.

The behaviour of Antarctic SIE or SIA in the global warming scenario has been quite different and more puzzling than the situation of Arctic sea ice, which is showing a decreased trend. The SIE in the Antarctic has shown an increasing trend since 1978 (Zwally et al., 2002; Parkinson and Cavalieri, 2012; Simmonds, 2015; Turner et al., 2015; Hobbs et al., 2016). Many explanations have been suggested, some countering others. For example, (1) there are supports (Thompson and Solomon, 2002; Turner et al., 2009) and rejections (Sigmond and Fyfe, 2010; Bitz and Polvani, 2012) of the role of the ozone hole in increasing SIE and (ii) supports on ties to basal melt-water from the ice shelves to sea ice increase (Bintanja et al., 2013) and this has also been rejected (Swart and Fyfe, 2013). However, the links among changes in sea ice in different sectors of the Antarctic and El Niño–Southern Oscillation (ENSO) (Stammerjohn et al., 2008), the Interdecadal Pacific Oscillation (IPO) (Meehl et al., 2016), and/or the Amundsen Sea Low (ASL) (Turner et al., 2009; Meehl et al., 2016) are univocal.

The total Antarctic/Southern Ocean shows a prominent annual cycle having minimum monthly SIE in February always (for the 41 years of the dataset, 1979–2019) (Figure 7.5a). This also is seen in the case of sea ice area (SIA). However, except for two years, 2002 (August) and 2018 (October), the maximum monthly SIE occurred in September for all years. September 2014 has the maximum SIE (20.12×10^6 km^2) as well as SIA (15.67×10^6 km^2) (Figure 7.5a). September 1986 has the minimum winter SIE (17.98×10^6 km^2), whereas the minimum winter SIA was in August 2016 (13.89×10^6 km^2). Similarly, during the melt season, maximum SIE (SIA) occurred in February 2008 (3.95×10^6 km^2) (February 2003 [2.62×10^6 km^2]) and minimum observed in February 2018 (2.35×10^6 km^2) (February 1993 [1.33×10^6 km^2]). Vyas et al. (2003) used MSMR observations to study the SIE in the entire Southern Ocean for the period 1999–2001 and found that the monthly minimum SIE occurred in February and the maximum (12.634 million km^2 during 1999–2001) occurred in September. The above observations clearly show that the maximum/minimum SIE and SIA may not occur simultaneously. This may be attributed to the local dynamics, i.e. wind and ocean current, which can advect the sea ice to different regions or thicken the ice due to ridging and rafting.

The Weddell Sea, part of the South Atlantic Ocean, is one of the active regions of bottom-water formation. It is the region where sea ice stays for more than one melt season – multiyear sea ice. In the other words, multi-year ice is available in this region. February is the month of minimum sea ice cover, as for the Southern Ocean as a whole. However, typically the maximum monthly SIE occurred in September, while it was in August in 1992, 1994, 2004, and 2017 and in October in 1997, 2002, 2015, and 2018 (Figure 7.5b). Interestingly, in the 41 years of analysis, the largest monthly sea ice coverage occurred in this sector in September 1980 ($7.66 \; 10^6$ km^2) and the minimum occurred in February, 1988 ($0.83 \; 10^6$ km^2). The minimum peak winter ice covered in the Weddell Sea sector occurred in September, 1990 (6.12×10^6 km^2) and the maximum February ice cover occurred in 2014 (2.18×10^6 km^2) (Figure 7.5b). This clear depicts that the year of maximum winter ice cover is not followed by the year of below-average ice extent in the following summer. In the 41 years of analysis, the maximum SIA occurred in October for three years, in August for 13 years, and for the rest of the years in September. The winter maximum and minimum SIA occurred in September, 1980 (6.21×10^6 km^2) and August, 1990 (5.08×10^6 km^2), respectively (Figure 7.5b). Similarly, the maximum (minimum) February (month of minimum ice cover) SIA occurred in 2014 (1.635×10^6 km^2) (1981 [0.57×10^6 km^2]) (Figure 7.5b). High ice extents with low SIA could bring about particularly effective decay seasons; winds and ocean currents could transport more warm air and water to the region than normal, accelerating melting. This illustrates interannual variability

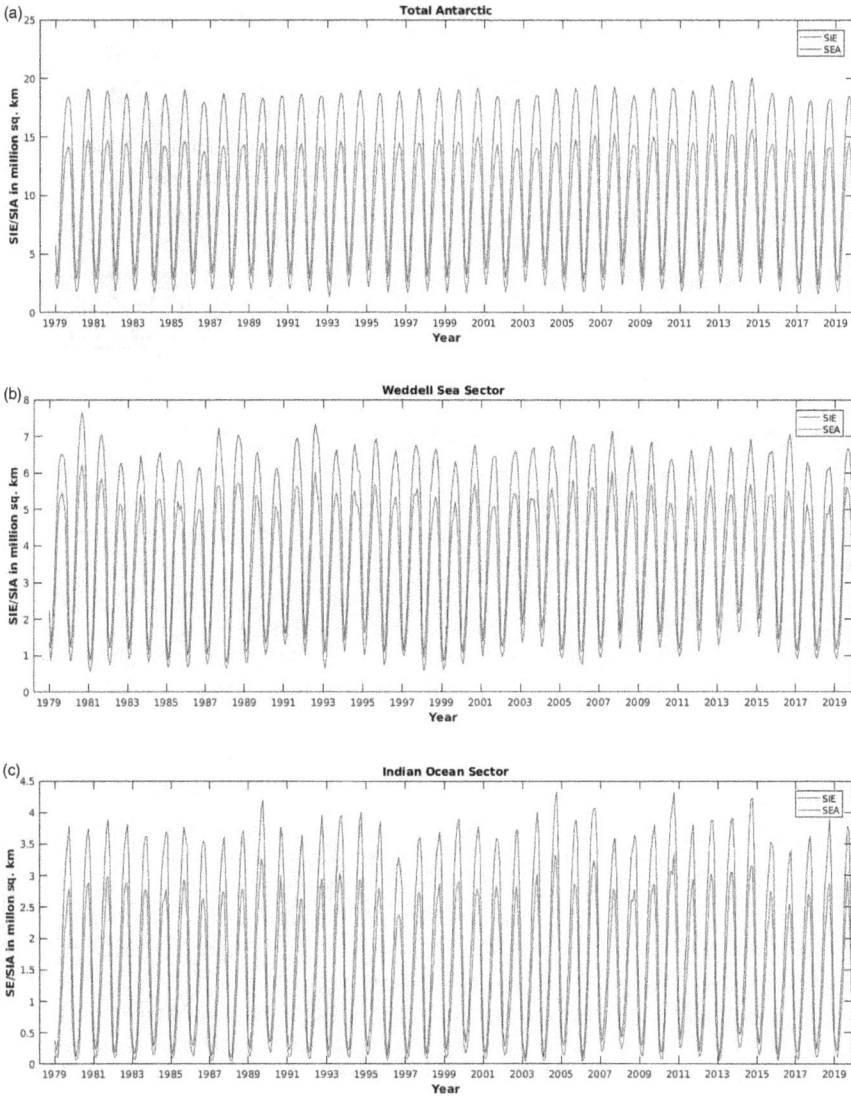

FIGURE 7.5 Monthly average sea ice extents and sea ice areas for (a) total Antarctic, (b) Weddell Sea sector, (c) Indian Ocean sector. The extents and areas are calculated using sea ice concentration data (NASA team algorithm) for January 1979–December 2019. 15% SIC is used as threshold.

and the difficulty of forecasting ice extent months in advance based simply on current ice extents.

Among the five sectors defined, the Indian Ocean sector is the only sector in which the monthly ice extent peaks in October rather than September. However, except for 13 years in the study period, the maximum SIA occurred in the month of September

(Figure 7.5c). This sector has the minimum summer ice extent and area among all sectors. Like other sectors, February is the month of minimum ice extent in this sector except for three years: 1986, 1988, and 2019 (Figure 7.5c). However, out of the 41 years analysed, the minimum monthly SIA occurred in February except for 10 years. The maximum (minimum) winter SIE occurred in October 2004 (4.32 × 10^6 km^2) (1996 September [3.28 × 10^6 km^2]), whereas the summer monthly minimum (maximum) SIE occurred in March 2003 (0.1 × 10^6 km^2) (February 2014 [0.48 × 10^6 km^2]). The maximum winter SIA was in October 2010 (3.33 × 10^6 km^2) and minimum summer SIA occurred in March 2003 (0.04 × 10^6 km^2). The minimum winter (maximum summer) SIA occurred in September 1996 (2.37 × 10^6 km^2) (February 2014 [0.27 × 10^6 km^2]). This clearly shows the interannual variability of sea ice cover in this sector.

Like other sectors and the Southern Ocean as a whole, the Western Pacific Ocean has minimum SIE in February except for 4 years. More than 50% of the year of study winter SIE maximum occurred either in October or August. The maximum winter SIE (SIA) was in September 1982 (2.57 × 10^6 km^2) (September 1982 [1.87 × 10^6 km^2]) (Figure 7.6a). Similarly winter minimum SIE (SIA) was in August 1989 (1.64 × 10^6 km^2) (July 1989 [1.14 × 10^6 km^2]). The maximum (minimum) end summer melt SIE in the study period was in February 2013 (0.77 × 10^6 km^2) (March 1986 [0.15 × 10^6 km^2]). The maximum (minimum) SIA at the end of the melting season was 0.51 × 10^6 km^2 in February 2013 (0.06 × 10^6 km^2 in March 1986) (Figure 7.6a). This is the only sector where the maximum and minimum SIA follow the SIE.

The SIE and SIA in the Ross Sea sector has a prominent, consistent monthly minimum in February but large variability appeared in its month of maximum, which was July in 2 years, August in 8 years, September in 18 years, and October in 13 years (Figure 7.6b). The record high monthly value SIE and SIA were in September 2007. However, the minimum winter ice extent (ice cover [i.e., area]) was observed in October 2017 (September 2017). The record low, almost total disappearance, of sea ice extent and sea ice cover (i.e., area) occurred in February 2017, with some rebounding in the following years (Figure 7.6b).

The Bellingshausen and Amundsen Seas sector shows more substantial interannual variability than the rest of the sectors and the Southern Ocean as a whole also, although interannual variability exists in all the sectors (Figure 7.6c). This sector is different from the other sectors in having a minimum SIE for 9 years in March rather than February, whereas no other sector had more than 4 years (western Pacific sector) with a minimum ice cover other than February. However, Parkinson (2019) reported that between 1979–2018, 11 years minimum SIE occurred in March in this sector. The minimum SIA occurred for 13 years in March, 2 years in January, and for the rest of the years in February (Figure 7.6c). The month of maximum winter ice extent is more in line with variability in the other sectors, being 3 years in July, 14 years in August, 18 years in September, 4 years in October, and 1 year in November. The maximum (minimum) winter ice extent was in September 2015 (August 1988). Summer ice cover was minimum in March 2010 and maximum in March 1979 (Figure 7.6c).

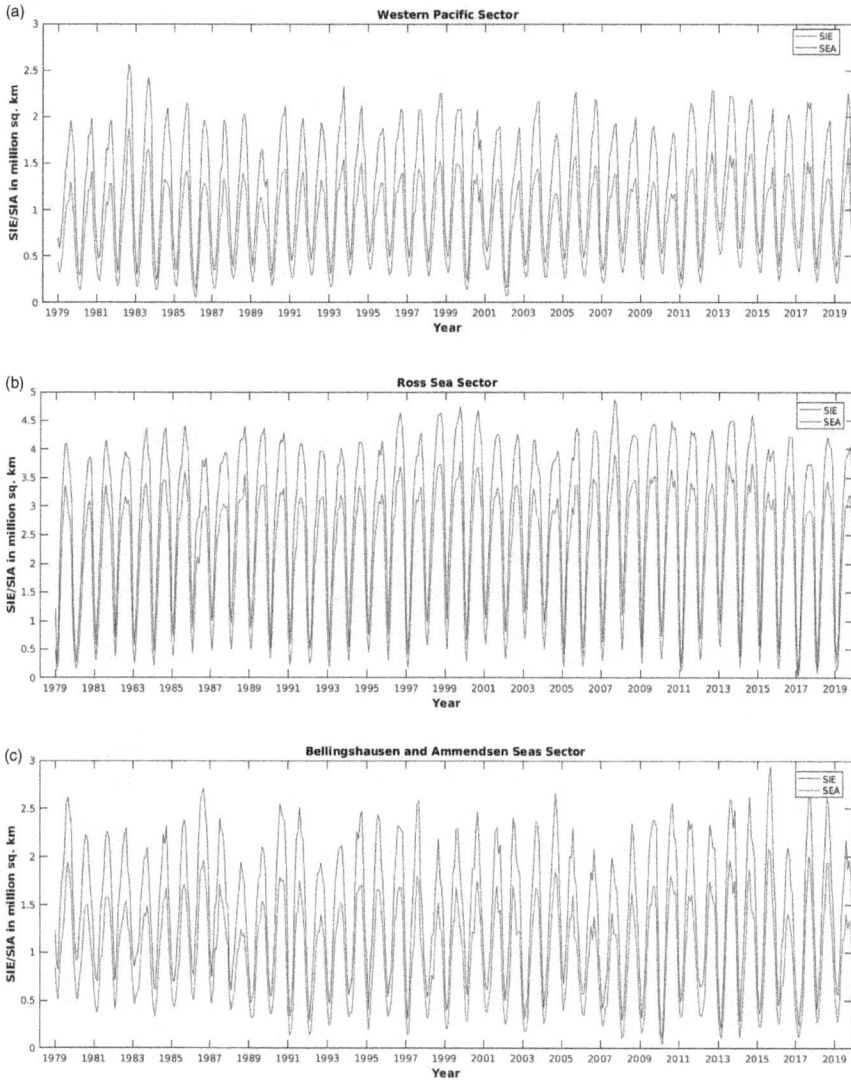

FIGURE 7.6 As Figure 7.5, but for (a) Western Pacific sector, (b) Ross Sea sector, (c) Bellingshausen and Amundsen Seas sector.

7.3.3 Sea Ice Duration

The number of continuous days in a year at a given location (i.e., grid/pixel) covered with sea ice is known as the sea ice duration (SID) of that position. Variation in SID impacts the insolation of the ocean surface, which affects the upper ocean heat content. Also, it possibly acts as a feedback and influences ice formation and advances in subsequent winter (Stammerjohn et al., 2012). Additionally, it also has foremost implications for local marine ecosystems (Massom and Stammerjohn, 2010). The first analysis of sea ice duration in the Southern Ocean was conducted by Parkinson in 1994. Parkinson (2002) analysed 21 years (1979–1999) of sea ice cover and reported that there exists a seesaw between the Ross Sea and Bellingshausen & Amundsen Seas sectors. Analysing 26 years of passive microwave data, Massom and Stammerjohn (2010) found that a significant positive trend (greater than 1.5 days per year) was observed in the Ross Sea, while the Bellingshausen and Amundsen Seas sector showed a predominantly negative trend (greater than 1.5 days per year) (Figure 7.7). Certain regions in the Bellingshausen and Amundsen Seas show an overwhelmingly negative trend greater than 2.5 days per year also. However, a detailed study by Stammerjohn et al. (2012) during 1979–2011 showed that sea ice duration in the Bellingshausen Sea was shortened by 3.3 months, whereas it was increased by 2.6 months in the western Ross Sea region (Figure 7.7b). They also reported that an increase in sea ice duration in the western Ross Sea was roughly divided equally between ice formation (i.e. early formation) and retreat (later retreat) period. Meanwhile, shortening of SID in the Bellingshausen Sea is attributed to a delayed trend in sea ice advance (i.e. 1.5 times the trend toward earlier retreat). Another independent study by Holland (2014) showed that intensification of the spring melt in the ABS leads to a decrease in SIC in the following summer, thus reducing the sea ice duration, whilst the reverse pattern emerges in the Ross Sea. Although a dominant pattern of enhance/retreat is found between the Ross Sea and BAS sectors, there exist complex patterns of change in the advance, retreat, and duration of sea ice within each sector in the west Antarctic (Massom et al., 2013; Stammerjohn et al., 2015).

On the other hand, the SID in the East Antarctic is considerably more complex than that of the West Antarctic and encompasses mixed signals spreading from the regional to local scale. Massom et al. (2013) analysed radiometric observations of sea ice over the Antarctic (region shown by purple lines in Figure 7.7a) from 1979/80

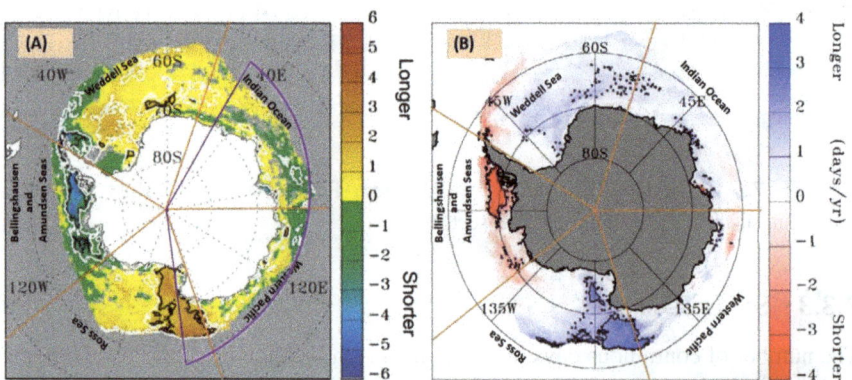

FIGURE 7.7   Sea ice duration in days/year (A) for years 1979–2004. (Adapted and modified from Massom and Stammerjohn [2010].) The purple colour counter represents the east Antarctic region studied by Massom et al. (2013) (B) for years 1979–80 to 2010–11. (Adapted and modified from Stammerjohn et al. [2012].)

to 2009/10 and found that strong positive and negative trends in sea ice seasonality occur in close juxtaposition in regions like polynyas, marginal ice zones, fast ice, etc. Shorter sea-ice duration trend (of 1–3 days per annum) is noticed in isolated pockets in the outer pack spread between 95–110°E, and in certain near-coastal areas between the Amery and West Ice Shelves and close to Davis Station (an Australian station) (Massom et al., 2013). Moreover, positive trends in SID are more extensive, and extend in the western Ross Sea sector (160–170°E) and near the coastal zone between 40–100°E (Massom et al., 2013).

7.3.4 Ice Thickness/Volume

Although SIC provides information about the spatial distribution of the ice cove, the changes in sea ice volume set the salinity at the ocean surface, hence there is convective mixing at the upper surface. It also controls the surface freshening during the melt season. Nearer to the coast, it controls the formation of Antarctic Bottom Water (AAB) through shelf water formation. However, the availability of reliable sea ice thickness (SIT) data in the Southern Ocean is limited and restricted to a few dedicated research stations. SIT data are available from a few in situ observations (Hass et al., 2009), underway observations from icebreakers (Worby et al., 2008), and recently from autonomous underwater vehicles (Williams et al., 2015). All these data acquired are temporally and spatially sparse and also have many uncertainties. However, regular monitoring and estimation of SIT from space is of high priority and an important area of research. In principle, it is possible to retrieve the height of ice or snow surface using satellite laser altimeter measurements using the concept of hydrostatic equilibrium (Kurtz and Markus, 2012). Practically, extracting the SIT from satellite laser altimeter observations in the Antarctic is challenging due to:

(a) Small freeboard of Antarctic sea ice compared to Arctic;
(b) Lack of independent and accurate knowledge of coexisting snow density, snow cover thickness over the sea ice in a particular time and space. Further, redistribution of snow by wind (Leonard and Maksym, 2011) and extensive conversion of snow into sea ice make the processes more complex.

Besides, some success is achieved using alternate methods to overcome these issues, such as:

(i) An empirical relation is designed considering the thickness of snow plus sea ice as one unit above the ocean surface (Xie et al. 2013);
(ii) Assuming freeboard of sea ice consists of only snow (Kurtz and Markus, 2012).

However, significant challenges remain in getting a long-term record of the Southern Ocean sea ice volume as there exists an interdependency between spatial and temporal (seasonal) variations of snow cover and snow and ice density (Kwok and Maksym, 2014). Therefore, the obvious question arises, is SIE/SIA a robust proxy for the SIT? The evidence is somewhat contradicting. Coupled sea ice–ocean models forced by realistic winds showed that SIT and SIC trends counter each other's contribution in the Bellingshausen Sea, whereas they complement each other in the Weddell, Ross, and Amundsen Seas (Zhang, 2014). Therefore, the spatial dependency of SIT or sea ice volume is evident.

Hence, a reliable long-term (climatic scale) record of SIT remains a challenge and creates a significant gap in Southern Ocean observations. Nevertheless, data assimilation to the coupled models demonstrated the potential to predict the climatology of SIT along with the SIC trend and created good hope for estimating the time-varying SIT (Massonnet et al., 2013; Holland et al., 2014; Zhang, 2014). However, the lack of reliable in situ data for validation of SIT indicates that it is likely to remain highly uncertain for some more years and required dedicated research.

7.4   SEA ICE TREND IN DIFFERENT SECTORS OF THE ANTARCTIC

The increasing trend in sea ice extent/area for last three and half decades since 1978 in the Antarctic (Zwally et al., 2002; Parkinson and Cavalieri, 2012; Simmonds, 2015; Turner et al., 2015; Hobbs et al., 2016) has been quite different and puzzling from that of the Arctic sea ice decrease. There are many suggestions and rejections on the explanation for this. The drivers, the El Niño–Southern Oscillation (ENSO) (Stammerjohn et al., 2008), the Interdecadal Pacific Oscillation (Meehl et al., 2016), and/or the Amundsen Sea Low (Turner et al., 2009; Meehl et al., 2016) to sea ice increasing trend are well accepted by the scientific community, whereas ties to the ozone hole and basal meltwater from the ice shelves are rejected by some researchers (Sigmond and Fyfe, 2010; Bitz and Polvani, 2012; Swart and Fyfe, 2013). However, researchers have not yet arrived at a consensus view of why the decade-long overall Antarctic sea ice increases occurred. It is to be noted that during the SMMR period almost no trend in SIE was observed. However, combined analysis of SMMR and MSMR estimates of the sea ice extents over the entire Southern Ocean indicate a steepening in the trend (Vyas et al., 2003). In the meantime, Antarctic SIE has taken a dramatic turn and showed a rapid decrease in 2017 and 2018 to the lowest value in the entire 1979–2018 record, wiping out the 35 years of increasing trend. It will be very interesting to monitor, examine, and explore the SIE and SIA trends in different sectors of the Antarctic and the Southern Ocean as a whole, including following years to have a valuable insight to test and validate the earlier suggested explanations for the long-term sea ice increase.

Figures 7.8–7.13 present a secular trend analysis of monthly SIE and SIA anomalies for different sectors of the Antarctic (as described in Figure 7.3) and Southern Ocean as a whole. Table 7.1 provides details on the yearly average trends in SIE and SIA and includes values for 1979–2018 by different researchers from time to time.

7.4.1   WEDDELL SEA SECTOR

The Weddell Sea is one of the active regions for bottom water formation and is also dynamic due to the presence of the Weddell Gyre. This region acts as a pole to the Antarctic Oscillation. It feeds sea ice into the Indian Ocean and influences the

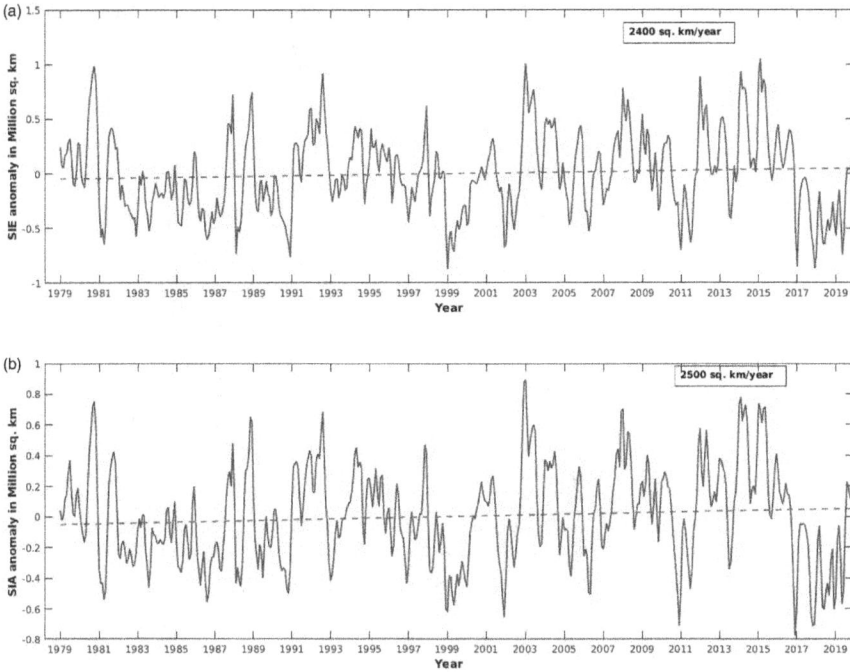

FIGURE 7.8 Trend in sea ice cover for the Weddell Sea sector using 41 years (January 1979–December 2019) of passive microwave observations: (a) monthly sea ice extent anomalyand (b) monthly sea ice area anomaly. The anomalies are smoothed using a 2-month running mean.

sea ice cover in the Bellingshausen and Amundsen Seas sector. The variations in SIE and SIA in this sector can influence the local and Bellingshausen–Amundsen temperatures (De Santis et al., 2017). SIE in this sector shows a continuous increasing trend, although the rate of expansion decreases due to a sudden decrease in ice cover in the last few years (Parkinson, 2019). Using the monthly sea ice information from 1979–2005, Comiso and Nishio (2008) found an increasing trend of $3.12 \pm 2.36 \times 10^3$ km²/year. Further, adding another two more years of data (i.e. from November 1978 through 2008) Comiso et al. (2011) found that the SIE trend increased to $5.53 \pm 2.14 \times 10^3$ km²/year and the uncertainty reduces slightly. Analysing 38 years of SIE (January 1979 to December 2016), De Santis et al. (2017), reported a further increase in the trend to $1.7 \pm 0.8\%$ change per decade. All these studies project a slow accelerating trend in SIE over the Weddell Sea (Table 7.1). However, in 2017, a sudden decrease in SIE is noticed which continued to oscillate at its minimum level for almost 36 months before it reached its monthly climatological value during the month of October–November 2019 (Figure 7.8a). This continuous 3-year low (compared with the monthly climatology) in SIE reduced the trend to 2.4×10^3 km²/year (Figure 7.8a). Using 40 years of SIE data (January 1979–December 2018), Parkinson (2019) reported a trend of $4.0 \pm 3.5 \times 10^3$ km²/year in this sector. Also,

TABLE 7.1
Linear Trends in Sea Ice Extent (SIE) and Sea Ice Area (SIA) Reported by Different Researchers

Sector in the Antarctic	Cavalierie and Parkinson (2008), (Nov. 1978–Dec. 2006)		Parkinson and Cavalierie (2012), (Nov. 1978–Dec. 2010)		Comiso et al, (2017) (Nov. 1978– Dec. 2015)	Comiso and Nishio (2008) (Nov. 1978– Dec. 2006)	Comiso et al. (2011) (1979–2008)	De Santis et al. (2017) (1979–2016)	Parkinson (2019) (1979–2018)
	SIE	SIA	SIE	SIA	SIA	SIE	SIE	SIE	SIE
Weddell Sea	3.30 ± 2.33	2.08 ± 2.03	5.20 ± 1.90	4.30 ± 1.70	8.90 ± 1.40	3.12 ± 2.36	5.53 ± 2.14	8.80 ± 1.00	4.00 ± 3.50
Indian Ocean	3.66 ± 1.21	2.33 ± 1.03	5.90 ± 1.10	4.20 ± 0.90	4.80 ± 0.80	3.57 ± 1.20	3.98 ± 1.08	4.80 ± 0.60	2.60 ± 1.80
Western Pacific Ocean	1.40 ± 1.02	2.24 ± 0.79	0.40 ± 0.80	1.40 ± 0.60	3.40 ± 0.60	1.38 ± 1.10	1.04 ± 0.89	2.70 ± 0.40	2.60 ± 1.30
Ross Sea	11.51 ± 1.86	8.93 ± 1.55	13.70 ± 1.50	10.90 ± 1.30	10.20 ± 1.10	11.50 ± 1.88	13.9 ± 1.69	10.30 ± 0.80	5.80 ± 2.90
Bellingshausen/ Amundsen Seas	-8.76 ± 1.43	-6.01 ± 1.10	-8.20 ± 1.20	-5.90 ± 0.90	-3.30 ± 0.90	-8.35 ± 1.44	-10.3 ± 1.29	-3.40 ± 0.60	-3.70 ± 1.80
Total Antarctic	11.11 ± 2.62	9.76 ± 2.41	17.10 ± 2.30	14.90 ± 2.10	24.0 ± 1.80	11.20 ± 2.68	14.1 ± 2.52	24.10 ± 1.20	11.30 ± 5.30

Note: The trend is expressed in 10^3 km² per year. Data used for the analysis are presented in bold.

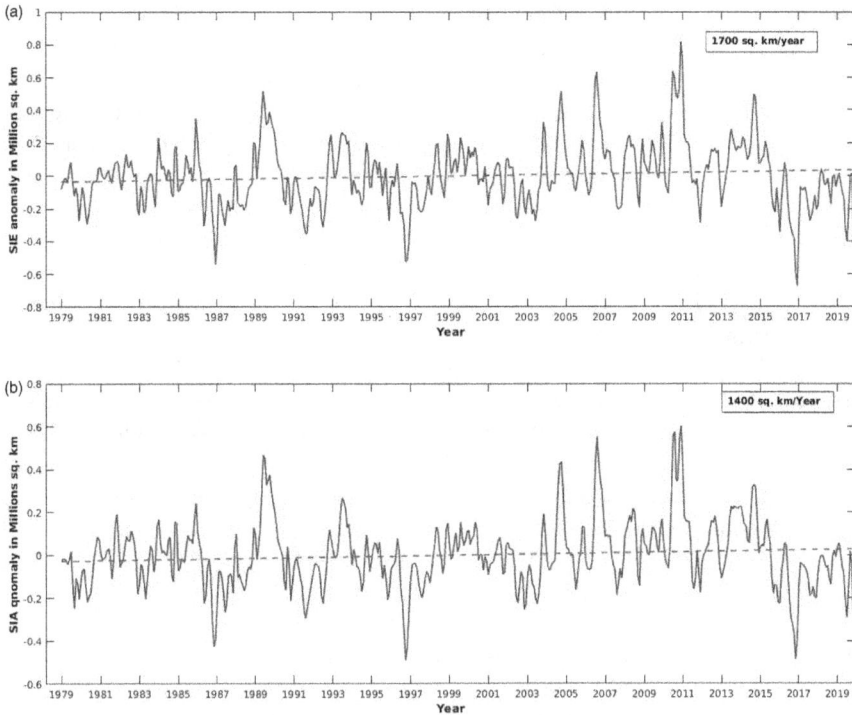

FIGURE 7.9 Trend in sea ice cover for the Indian Ocean sector of the Antarctic using 41 years (January 1979–December 2019) of passive microwave observations: (a) monthly sea ice extent anomaly and (b) monthly sea ice area anomaly. The anomalies are smoothed using a 2-month running mean.

they found that the SIA in this sector increases at a much faster rate ($7.0 \pm 3.7 \times 10^3$ km²/year) than the SIE. On the same line the analysis of monthly SIA anomaly from January 1979 to December 2019 shows a positive trend of 2.5×10^3 km²/year (Figure 7.8b) in this sector. The comparison between trends of SIE and SIA shows that the ice concentration is increased in this sector.

7.4.2 INDIAN OCEAN SECTOR

The asymmetry in growth (8 months) and decay (4 months) period of sea ice in the Indian Ocean sector of the Antarctic is most prominent. The Weddell Gyre pumps sea ice into this sector from the west, and the sea ice anomaly in the western Indian Ocean has a strong correlation with the tropical Indian Ocean sea surface temperature (Yuan and Martinson, 2000). Further, the SIE anomaly in this sector during austral winter (i.e. from April to October) following ENSO events and the Southern Oscillation Index in the previous 12 months (Simmonds and Jacka, 1995) are linked. Moreover, Deb et al. (2014) found that an inverse relation exists between ENSO and SIE in this sector. The SIE in this sector shows an accelerating trend from November

1978 through 2008 (1.9 ± 0.5% change per decade between 1979–2005) (Comiso and Nishio, 2008) and the trend increased to 2.1 ± 0.6 % change per decade between November 1978 through 2008 (Comiso et al., 2011). Analysis of longer SIE time series (January 1979 through December 2016) shows a reduction in the increasing trend to 1.7 ± 0.9% change per decade, which suggests that in the latter 8 years there is a decrease in SIE compared with the previous two decades (De Santis et al., 2017). The addition of another two years (i.e., 2017 and 2018) of the SIE anomaly to the De Santis et al. (2017) analysis, reduced the trend further to 1.4 ± 0.9% change per decade (i.e. 2.6 ± 1.8 × 10^3 km²/year) (Parkinson, 2019) (Table 7.1). The analysis of 41 years (1979–2019) of the monthly SIE anomaly, as observed by radiometers onboard SMMR, SSM/I, and SSMIS, shows that the SIE is increasing at a rate of 1.7 × 10^3 km²/year (Figure 7.9a). This clearly shows that until the first decade of the 21st century the SIE in this sector accelerated then the rate of expansion decreased. Using 40 years of sea ice data, Parkinson (2019) found an increasing trend in SIA in the order of 5.9 ± 1.8 × 10^3 km²/year. However, the trend of monthly SIA anomaly for 41 years (1979–2019) shows that the SIA increases at a rate of 1400 km²/year. Like the Weddell Sea in this sector, the secular trend in SIA is higher than that of SIE.

7.4.3 WESTERN PACIFIC SECTOR

The Western Pacific sector of the Antarctic shows maximum and minimum sea ice extents in September and February, unlike the Southern Ocean as a whole. However, there exists a very small difference in area between October and September SIEs and the August ice cover is not far behind. The monthly SIE anomaly shows a decreasing trend from 2006 through 2010 and then it suddenly peaked up to 2014 before it shows an oscillating behaviour till the end of 2017 (Figure 7.10a). The SIE anomaly showed an increasing trend of 1.38 ± 1.01 × 10^3 km²/year for the period 1979–2005 (Comiso and Nishio, 2008). However, the analysis of a longer time series (i.e. from November 1978 through 2008) of the SIE anomaly shows a reduction in the increasing trend to 1.04 ± 0.89 × 10^3 km²/year. This shows that a decreasing trend in the SIE in subsequent years after 2005 (Figure 7.10a). During the same time the SIA anomaly also shows a negative trend (Figure 7.10b). However, in the next half decade the sea ice cover regains the accelerating trend and the SIE anomaly shows an increasing trend of 1.8 ± 1.2% per decade (De Santis et al., 2017). This trend further picked up and reached a value of 2.3 ± 1.2% per decade during 1979 to 2019 (Parkinson, 2019). During the same period, the SIA anomaly increased at a slightly higher rate, 2.8 ± 1.4% per decade. The analysis of last 41 years (1979–2019) of the monthly sea ice anomaly shows an increasing trend of 2.0 × 10^3 km²/year in SIE and 2.7 × 10^3 km²/year in SIA (Figure 7.10). This sector shows an almost 30% more positive trend in SIA than SIE.

7.4.4 ROSS SEA SECTOR

Ross Sea, Antarctica, is another important sector for the formation of bottom water and has the second highest sea ice coverage after Weddell Sea. This sector has a see-saw effect with the Bellingshausen and Amundsen Seas sector (Massom and

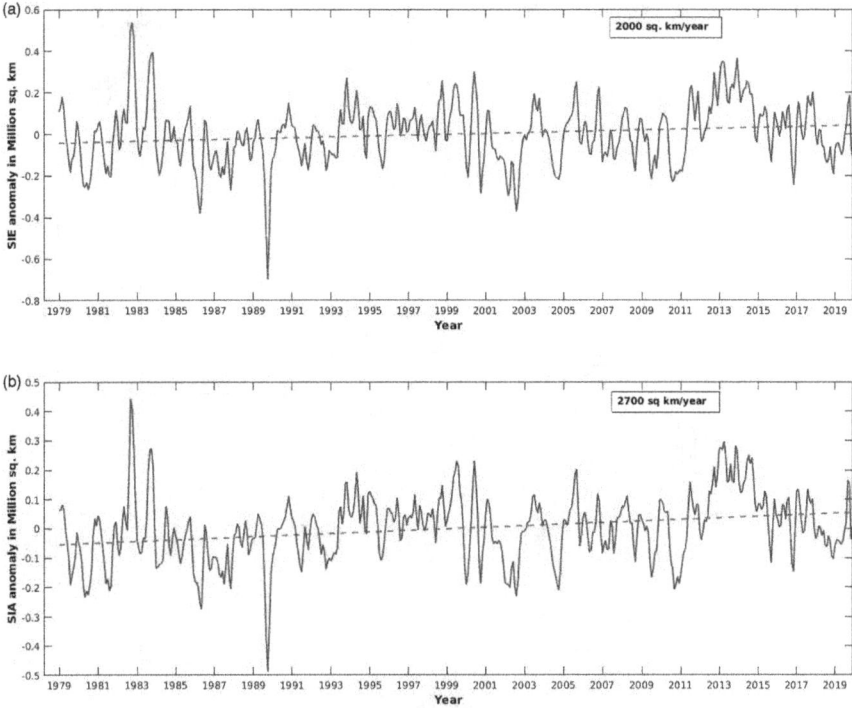

FIGURE 7.10 Trend in sea ice cover for the Western Pacific sector of the Antarctic using 41 years (January 1979–December 2019) of passive microwave observations: (a) monthly sea ice extent anomaly and (b) monthly sea ice area anomaly. The anomalies are smoothed using a 2-month running mean.

Stammerjohn, 2010; Stammerjohn et al., 2012; Massom et al., 2013). The first three decades of radiometric observations (November 1978 through 2008) showed a steep increase in SIE. For example: SIE from 1979–2005 shows an increasing trend of 11.5 $\pm$ 1.88 $\times$ 10^3 km^2/year (Comiso and Nishio, 2008). Analysis of a longer time series data (November 1978 through 2008) shows a positive trend of 13.9 $\pm$ 1.69 $\times$ 10^3 km^2/ year (Comiso et al., 2011). After 2008, this region experienced a downward trend in the sea ice cover (Figure 7.11a) and the resultant rate of increase of SIE from January 1979 to December 2016 reduced to 3.3 $\pm$ 0.9% per decade (de Santis et al., 2017). The decrease in SIE after 2014 reduced the SIE trend to 2.1 $\pm$ 1.1% per decade (5.8 $\pm$ 2.9 $\times$ 10^3 km^2/year) (Parkinson 2019). Analysis of 41 years (January 1979 to December 2019) of monthly SIE anomaly shows that the SIE increases at a rate of 4.7 $\times$ 10^3 km^2/ year (Figure 7.11a). The monthly SIA anomaly also shows an increasing trend of 3.9 $\times$ 10^3 km^2/year (Figure 7.11b). This is the only sector where the SIA trend is less than that of the SIE. This signifies that the SIC is decreasing in this sector.

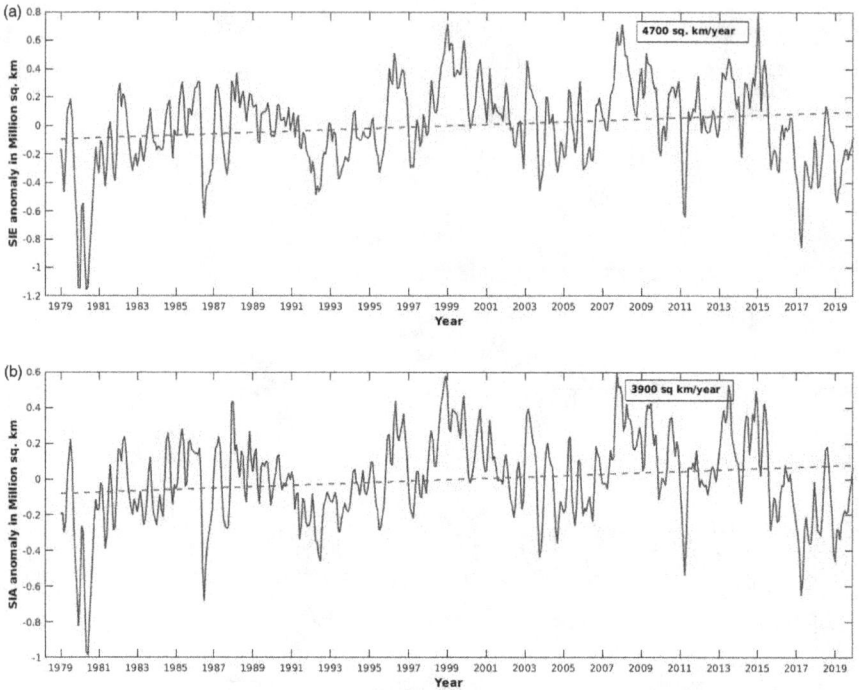

FIGURE 7.11 Trend in sea ice cover for the Ross Sea using 41 years (January 1979–December 2019) of passive microwave observations: (a) monthly sea ice extent anomaly and (b) monthly sea ice area anomaly. The anomalies are smoothed using a 2-month running mean.

7.4.5 BELLINGSHAUSEN AND AMUNDSEN SEAS SECTOR

The Bellingshausen and Amundsen Seas sector in the Antarctic is the most out of line with the rest of the Southern Ocean and other sectors. Sea ice cover in this sector is influenced by the processes in both the Weddell Sea and Ross Sea sectors, and vice versa. The sea ice cover in this sector is also influenced by the weather and climate change occurring in the Antarctic Peninsula. During the late 1970s through the first decade of the 21st century the SIE in this sector showed a more rapid decrease than in recent decades. Since 1979–2005 SIE has reduced at a rate of $-8.35 \pm 1.44 \times 10^3$ km²/year (Comiso and Nishio, 2008). In the subsequent three years the decrease in SIE was more rapid than the previous two and half decades and the trend was $-10.3 \pm 1.29 \times 10^3$ km²/year (Comiso et al., 2011). It is interesting to note that during the same period Ross Sea sector showed a higher positive trend in SIE. In the latter two decades the sea ice cover recovered and by 2016 the decreasing trend reduced to $-2.9 \pm 1.4\%$ per decade (de Santis et al., 2017). Further, this trend shows a slight improvement and 40 years of analysis by passive microwave observation of SIE shows that the sea ice cover was further regained and the trend reached a value $-2.5 \pm 1.2\%$ per decade ($-3.7 \pm 1.8 \times 10^3$ km²/year) (Parkinson, 2019). Forty-one years of monthly SIE anomalies showed a decreasing trend of -3.6×10^3 km²/year (Figure 7.12a).

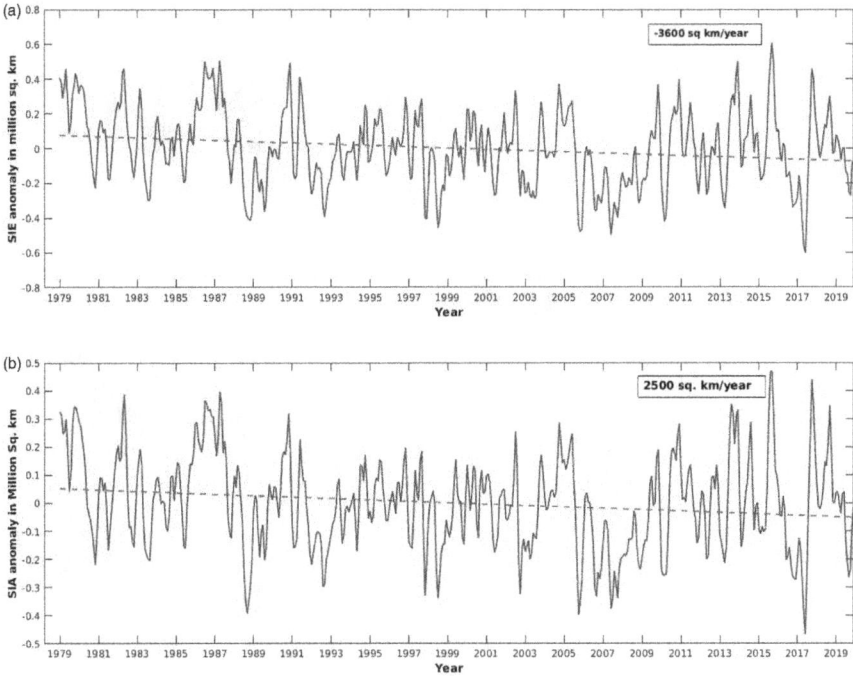

FIGURE 7.12 Trend in sea ice cover for the Bellingshausen and Amundsen Seas sector using 41 years (January 1979–December 2019) of passive microwave observations: (a) monthly sea ice extent anomaly and (b) monthly sea ice area anomaly. The anomalies are smoothed using a 2-month running mean.

The sea ice area decreased at a rate of $-4.9 \pm 2.1 \times 10^3$ km²/year during 1979–2018 (Parkinson, 2019). It can be inferred from Parkinson (2019) that the sea ice concentration in this sector is decreasing at a faster rate. However, 41 years of monthly SIA anomalies showed a reduction in the negative trend in SIA to -2.5×10^3 km²/year (Figure 7.12b). This contradicts Parkinson (2019) and shows that the rate of decrease in SIA is less than that of the SIE. This aligns well with other sectors, except the Ross Sea sector. Nevertheless, the sea ice cover in the Ross Sea sector behaved the opposite to that of the Bellingshausen and Amundsen Seas sector.

7.4.6 TOTAL ANTARCTIC

The SIE in the Southern Ocean or total Antarctic shows a prominent annual cycle and has monthly minima in February and maxima in September. Meanwhile, the SIE shows interannual variations, and the monthly deviations depict clearly the overall upward trend. This is also true with SIA. The monthly anomaly of SIE and SIA shows a marked decrease since 2015 and remains below its monthly climatology since 2017 (Figure 7.13a,b). Despite the marked decreases in ice extent following the 2014 record high, the secular trends remain positive, although at roughly half

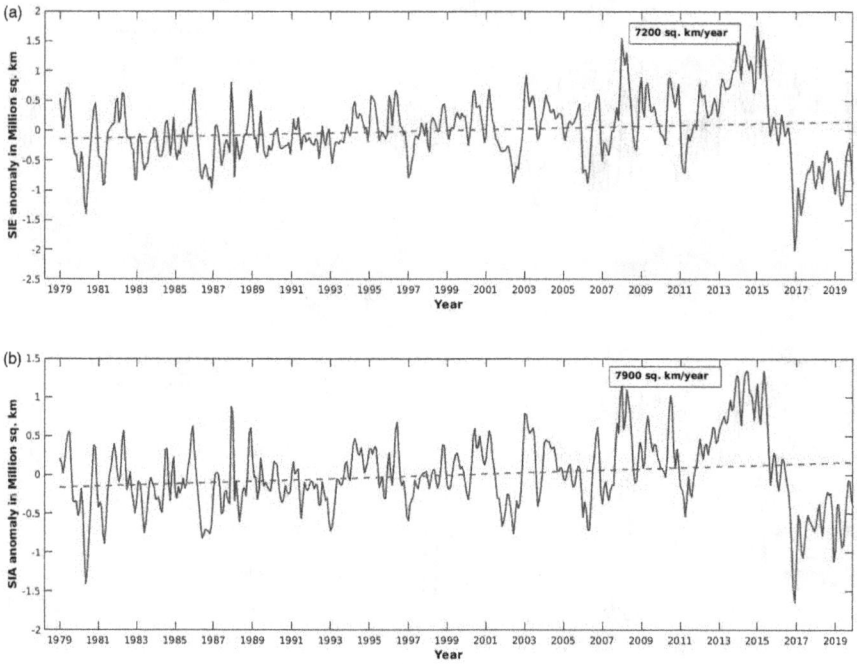

FIGURE 7.13   Trend in sea ice cover for the southern Hemisphere using 41 years (January 1979–December 2019) of passive microwave observations: (a) monthly sea ice extent anomaly and (b) monthly sea ice area anomaly. The anomalies are smoothed using a 2-month running mean.

the magnitude of the 1979–2014 trends (Table 7.1). Specifically, 40 years (1979–2018) of monthly deviations ($11.3 \pm 5.3 \times 10^3$ km²/year) is only 50.7% of the trend ($22.100 \pm 2 \times 10^3$ km²/year) in 1979–2014 (Parkinson, 2019), and similarly the yearly averaged sea ice trend for 1979–2018 (11.3×10^3 km²/year) is only 50.4% of the trend for 1979–2014 (Parkinson, 2019). Moreover, the SIE shows a systematic and continuous increasing trend from 1979 to 2016. The monthly sea ice deviation from 1979 to 2005 shows an increasing trend of $11.2 \pm 2.68 \times 10^3$ km²/year (Comiso and Nishio, 2008). An addition roughly 3 years of observation showed a strengthening in the trend to $14.1 \pm 2.52 \times 10^3$ km²/year (Comiso et al., 2011). Further, this positive trend continues to grow slowly and was found to be $1.6 \pm 0.4\%$ change per decade by the end of 2016 (de Santis et al., 2017). A drastic decrease in sea ice cover in 2017 being maintained below the monthly climatology for a considerable period reduced the increasing trend (Figure 7.13a). Forty years of radiometric observation shows that the SIE increased at a rate of $1.0 \pm 0.5\%$ per decade, which is less than its trend since 2016 (Parkinson, 2019). Moreover, at the same time the SIA has been increasing at a much faster rate (almost double) than the SIE (Table 7.1) (Parkinson 2019). The radiometric observations from January 1979 to December 2019 show a positive trend of $7.2 \ 10^3$ km²/year (Figure 7.13a) over the whole Southern Ocean. The SEA was also found to be increasing at a rate of 7.9×10^3 km²/year (Figure 7.13b).

This shows that the whole Antarctic SIE trend behaves in the same way as that of other sectors except the Bellingshausen and Amundsen Seas sector, whereas the SIA behaves in a similar manner to other sectors except the Ross Sea and Bellingshausen and Amundsen Seas sectors.

7.5   SEA ICE VARIABILITY, PHYSICAL PROCESSES AND TELECONNECTIONS

The SIC, hence the SIE and SIA, in the Antarctic is influenced by both atmospheric and oceanic processes that occur locally and remotely. For example, extreme sea ice retreat in late 2016 has been related to (i) rapid sea ice retreat in the Weddell Sea caused by a strong northerly atmospheric flow (Turner et al., 2017); (ii) reduction in SIE in the Indian and Pacific Oceans was a result of the weakening of circumpolar westerlies caused by a weakened polar stratospheric vortex (Wang et al., 2019); (iii) the presence of a zonal wave 3 atmospheric pattern around Antarctica for a longer period resulted in a reduction in sea ice cover in the Indian Ocean, Ross Sea, Bellingshausen Sea, and western Weddell Sea (Schlosser et al., 2018; Wang et al., 2019); the (iv) occurrence of an unusually negative southern annular mode (SAM) in November 2016 caused rapid ice retreat in the Ross Sea and elsewhere (Turner et al., 2017; Schlosser et al., 2018; Wang et al., 2019; Meeh et al., 2019); and (v) the extreme El Niño that peaked months earlier, in December 2015 through February 2016, contributed to unusually warm ocean waters in the Bellingshausen, Amundsen, and eastern Ross Seas, with anomalous warmth that persisted into the austral spring (Stuecker et al., 2017). Thus, sea ice in the Antarctic responds to many different interactive processes in the atmosphere and ocean (Hobbs et al., 2016). These include the horizontal and vertical advection of heat, the rigging and rafting of sea ice by wind and ocean currents; meltwater from icebergs, ice shelves, and marine-terminating glaciers, etc. Potential driving mechanisms for these changes include changes in surface temperature (Comiso et al., 2017), wind direction (Hobbs et al., 2016; Holland & Kwok, 2012; Turner et al., 2016), sea ice drift (Pope et al., 2017), precipitation (Liu & Curry, 2010), ice-ocean feedbacks (Goosse & Zunz, 2014), changes in the entrainment of warm deep water into the winter mixed layer (Zhang, 2007), and meltwater from the Antarctic ice sheet (Bronselaer et al., 2018). These drivers are schematically illustrated in Figure 7.14.

7.5.1   ATMOSPHERIC DRIVERS

Near-surface winds have a very high impact on advection of sea ice and are a dominant driver of SIC trends around much of West Antarctica (Holland and Kowk, 2012). In contrast to the SIC trend, thermodynamic changes caused by wind play a major role elsewhere, and autumn SIE trends in the Bellingshausen Sea oppose those in near-surface winds (Holland, 2014). The surface winds that drive sea ice motion and variability in SIC, hence also the sea ice trend around Antarctica, are strongly linked to synoptic weather systems, which in turn are connected to larger-scale climate variability in the Southern Hemisphere (Hobbs et al., 2016; Turner et al., 2016; Schemm, 2018).

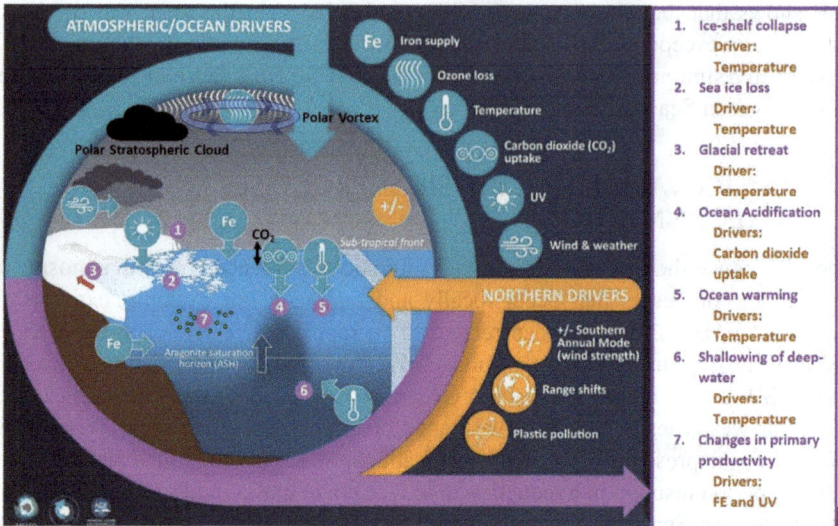

FIGURE 7.14 Pictorial representation of the global drivers that are affecting the Southern Ocean. Northern drivers are global drivers whose influence reaches from north of the Southern Ocean. (Adapted and modified from Morley et al. [2020].)

A Baroclinic region, known as the circumpolar trough (low-pressure belt), present between 60–70°S is known to be responsible for affecting the ice cover in the Antarctica. There exist three climatological low-pressure centres, located approximately at 20°E, 90°E and 150°W, within the circumpolar trough, associated with the atmospheric wavenumber 3 pattern around the continent (Raphael, 2004). Further, this zone is associated with the propagation of a large number of storms from mid-latitudes to the south or development of storms in the area. The low-pressure centre present off West Antarctica at approximately 150°W is quasi-stationary and known as the Amundsen Sea Low (ASL) (Turner et al., 2012; Hosking et al., 2013). Climatologically, the ASL is positioned at about 220°E in winter and shifts to 250°E in summer (Hosking et al., 2013). The variability in pressure in the ASL region has a major influence on spreading the climate signal from the Antarctic Peninsula to the Ross Sea. Further, the ASL region has the strongest teleconnection with the tropical Pacific (Ding et al., 2011). During the positive (negative) phase of ENSO, the mean sea level pressure in the ASL region is higher (lower) than the normal (Turner, 2004). This pressure change influences the wind field and therefore the sea ice distribution between the Antarctic Peninsula and the Ross Sea. Also, the sea surface temperature trend across the Atlantic Ocean has been linked to mean sea level pressure (MSLP) and sea ice changes in this sector (Li et al., 2014). Additionally, the atmospheric polar front jet (PFJ) in the Pacific immensely impacts the SIE in the Ross Sea (Turner et al., 2015).

The Southern annular mode (SAM) or the Antarctic oscillation is another primary mode of climate variability at the southern high latitude (Thompson and Wallace, 2000). The SAM is a low-frequency atmospheric mode, characterized as an oscillation of mass between mid- and high-latitude areas of the Southern Hemisphere. The positive phase of the SAM is characterized by relatively low and high pressure over the Antarctic and the mid-latitudes, respectively. The negative phase has an opposite feature to that of the positive phase. Different phases of the SAM have intrinsic variability, but are also influenced by greenhouse gases concentration, volcanic aerosols, and the loss of stratospheric ozone over the Antarctic (Arblaster and Meehl, 2006). The positive SAM generates more cyclonic conditions over ASL, thus driving warm (cold) winds, poleward (equatorward) into the Antarctic Peninsula/Bellingshausen Sea (Ross Sea) region, which in turn reduces (enhances) SIE in the sector. Experiments with atmospheric model suggest that the ozone hole would deepen the ASL, so also moving the SAM into its positive phase and strengthening the winds around the Antarctic resulting in more sea ice formation in the Ross Sea (Holland & Kwok, 2012; Hosking et al., 2013; Turner et al., 2009; Turner et al., 2015). Further, Marshall et al.(2014) suggested that stratospheric ozone loss over the Antarctic would lead to a decrease in SSTs around the continent, which would enhance sea ice cover. However, the role of stratospheric ozone depletion in the recent increase in Antarctic SIE was questioned by Sigmond and Fyfe (2010).

A differing pattern of correlation is observed between inter-annual sea ice variability and atmospheric circulation in different sectors of the Antarctic (Turner et al., 2015). For example, SIE in the Bellingshausen and Amundsen Seas sector is correlated with mean sea level pressure in a similar way to that of the Weddell Sea (Turner et al., 2015). Further, SIE variability in a limited area of the Ross Sea, just to the west of 180°W, shows a dependency on atmospheric circulation. Moreover, the influence of atmospheric circulation on SIE variability in the Indian Ocean sector behaves quite differently from that of other sectors of the Antarctic. Using a coupled ocean–sea ice model, Deb et al. (2017) showed that dynamical SAM forcing leads to a non-annular response in sea ice cover over the Indian Ocean sector of the Southern Ocean (IOS). In positive SAM years, westerlies are stronger over the Indian Ocean, resulting in advection of sea ice from the Weddell Sea to the western part of the IOS, increasing SIC and thickness over the region. In contrary, in the east IOS, strong sea ice drift along the south-eastward direction pushes the sea ice edge towards the coast and creates a divergence of sea ice near the edge (Deb et al., 2017). The reasons for the different patterns of correlation between SIE and MSLP in the five sectors around the Southern Ocean may well be related to the shape of the continent, as well as different forcing from areas outside the Antarctic. However, further research is needed to gain a better understanding of the relationships.

Antarctic sea ice has a strong response to changes in the Pacific Ocean sea surface temperature (SST), i.e. El Niño Southern Oscillation (ENSO). The tropical Pacific convection excites an atmospheric disturbance that propagates in a south-eastward direction to the south-eastern Pacific in the Antarctic as the Rossby wave train (Yu et al., 2011). This type of mode is commonly known as the Pacific South American mode (Mo, 2000; Mo & Higgins, 1998; Mo & Paegle, 2001). During El Niño, an anomalous high-pressure system develops in the eastern Pacific, centred at about 90°W and 55°S, which weakens the ASL (Yuan, 2004; Yuan & Li, 2008). Deb et al. (2014) showed that canonical and central Pacific ENSOs impact differently the sea ice cover in the IOS. They showed that winters following La Niña years are associated

with more SIA compared to that of ENSO-neutral years. In winter following La
Niña years, the sea level pressure gradient between the Antarctic land mass and the
subpolar region increases, enhancing the southerly wind, with a reduction in surface
air temperature (SAT) and enhancing northward advection of sea ice in the IOS. The
in-phase relation among SAT, SST and sea ice advection results in an increase in
SIA. However, in the winters following El Niño years, anomalous easterlies, wind-
induced sea ice motion, SAT anomalies and heat transport by the regional ferrel cell
increase (decrease) the SIA in the western (eastern) part of the IOS. During El Niño
Modoki years, an increase in SST and the presence of warmer surface air reduce SIA
during summer as well as the winter following it (Deb et al., 2014). Additionally,
Dash et al. (2013) showed that teleconnection between tropical Pacific and SIE in the
Weddell Sea (South Atlantic) and the Bellingshausen–Amundsen Sea (South-Eastern
Pacific) sectors of the Southern Ocean flipped around 1992. They attributed this phase
shift in the tropical and polar teleconnection to the modulation of the regional Ferrell
Cell. Thus, there exists significant asymmetry between the responses of El Niño and
La Niña to the SIE in the Antarctic (Houseago-Stokes & Mcgregor, 2000; Simpkins
et al., 2013; Yuan, 2004). This shows the inconstancy of tropical and Southern Ocean
teleconnection, and is attributed to different phases of SAM (Fogt & Bromwich, 2006;
Fogt et al., 2011; Stammerjohn et al., 2008). The ENSO and high-latitude Southern
Hemisphere teleconnection is significant with El Niño (La Niña) and is associated
with the negative (positive) phase of SAM (Fogt & Bromwich, 2006; Fogt et al.,
2011; Stammerjohn et al., 2008). Although, the contribution of ENSO is small in the
overall trend of Antarctic sea ice, its importance on interannual variability cannot
be ruled out (Kohyama & Hartmann, 2016; Simpkins et al., 2013; Yu et al., 2011).
Additionally, Ding and Steig (2013) reported that a widespread warming trend over
the Antarctic Peninsula in fall is forced by the Rossby wave trend emanating from
the tropical Pacific associated with the SST anomaly over the region. Also, Clem
and Fogt (2015) reported that statistically significant warming after 1979 across the
Western Antarctic Peninsula corresponds to increasing pressure in the South Atlantic
associated with more La Niña-like conditions in the tropical Pacific and an associated
Rossby wave train. Meanwhile, the warming trend in western West Antarctica is
attributed to deepening in pressure in the ASL present near the eastern Ross Sea,
which is tied to the persistence of a negative phase of Pacific Decadal Oscillation
(PDO) since the 1990s. Nevertheless, no widely established mechanism exists to link
the ENSO and SAM (Ciasto et al., 2015).

Multidecadal variability in convective heating over the tropical Atlantic produced
due to the variation in sea surface temperature drives the anomalous Rossby wave
and affects the sea ice in the west Antarctic (Li et al., 2014; Simpkins et al., 2014).
With a warming trend in tropical Atlantic since 1979, the ASL deepens, driving
Bellingshausen and eastern Ross Sea sea-ice changes in winter (Li et al., 2014) and
spring (Simpkins et al., 2014). Similar results were also reported by Hobbs et al.
(2015). Antarctic Sea ice teleconnections with multidecadal SST variability over
tropical Pacific and Atlantic have been suggested as a contributing factor, but the
magnitude of their influence on trends of the former is not well-established; while an
initial analysis by Hobbs et al. (2016) indicated it to be modest.

In summary, the variability in the sea ice cover in the Antarctic is controlled by the local, regional and remote (tropical) atmospheric forcings. Variability in different atmospheric modes (such as SAM, atmospheric wavenumber 3 pattern, ASL, etc.) in the Southern Hemisphere impacts the interannual variability in ice cover in the Antarctic through pronounced changes in wind stress over the region and thermodynamical forcing. However, these regional atmospheric changes are not solely able to explain the observed sea ice trends, which is especially true for the Ross Sea sector (Parkinson & Cavalieri, 2012; Turner et al., 2015). A coupled ocean–sea ice model study showed that the sea ice variability during autumn and winter is mainly driven by thermodynamic forcing, whereas wind stress plays a dominant role during summer (Kusahara et al., 2018). Further, dynamic-thermodynamic forcings evidently impact the SIE in the melt season. However, regionally SIE variability in the Ross Sea is mainly controlled by wind stress, while thermodynamic forcing plays a greater role in other sectors. However, rigorous experiments and analysis are required to comprehend the contributing factors and quantify their influence on the trends of SIE in different sectors of the Southern Ocean.

7.5.2 Oceanic Drivers

Moderately slow warming at southern polar latitudes in a global warming scenario could be attributed to anomalous heat uptake into the Southern Ocean. However, limited and scattered in situ observations around the continent and Southern Ocean restrict the study to gather concrete evidence that the ocean is a primary driver of interannual variability and trend of SIE in the Antarctic. The increasing trend of Antarctic sea ice can be explained using ocean–atmosphere feedback (Zhang 2007). Using a sea ice–ocean coupled model, Zhang (2007) argued that in a global warming scenario, with an increase in air temperature, downward long-wave radiation increases, which increases the upper-ocean temperature and decreases sea ice growth. The decrease in sea ice growth leads to a reduction in brine rejection, which decreases the upper ocean salinity and density, thus increasing the stratification of the upper ocean. This suppresses the convective overturning, leading to a reduction in upward ocean heat transport and availability of ocean heat flux to sea ice melt, which increases the overall SIE in the Antarctic. Other arguments suggest that the atmosphere can have a two-timescale response in the Southern Ocean (Holland et al., 2017; Kostov et al., 2017; Ferreira et al., 2015). For example, strengthening of westerlies enhances the Ekman drift and increases the Antarctic SIE due to direct wind forcing and colder sea surface temperature in the short term. Nonetheless, the enhanced upwelling can bring warmer subsurface water to the surface and cause a reduction in ice growth/enhancement of the ice melting in the long run. A recent reduction in sea ice cover, that began in 2016, could be attributed partly to this two-timescale response of the atmosphere (Meehl et al., 2019). The coexistence of a positive phase of SAM and negative phase of interdecadal Pacific oscillation (IPO) over the 2000s, enhances the wind stress curl that leads to upwelling of warm water and results in a reduction in sea ice cover in the epoch that began in 2016 (Meehl et al., 2019). Another, two-timescale response suggested is the ozone-induced increase in polarity of the SAM that impacts the surface temperature and sea ice in the Southern Ocean (Marshall et al., 2014). However, there is considerable uncertainty over the length of transition between the two timescales, which ranges from a few years to a few decades. Extensive research into this/these mechanism(s) could potentially help to explain partially/totally the observed trend of SIE in the Southern Hemisphere.

On the other hand, the climatological pattern of sea ice drift around the Antarctic depicts the importance of sub-polar gyres contributing to the sea ice cover of different sectors of the Antarctic. For example, the Ross Sea gyre shows coastal inflow of sea ice into the Ross Sea that moves southward along the coast, with a strong northward outflow in the west (Kwok et al., 2017). The Weddell gyre is associated with a cyclonic drift pattern that dominates the ice drift during March through November (Kwok et al., 2017).

Sea ice–ocean feedback influences the sea ice anomaly from one season to the next (Holland, 2014; Stammerjohn et al., 2012), However, these interactions and feedbacks show strong seasonal and regional dependency (Maksym et al., 2012). The timing and rate of melting of sea ice appears to have a significant influence on the timing of the advance of ice cover in subsequent seasons, whereas the reverse is weakly related (Stammerjohn et al., 2012). This suggests that the acceleration or deceleration of the onset of autumn freezing is strongly related to ocean thermal feedback. The acceleration of sea ice retreat in the Bellingshausen Sea is attributed to the positive feedback of wind-driven sea ice retreat that can lead to upper-ocean warming, which favours its cause, resulting in a positive feedback (Meredith & King, 2005). On the other hand, strengthening of westerlies can enhance the ice divergence, with more sea ice production and enhancement in the convective overturning circulation and upward heat flux that inhibits ice growth, which is a negative feedback (Sigmond & Fyfe, 2010). These feedback mechanisms are critical to understanding the present and predicting the future sea ice cover in the Antarctic.

In the Antarctic coastal region, melt water from the ice-shelf has the potential to affect the sea ice growth. Fresh water input from the melting of the Antarctic ice-sheet can increase stratification and shield the surface ocean from the warm deeper waters which could reduce SSTs and cause subsurface ocean warming (Fogwill et al., 2015; Park & Latif, 2018). This has been noticed in the Ross Sea (Jacobs et al., 2002). However, the model experiment found that ice-sheet melting has a very minimal impact in explaining the observed sea ice trend. In another modelling study, Bronselaer et al. (2018) found evidence that by the end of the century the influx of ice sheet meltwater may cause a substantial increase in mean winter SIA in the Antarctic.

7.6 ANTHROPOGENIC FORCING AND FUTURE PROJECTIONS

There is a clear indication of anthropogenic emissions impacts on the recent decline of Arctic sea ice (Notz & Stroeve, 2016). However, the link between the Antarctic sea ice trends and human-induced greenhouse gas emissions is currently not understood. The consequences of anthropogenic greenhouse gas emissions on Antarctic sea ice is challenging because (i) there has been a relatively short record of sea ice observation, (ii) the response may be small compared to the natural climate variability of the Southern Ocean, which dominated the recent SIE and its decadal trend (Crosta et al., 2021; Turner et al., 2012), and (iii) coupled global climate models are unable to faithfully reproduce the complex Antarctic climate system (Turner et al., 2013; Hobbs et al., 2016). Furthermore, some of the climate modes, such as SAM, ENSO, etc. also are influenced anthropogenically.

The Coupled Model Intercomparison Project (CMIP) is a standard protocol and initiative of the World Climate Research Programme (WCRP) to bring together and study the outputs of around 50 state-of-the-art coupled atmosphere–ocean–sea ice climate models which are independently run for past conditions and a range of future scenarios. The CMIP was established in 1985. Phase 5 (CMIP5) is the most complete and extensive of the CMIPs, and the sixth phase is currently underway.

Three types of coupled model runs are of particular interest:

(1) Control runs with fixed pre-industrial concentrations of greenhouse gases, and fixed stratospheric ozone, aerosols and solar forcing.
(2) Historical runs from the mid-19th century to 2005 with observed concentrations of greenhouse gases and development of the ozone hole from about 1980.
(3) Projections for the coming decades and centuries run with a range of Representative Concentration Pathways (RCPs).

These CMIP5 model simulations are performed independently and many models include different physics and parameterization schemes and have a range of complexity in sea ice components (Roach et al., 2018). Many of the CMIP5 models simulate a decreasing sea ice trend in the Antarctic, which points to the inadequate representation of key processes in the model or/and internal variability in SIE which the models are unable to capture (Shu et al., 2015).

Turner et al. (2013) performed an assessment of annual cycle and trends (over 1979–2005) of Antarctic sea ice in the historic run of 18 CMIP5 models, and reported that many models failed to simulate the annual cycle of SIE. They reported that the maximum root mean square error in these CMIPs simulated SIE as compared to satellite radiometric observations is 9.2×10^6 km^2, which is half the maximum ice extent observed by the latter and three-quarters of the mean satellite SIE. The mean root mean square error is 3.8×10^6 km^2, which is one-third of the mean satellite SIE. Gross errors in some models were linked to large warm or cold biases in ocean temperatures. Although CMIP5 models have been unable to capture the magnitude of the seasonal cycle, the asymmetric pattern (i.e. the length of formation and melt seasons) and its timing are largely captured by them (Eayrs et al., 2019; Turner et al., 2013). This suggests that climate models include the physics/mechanism that drive the asymmetric cycle of Antarctic sea ice. Moreover, Polvani and Smith (2013) suitably selected four CMIP5 models which showed that the sea ice trend observed in the Antarctic since 1979 falls well within the natural distribution of trends arising in the coupled atmosphere–ocean–sea ice system. Additionally, it was found that a few CMIP5 models correctly simulate the negative phase of the Interdecadal Pacific Oscillation (IPO) and show an increasing trend in SIE (Meehl et al., 2016). In contrast, Zunz et al. (2013) examined Antarctic sea ice in the historical and hindcast experiments of 24 CMIP5 models and concluded that no CMIP5 model could reasonably reproduce Antarctic sea ice cover which agrees with satellite observations, and all models overestimate the interannual variability of SIE, particularly in the austral winter. Roach et al. (2018) pointed out that in spite of different sea ice physics adopted by the models, these tendencies are broadly consistent across the CMIP5 simulations. However, the multi-model mean of CMIP5 simulations improved the prediction of the growth season (7 months) followed by the melt season (5 months) that matches with the observations (Rosenblum and Eisenman, 2017).

It is a general view that CMIP5 models poorly simulate the mean and variability of Antarctic sea ice (Collins et al., 2013). Thus, the Intergovernmental Panel on Climate Change (IPCC) Fifth Assessment Report states low confidence in projected changes of ice cover in the Antarctic. In contrast, Holmes et al. (2019) revisited the CMIP5 models and stated that a subset of good models were able to reproduce the seasonal cycle of Antarctic sea ice. However, they (the models) greatly differ in their skill in simulating ice processes and sea ice concentration budget terms. Further, Holmes et al. (2019) reported that ACCESS climate models appear to be an excellent candidate in representing sea ice budget terms, and argued reported overcompensation between model dynamics and thermodynamics (Schroeter et al., 2018; Uotila et al., 2014) which is largely an artefact of incomplete observational sampling. Additionally, they showed that key to faithful simulation of winter maximum of Antarctic sea ice requires accurate representation of the circumpolar pressure trough and its climatological lows. A study with the Community Earth System Model (CESM) revealed that accurate representation of sea ice drift velocity is crucial for accurate prediction and projection of sea ice in the Antarctic (Sun & Eisenman, 2021). The appropriate choice of existing coupled climate models, as well as improved coupling of dynamics-thermodynamics among atmosphere–ocean–sea ice systems is expected to provide more reliable projections of future change in Antarctic sea ice.

7.7 SUMMARY

The annual growth and melt cycle of sea ice in the Antarctic is the largest cryospheric seasonal variation on Earth. Sea ice plays a key role in inhibiting the exchange of mass, momentum and energy between the atmosphere and ocean in the polar regions. It is an important habitat and plays a key role also in maintaining the diverse and unique Southern Ocean ecosystem. Therefore, exploration, exploitation, understanding and modelling these phenomena/processes are of broad interest. This chapter briefly explains the formation and decay of SIE in the Antarctic. Also, different methods regularly used to map and monitor the sea ice are discussed. Different factors affecting the sea ice growth/decay and trend in different sectors of the Antarctic are explained. Satellite observation since 1979 shows a modest increasing trend in Antarctic SIE and SIA, albeit there has been a decrease in sea ice cover from 2014 with a record low in 2017. The increasing trend of sea ice cover is a combination of substantial regional changes, with a positive trend in the Ross Sea and Weddell Sea sectors and a decreasing trend in the Amundsen and Bellingshausen Seas sector. The SIA trend is relatively higher than the SIE except for the Ross Sea sector. The sea ice cover in Ross Sea and the Bellingshausen and Amundsen Seas sectors behaves like a dipole. The increase of ice in the Ross Sea is closely linked to a deepening of the ASL, which is part of a large atmospheric variability known as wave number 3 pattern. Rather, modelling studies advocate that the sea ice trend observed in the last four decades is within the bounds of internal variability of the atmosphere–ocean–ice system. The SIEs in different sectors of the Antarctic and Southern Ocean as a whole show extreme inter-annual variability, but yet retain asymmetry (slow growth and fast melt) in annual evolution of sea ice cover. The regional or seasonal variability in SIC/SIE/SIA depends on different drivers, be it atmospheric or oceanic or melting of ice sheet, but above all associated surface wind plays a key role. The divergence created by wind in spring facilitates rapid sea ice melting through ice–ocean feedback. Since smaller floes melt faster, this adds to the above feedback processes.

Almost all CMIP5 models poorly represent Antarctic sea ice cover and trend. Although the multi-model mean could simulate the asymmetry in the annual cycle, a large difference exists in SIE compared with the satellite data. However, some of the uncertainties are the results of large biases in the simulation of ocean temperatures, sea ice drift, sea ice volume, different drivers of sea ice changes, etc. This highlights the necessity of correct representation of atmospheric and oceanic conditions in the coupled model to simulate realistic sea ice conditions over the Antarctic. Another major difficulty is that in situ observations are sparse and the variabilities are large and strong. Finally, in order to understand sea ice variability across different sectors of the Antarctic, influences of various contributing processes/factors such as surface winds, the dynamic and thermodynamic processes in the ice pack and its interaction with the Southern Ocean, the variation of the hydrological cycle at the local and global scale, and interactions with various tropical (ENSO, IPO, PDO) and polar (SAM, wave number 3 pattern) atmospheric indices/drivers need to be considered. Hopefully, future developments will improve the model skill and in situ observation of different atmospheric, oceanic and sea ice parameters to better understand, simulate, validate and predict Antarctic sea ice cover.

ACKNOWLEDGEMENTS

The author is thankful to the editor for providing the opportunity to contribute a chapter in this book. The author also thanks Mr. Balaji Senapati for helping to draw some of the plots. NSIDC is greatly acknowledged for making the sea ice data publicly available for analysis. Thanks are due to the authors and journals of the respective articles from which figures have been adopted and modified.

REFERENCES

Abram, N.J., Thomas, E.R., McConnell, J.R., Mulvaney, R., Bracegirdle, T.J., Sime, L.C., Aristarain, A.J. (2010) Ice core evidence for a 20th century decline of sea ice in the Bellingshausen Sea, Antarctica. *Journal of Geophysical Research: Atmospheres* 115: D23101.

Ackley, S., Wadhams, P., Comiso, J.C., Worby, A.P. (2003) Decadal decrease of Antarctic sea ice extent inferred from whaling records revisited on the basis of historical and modern sea ice records. *Polar Research* 22: 19–25.

Andersen, S., Tonboe, R.T., Kern, S., Schyberg, H. (2006) Improved retrieval of sea ice total concentration from spaceborne passive microwave observations using numerical weather prediction model fields: An intercomparison of nine algorithms. *Remote Sensing of the Environment* 104(4): 374–392.

Arblaster, J., Meehl, G.A. (2006) Contributions of external forcings to Southern Annular Mode trends. *Journal of Climate* 19: 2896–2905.

Arrigo, K.R., van Dijken, G.L., Bushinsky, S. (2008) Primary production in the Southern Ocean, 1997–2006. Journal of Geophysical Research 113: C08004.

Barrett, P.J. (2013) Resolving views on Antarctic Neogene glacial history – the Sirius debate. *Earth and Environmental Science Transactions of the Royal Society of Edinburgh* 104: 31–53.

Bhandari S.M., Vyas, N.K., Dash, M., Khanolkar, A., Sharma, N., Khare, N., Pandey, P.C. (2005) Simultaneous MSMR and SSM/I observations and analysis of sea-ice characteristics over the Antarctic region. *International Journal of Remote Sensing* 26(15): 3123–3136.

Bintanja, R., van Oldenborgh, G.J., Drijfhout, S.S., Wouters, B., Katsman, C.A., (2013) Important role for ocean warming and increased ice-shelf melt in Antarctic sea-ice expansion. *Nature Geoscience* 6: 376–379.

Bitz, C.M., Polvani, L.M. (2012) Antarctic climate response to stratospheric ozone depletion in a fine resolution ocean climate model. *Geophysical Research Letters* 39: L20705.

Bitz, C.M., Lipscom, W.H. (1999) An energy-conserving thermodynamic model of sea ice. *Journal of Geophysical Research* 104(C7): 15,669–15,677.

Bouillon S., Morales Maqueda, M.A., Legat, V., Fichefet, T. (2009) An elastic-viscous-plastic sea ice model formulated on Arakawa B and C grids. *Ocean Modelling* 27:174–184.

Bronselaer, B., Winton, M., Griffies, S.M., Hurlin, W.J., Rodgers, K.B., Sergienko, O.V., Stouffer, R.J., Russell, J.L. (2018) Change in future climate due to Antarctic meltwater. *Nature* 564: 53–58.

Cavalieri, D.J., Parkinson, C.L., DiGirolamo, N., Ivanoff, A. (2012) Inter-sensor calibration between F13 SSMI and F17 SSMIS for global sea ice data records. *IEEE Geoscience Remote Sensing Letters* 9: 233–236.

Cavalieri, D.J., Parkinson, C.L., Gloersen, P., Comiso, J.C., Zwally, H.J. (1999) Deriving long term time series of sea ice cover from satellite passive-microwave multisensor data sets. *Journal of Geophysical Research* 104: 15803–15814.

Cavalieri, D.J., Gloersen, P., Campbell, W.J. (1984) Determination of sea ice parameters with the NIMBUS 7 scanning multichannel microwave radiometer. *Journal of Geophysical Research* 89: 5355–5369.

Ciasto, L.M., Simpkins, G.R., England, M.H. (2015) Teleconnections between tropical Pacific SST anomalies and extratropical Southern Hemisphere climate. *Journal of Climate* 28: 56–65.

Clem, K.R., Fogt, R.L. (2015) South Pacific circulation changes and their connection to the tropics and regional Antarctic warming in austral spring, 1979–2012. *Journal of Geophysical Research: Atmospheres* 120: 2773–2792.

Collins, M., Knutti, R., Arblaster, J., Dufresne, J.L., Fichefet, T. Friedlingstein, P., et al. (2013) Long-term climate change: Projections, commitments and irreversibility. In: T. F. Stocker, et al. (Eds.) *Climate change 2013: The physical science basis. Contribution of Working Group I to the Fifth Assessment Report of the Intergovernmental Panel on Climate Change*, pp. 1029–1136. Cambridge: Cambridge University Press.

Comiso, J.C. (1995) SSM/I sea ice concentrations using the bootstrap algorithm. *NASA Reference Publications* 1380: 49

Comiso, J.C. (2000) *Bootstrap sea ice concentrations from Nimbus-7 SMMR and DMSP SSM/I-SSMIS, version 2 [1992–2010 used]*; National Snow and Ice Data Center, Boulder, CO, digital media. Available online at: http://nsidc.org/data/nsidc-007/.

Comiso, J.C., Nishio, F. (2008) Trends in the sea ice cover using enhanced and compatible AMSR-E, SSM/I, and SMMR data. *Journal of Geophysical Research* 113: C02S07.

Comiso, J.C., Cavalieri, D.J., Parkinson, C.L., Gloersen, P. (1997) Passive microwave algorithms for sea ice concentration: A comparison of two techniques. *Remote Sensing of the Environment* 60(3): 357–384.

Comiso, J.C., Kwok, R., Martin, S., Gordon, A.L. (2011) Variability and trends in sea ice extent and ice production in the Ross Sea. *Journal of Geophysical Research* 116: C04021.

Comiso, J.C., Gersten, R.A., Stock, L.V., Turner, J., Perez, G.J., Cho, K. (2017) Positive trend in the Antarctic sea ice cover and associated changes in surface temperature. *Journal of Climate* 30: 2251–2267.

Cotte, C., Guinet, C. (2007) Historical whaling records reveal major regional retreat of Antarctic sea ice. *Deep-Sea Research Part I* 54: 243–252.

Crosta, X., Etourneau, J., Orme, L.C., Dalaiden, Q., Campagne, P., Swingedouw, D., Goosse, H., Massé, G., Miettinen, A., McKay, R.M., Dunbar, R.B., Escutia, C., Ikehara, M. (2021) Multi-decadal trends in Antarctic sea-ice extent driven by ENSO–SAM over the last 2,000 years. *Nature Geoscience* 14: 156–160.

Curran, M.A., van Ommen, T.D., Morgan, V.I., Phillips, K.L., Palmer, A.S. (2003) Ice core evidence for Antarctic sea ice decline since the 1950s. *Science* 302: 1203–1206.

Dash, M.K., Pandey, P.C., Vyas, N.K., Turner, J. (2013) Variability in the ENSO-induced southern hemispheric circulation and Antarctic sea ice extent. *International Journal of Climatology* 33: 778–783.

Dash, M.K., Bhandari, S.M., Vyas, N.K., Khare, N., Mitra, A., Pandey, P.C. (2001) Oceansat-MSMR imaging of the Antarctic and the Southern Polar Ocean. *International Journal of Remote Sensing* 22(16): 3253–3259.

Deb, P., Mihir, K. Dash, S., Dey, P., Pandey, P.C. (2017) Non-annular response of sea ice cover in the Indian sector of the Antarctic during extreme SAM events. *International Journal of Climatology* 37: 648–656.

Deb, P., Dash, M.K., Pandey, P.C. (2014) Effect of Pacific warm and cold events on the sea ice behavior in the Indian sector of the Southern Ocean. *Deep-Sea Research – I* 84: 59–72.

de la Mare, W.K. (1997) Abrupt mid-twentieth-century decline in Antarctic sea-ice extent from whaling records. *Nature* 389 57–60.

de la Mare. W.K. (2009) Changes in Antarctic sea-ice extent from direct historical observations and whaling records. *Climate Change* 92: 461–493.

Delille, B., Jourdain, B., Borges, A.V., Tison, J.-L., Delille, D. (2007) Biogas (CO_2, O_2, dimethylsulfide) dynamics in spring Antarctic fast ice. *Limnology and Oceanography* 52(4): 1367–1379.

De Santis, A., Maier, E., Gomez, R., Gonzalez, I. (2017) Antarctica, 1979–2016 sea ice extent: total *versus* regional trends, anomalies, and correlation with climatological variables. *International Journal of Remote Sensing* 38(24): 7566–7584.

Dieckmann, G.S., Nehrke, G., Papadimitriou, S., Go ̈ttlicher, J., Steininger, R., Kennedy, H., Wolf-Gladrow, D., Thomas, D.N. (2008) Calcium carbonate as ikaite crystals in Antarctic sea ice. *Geophysical Research Letters 35*: L08501.

Ding, Q., Steig, E.J. (2013) Temperature change on the Antarctic Peninsula linked to the tropical Pacific. *Journal of Climate* 26: 7570–7585.

Ding, Q., Steig, E.J., Battisti, D.S., Kuttel, M. (2011) Winter warming in West Antarctica caused by central tropical Pacific warming. *Nature Geoscience 4*: 398–403.

Eayrs, C., Holland, D.M., Francis, D., Wagner, T.J.W., Kumar, R., Li, X. (2019) Understanding the seasonal cycle of Antarctic sea ice extent in the context of longer-term variability. *Reviews of Geophysics* 57: 1037–1064.

Fabry V.J., McClintock, J.B., Mathis, J.T., Grebmeier, J.C. (2009) Ocean acidification at high latitudes: the bellwether. *Oceanography* 22(4): 160–171.

Ferreira, D., Marshall, J., Bitz, C.M., Solomon, S., Plumb, A. (2015) Antarctic Ocean and sea ice response to ozone depletion: A two-time-scale problem *Journal of Climate* 28: 1206–1226.

Fogt, R.L., Bromwich, D.H. (2006) Decadal variability of the ENSO teleconnection to the high-latitude South Pacific governed by coupling with the Southern Annular Mode. *Journal of Climate* 19: 979–997.

Fogt, R.L., Bromwich, D.H., Hines, K.M. (2011) Understanding the SAM influence on the South Pacific ENSO teleconnection. *Climate Dynamics* 36: 1555–1576.

Fogwill, C.J., Phipps, S.J., Turney, C.S.M., Golledge, N.R. (2015) Sensitivity of the Southern Ocean to enhanced regional Antarctic ice sheet meltwater input. *Earth's Future* 3: 317–329.

Goosse, H., Zunz, V. (2014) Decadal trends in the Antarctic sea ice extent ultimately controlled by ice-ocean feedback. *Cryosphere* 8: 453–470.

Gloersen, P., Campbell, W.J., Cavalier, D.J., Comiso, J.C., Parkinson, C.L., Zwally, H.J. (1992) *Arctic and Antarctic Sea Ice: 1978–1987*. NASA- SP-511 (1992).

Gloersen, P., Cavalieri, D.J. (1986) Reduction of weather effects in the calculation of sea ice concentration from microwave radiances. *Journal of Geophysical Research* 91: 3913–3919.

Haas, C., Lobach, J., Hendricks, S., Rabenstein, L., Pfaffling, A. (2009) Helicopter-borne measurements of sea ice thickness, using a small and lightweight, digital EM system. *Journal of Applied Geophysics* 67: 234–241.

Heygster, G., Wiebe, H., Spreen, G., Kaleschke, L. (2009) AMSR-E geolocation and validation of sea ice concentrations based on 89 GHz data. *Journal of the Remote Sensing Society of Japan* 29: 226–235.

Hobbs, W.R., Bindoff, N.L., Raphael, M.N. (2015) New perspectives on observed and simulated Antarctic sea ice extent trends using optimal fingerprinting techniques. *Journal of Climate* 28: 1543–1560.

Hobbs, W.R., Massom, R., Stammerjohn, S., Reid, P.P., Williams, G.D., Meier, W. (2016) A review of recent changes in Southern Ocean sea ice, their drivers and forcings. *Global and Planetary Change* 143: 228–250.

Holland, M.M., Landrum, L., Kostov, Y., Marshall, J. (2017) Sensitivity of Antarctic sea ice to the Southern Annular Mode in coupled climate models. *Climate Dynamics* 49: 1813–1831.

Holland, M.M. (2014) The seasonality of Antarctic sea ice trends. *Geophysical Research Letters* 41: 4230–4237.

Holland, P.R., Bruneau, N., Enright, C., Losch, M., Kurtz, N.T., Kwok, R. (2014) Modeled trends in Antarctic sea ice thickness. *Journal of Climate* 27: 3784–3801.

Holland, P., Kwok, R. (2012) Wind-driven trends in Antarctic sea-ice drift. *Nature Geoscience* 5(12): 872–875.

Holmes, C.R., Holland, P.R., Bracegirdle, T.J. (2019) Compensating biases and a noteworthy success in the CMIP5 representation of Antarctic sea ice processes. *Geophysical Research Letters* 46: 4299–4307.

Hosking, J.S., Orr, A., Marshall, G.J., Turner, J., Phillips, T. (2013) The influence of the Amundsen-Bellingshausen Seas Low on the climate of West Antarctica and its representation in coupled climate model simulations. *Journal of Climate* 26: 6633–6648.

Houseago-Stokes, R.E., Mcgregor, G.R. (2000) Spatial and temporal patterns linking southern low and high latitudes during South Pacific warm and cold events. *International Journal of Climatology: A Journal of the Royal Meteorological Society* 20(7): 793–801.

Hunke, E.C., Dukowicz, J.K. (1997) An elastic-viscous-plastic model for sea ice dynamics. *Journal of Physical Oceanography* 27: 1849–1867

Jacobs, S.S., Giulivi, C.F., Mele, P.A. (2002) Freshening of the Ross Sea during the late 20th century. *Science* 297: 386–389.

Kohyama, T., Hartmann, D.L. (2016) Antarctic sea ice response to weather and climate modes of variability. *Journal of Climate* 29(2): 721–741.

Kostov, Y., Marshall, J., Hausmann, U., Armour, K.C., Ferreira, D., Holland, M.M. (2017) Fast and slow responses of Southern Ocean sea surface temperature to SAM in coupled climate models. *Climate Dynamics* 48(5); 1595–1609.

Kurtz, N.T., Markus, T. (2012) Satellite observations of Antarctic sea ice thickness and volume. Journal of Geophysical Research 117: C08025.

Kusahara, K., Williams, G.D., Massom, R., Reid, P., Hasumi, H. (2018) Spatiotemporal dependence of Antarctic sea ice variability to dynamic and thermodynamic forcing: A coupled ocean-sea ice model study. *Climate Dynamics* 52(7–8): 3791–3807.

Kwok, R., Maksym, T. (2014) Snow depth of the Weddell and Bellingshausen sea ice covers from Ice Bridge surveys in 2010 and 2011: An examination. *Journal of Geophysical Research: Oceans* 119: 4141–4167.

Kwok, R., Pang, S.S., Kacimi, S. (2017) Sea ice drift in the Southern Ocean: Regional patterns, variability, and trends. *Elementa: Science of the Anthropocene* 5: 32.

Lebedev, V.V. (1938) Rost l'da v arkticheskikh rekakh i moriakh v zavisimosti ot otritsatel'nykh temperatur vozdukha. *Problemy arktiki* 5–6: 9–25.

Leonard, K.C., Maksym, T. (2011) The importance of wind-blown snow redistribution to snow accumulation on and mass balance of Bellingshausen Sea ice. Annals of Glaciology 52(57): 271–278.

Leppäranta, M. (1993) A review of analytical models of sea-ice growth. *Atmosphere-Ocean* 31(1): 123–138.

Leppäranta, M. (2005) *The Drift of Sea Ice*, 266 pp. Berlin and Heidelberg: Springer-Verlag.

Li, X.C., Holland, D.M., Gerber, E.P., Yoo, C. (2014) Impacts of the north and tropical Atlantic Ocean on the Antarctic Peninsula and sea ice. *Nature* 505: 538–542.

Liu, J., Curry, J.A. (2010) Accelerated warming of the Southern Ocean and its impacts on the hydrological cycle and sea ice. *Proceedings of the National Academy of Sciences* 107(34): 14,987–14,992.

Lubin, D., Massom, R. (2006) *Polar Remote Sensing, Volume 1 Atmosphere and Oceans*, 759 pp. Springer-Praxis.

Maksym, T., Sharon, E., Stammerjohn, S.A., Massom, B. (2012) Antarctic sea ice – a polar opposite? *Oceanography* 25(3): 140–151.

Massonnet, F., Mathiot, P., Fichefet, T., Goosse, H., Beatty, C.K., Vancoppenolle, M., Lavergne, T. (2013) A model reconstruction of the Antarctic sea ice thickness and volume changes over 1980–2008 using data assimilation. *Ocean Modelling* 64: 67–75.

Marshall J., Armour, K.C., Scott, J.R., Kostov, Y., Hausmann, U., Ferreira, D., Shepherd, T.G., Bitz, C.M. (2014) The ocean's role in polar climate change: asymmetric Arctic and Antarctic responses to greenhouse gas and ozone forcing. *Philosophical Transactions of the Royal Society A* 372: 20130040.

Massom, R.A., Stammerjohn, S.E. (2010) Antarctic sea ice change and variability—Physical and ecological implications. *Polar Science* 4: 149–186.

Massom, R., Reid, P., Stammerjohn, S., Raymond, B., Fraser, A., Ushio, S. (2013) Change and variability in East Antarctic sea ice seasonality, 1979/80–2009/10. *PLoS ONE* 8(5): e64756.

Meehl, G.A., Arblaster, J.M., Chung, C.T.Y., Holland, M.M., DuVivier, A., Thompson, L., Yang, D., Bitz, C.M. (2019) Sustained ocean changes contributed to sudden Antarctic sea ice retreat in late 2016. *Nature Communications* 10: 14.

Meehl, G.A., Arblaster, J.M., Bitz, C.M., Chung, C.T.Y., Teng, H. (2016) Antarctic sea-ice expansion between 2000 and 2014 driven by tropical Pacific decadal climate variability. *Nature Geoscience* 9(8): 590–595.

Meier, W.N., Ge Peng, Scott, D.J., Savoie, M.H. (2014) Verification of a new NOAA/NSIDC passive microwave sea-ice concentration climate record. *Polar Research* 33(1): 21004.

Meredith, M.P., King, J.C. (2005) Rapid climate change in the ocean west of the Antarctic Peninsula during the second half of the 20th century. *Geophysical Research Letters* 32: L19604.

Mo, K.C. (2000) Relationships between low-frequency variability in the Southern Hemisphere and sea surface temperature anomalies. *Journal of Climate* 13(20): 3599–3610.

Mo, K.C., Paegle, J.N. (2001) The Pacific–South American modes and their downstream effects. *International Journal of Climatology: A Journal of the Royal Meteorological Society* 21(10): 1211–1229.

Mo, K.C., Higgins, R.W. (1998) The Pacific–South American modes and tropical convection during the Southern Hemisphere winter. *Monthly Weather Review* 126(6): 1581–1596.

Moline, M.A., Karnovsky, N.J., Brown, Z., Divoky, G.J., Frazer, T.R., Jacoby, C.A., Torres, J.J., Fraser, W.R. (2008) High latitude changes in ice dynamics and their impact on polar marine ecosystems. *Annals of the New York Academy of Science* 1134: 267e319.

Morley, S.A., Abele, D., Barnes, D.K.A., Cárdenas, C.A., Cotté, C., Gutt, J., Henley, S.F., Höfer, J., Hughes, K.A., Martin, S.M., Moffat, C., Raphael, M., Stammerjohn, S.E., Suckling, C.C., Tulloch, V.J.D., Waller, C.L., Constable, A.J. (2020) Global drivers on Southern Ocean ecosystems: Changing physical environments and anthropogenic pressures in an Earth system. *Frontiers in Marine Science* 7: 547188.

Notz, D., Stroeve, J. (2016) Observed Arctic sea-ice loss directly follows anthropogenic CO_2 emission. *Science* 354(6313): 747–750.

Oza, S.R., ajak, D.R., Dash, M.K., Bahuguna, I.M., Kumar, R. (2017) Advances in Antarctic sea ice studies in India. *Proceedings of the Indian National Science Academy* 83(2): 427–435.

Park, W., Latif, M. (2018) Ensemble global warming simulations with idealized Antarctic melt-water input. *Climate Dynamics* 52: 3223–3239.

Parkinson, C.L. (1994) Spatial patterns in the length of the sea ice season in the Southern Ocean. *Journal of Geophysical Research* 99: 16,327–16,339.

Parkinson, C.L. (2014) Global sea ice coverage from satellite data: Annual cycle and 35-yr trends. *Journal of Climate* 27: 9377–9382.

Parkinson, C.L. (2019) A 40-y record reveals gradual Antarctic sea ice increases followed by decreases at rates far exceeding the rates seen in the Arctic. *PANAS* 116(29): 14414–14423.

Parkinson, C.L., Cavalieri, D.J. (2012) Antarctic sea ice variability and trends, 1979–2010. *Cryosphere* 6(4): 871–880.

Parkinson, C.L. (2002) Trends in the length of the Southern Ocean sea ice season, 1979–1999. *Annals of Glaciology* 34: 435–440.

Polvani, L.M., Smith, K.L. (2013) Can natural variability explain observed Antarctic sea ice trends? New modelling evidence from CMIP5. *Geophysical Research Letters* 40: 3195–3199.

Pope J.O., Holland, P., Orr, A., Marshall, G.J., Phillips, T. (2017) The impacts of El Nino on the observed sea ice budget of West Antarctica. *Geophysical Research Letters* 44: 6200–6208.

Quetin, L.B., Ross, R.M. (2009) Life under Antarctic pack ice: a krill perspective. In: Krupnik, I., Lang, M.A., Miller, S.E. (Eds.) *Smithsonian at the Poles: Contributions to International Polar Year Science*, pp. 285–298. Smithsonian Inst.: Washington DC.

Raphael, M.N. (2004) A zonal wave 3 index for the Southern Hemisphere. *Geophysical Research Letters* 31: L23212.

Roach, L.A., Dean, S.M., Renwick, J.A. (2018) Consistent biases in Antarctic sea ice concentration simulated by climate models. *The Cryosphere* 12(1): 365–383.

Rosenblum, E., Eisenman, I. (2017) Sea ice trends in climate models only accurate in runs with biased global warming. *Journal of Climate* 30(16): 6265–6278.

Schemm, S. (2018) Regional trends in weather systems help explain Antarctic sea ice trends. *Geophysical Research Letters* 45: 7165–7175.

Schlosser, E., Haumann, F.A., Raphael, M.N. (2018) Atmospheric influences on the anomalous 2016 Antarctic sea ice decay. *Cryosphere* 12: 1103–1119.

Schroeter, S., Hobbs, W., Bindoff, N.L., Massom, R., Matear, R. (2018) Drivers of Antarctic sea ice volume change in CMIP5 models. *Journal of Geophysical Research: Oceans* 123: 7914–7938.

Sharma, N., Dash, M.K., Vyas, N.K., Bhandari, S.M., Pandey, P.C., Khare, N. (**2011**) Signature of ice melt over the Greenland, derived from MSMR (OCEANSAT – 1) data. *Mausam* 62: 627–632.

Shu, Q., Song, Z., Qiao, F. (2015) Assessment of sea ice simulations in the CMIP5 models. *Cryosphere* 9(1): 399–409.

Sigmond, M., Fyfe, J.C. (2010) Has the ozone hole contributed to increased Antarctic sea ice extent? *Geophysical Research Letters* 37: L18502.

Simmonds, I. (2015) Comparing and contrasting the behaviour of Arctic and Antarctic sea ice over the 35-year period 1979–2013. *Annals of Glaciology* 56: 18–28.

Simpkins, G.R., Ciasto, L.M., England, M.H. (2013) Observed variations in multidecadal Antarctic sea ice trends during 1979–2012. *Geophysical Research Letters* 40: 3643–3648.

Simpkins, G.R., McGregor, S., Taschetto, A.S., Ciasto, L.M., England, M.H. (2014) Tropical connections to climatic change in the extratropical Southern Hemisphere: The role of Atlantic SST trends. *Journal of Climate* 27(13): 4923–4936.

Sinclair, K.E., Bertler, N.A.N., Bowen, M.M., Arrigo, K.R. (2014) Twentieth century sea-ice trends in the Ross Sea from a high-resolution, coastal ice-core record. *Geophysical Research Letters* 41: 3510–3516.

Smith, K.L., Polvani, L.M., Marsh, D.R. (2012) Mitigation of 21st century Antarctic sea ice loss by stratospheric ozone recovery. *Geophysical Research Letters* 39: L20701.

Smith Jr., W.O., Comiso, J.C. (2008) Influence of sea ice on primary production in the Southern Ocean: a satellite perspective. *Journal of Geophysical Research* 113: C05S93.

Stammerjohn, S.E., Maksym, T., Massom, R.A., Lowry, K.E., Arrigo, K.R., Yuan, X., Raphael, M., Randall-Goodwin, E., Sherrell, R.M., Yager, P.L. (2015) Seasonal sea ice changes in the Amundsen Sea, Antarctica, over the period of 1979–2014. *Elementa: Science of the Anthropocene* 3: 000055.

Stammerjohn, S.E., Martinson, D.G., Smith, R.C., Yuan, X., Rind, D. (2008) Trends in Antarctic annual sea ice retreat and advance and their relation to El Nino–Southern Oscillation and Southern Annular Mode variability. *Journal of Geophysical Research* 113: C03S90.

Stammerjohn, S., Massom, R., Rind, D., Martinson, D. (2012) Regions of rapid sea ice change: An interhemispheric seasonal comparison. *Geophysical Research Letters* 39: L06501.

Stuecker, M.F., Bitz, C.M., Armour, K.C. (2017) Conditions leading to the unprecedented low Antarctic sea ice extent during the 2016 austral spring season. *Geophysical Research Letters* 44: 9008–9019.

Sun, S., Eisenman, I. (2021) Observed Antarctic sea ice expansion reproduced in a climate model after correcting biases in sea ice drift velocity. *Nature Communications* 12: 1060.

Swart, N.C., Fyfe, J.C. (2013) The influence of recent Antarctic ice sheet retreat on simulated sea ice area trends. *Geophysical Research Letters* 40: 4328–4332.

Thomas, E.R., Abram, N.J. (2016) Ice core reconstruction of sea ice change in the Amundsen-Ross Seas since 1702 AD. *Geophysical Research Letters* 43: 5309–5317.

Thomas, E.R., Dennis, P.F., Bracegirdle, T.J., Franzke, C. (2009) Ice core evidence for significant 100-year regional warming on the Antarctic Peninsula. *Geophysical Research Letters* 36: L20704.

Thompson, D.W.J., Solomon, S. (2002) Interpretation of recent Southern Hemisphere climate change. *Science* 296: 895–899.

Thompson, D.W.J., Wallace, J.M. (2000) Annular modes in the extratropical circulation. Part I: month-to-month variability. *Journal of Climate* 13: 1000–1016.

Tre´guer, P., Pondaven, P. (2002) Climatic changes and the carbon cycle in the Southern Ocean: a step forward. *Deep Sea Research II* 49(9–10): 1597–1600.

Turner, J. (2004) The El Niño-Southern Oscillation and Antarctica. *International Journal of Climatology* 24: 1–31.

Turner J., Comiso, J.C., Marshall, G.J., Lachlan-Cope, T.A., Bracegirdle, T., Maksym, T., Meredith, M.P., Wang, Z., Orr, A (2009) Non-annular atmospheric circulation change induced by stratospheric ozone depletion and its role in the recent increase of Antarctic sea ice extent. *Geophysical Research Letters* 36: L08502.

Turner, J., Hosking, J.S., Marshall, G.J., Phillips, T., Bracegirdle, T.J. (2016) Antarctic sea ice increase consistent with intrinsic variability of the Amundsen sea low. *Climate Dynamics* 46(7–8): 2391–2402.

Turner J., Hosking, J.S., Bracegirdle, T.J., Marshall, G.J., Phillips, T. (2015) Recent changes in Antarctic sea ice. *Philosophical Transactions of the Royal Society A: Mathematical, Physical and Engineering Sciences* 373: 20140163.

Turner, J., Hosking, J.S., Phillips, T., Marshall, G.J. (2013) Temporal and spatial evolution of the Antarctic sea ice prior to the September 2012 record maximum extent. *Geophysical Research Letters* 40: 5894–5898.

Turner, J., Hosking, J.S., Bracegirdle, T.J., Marshall, G.J., Phillips, T. (2015) Recent changes in Antarctic sea ice. *Philosophical Transactions of the Royal Society A* 373: 20140163.

Turner, J., Bracegirdle, T.J., Phillips, T., Marshall, G.J., Hosking, J.S. (2012) An initial assessment of Antarctic sea ice extent in the CMIP5 models. *Journal of Climate* 26(5): 1473–1484.

Turner, J., Phillips, T., Marshall, G.J., Hosking, J.C., Pope, J.O., Bracegirdle, T.J., Deb, P. (2017) Unprecedented springtime retreat of Antarctic sea ice in 2016. *Geophysical Research Letters* 44: 6868–6875.

Turner, J., Phillips, T., Hosking, S., Marshall, G.J., Orr, A. (2012) The Amundsen Sea low. *International Journal of Climatology* 33: 1818–1829.

Turner, J., Bracegirdle, T.J., Phillips, T., Marshall, G.J., Hosking, J.S. (2013) An initial assessment of Antarctic sea ice extent in the CMIP5 models. *Journal of Climate* 26: 1473–1484.

Tynan, C.T., Ainley, D.G., Stirling, I. (2009) Sea ice: A critical habitat for polar marine mammals and birds. In: Thomas, D.N., Dieckmann, G.S. (Eds.) *Sea Ice*, second ed. Wiley-Blackwell: Oxford.

Untersteiner, N. (1961) On the mass and heat budget of Arctic sea ice. *Archives of Meteorology and Geophysics Bioklimatology Series A* 12: 151–182.

Untersteiner, N. (1964) Calculations of temperature regime and heat budget of sea ice in the Central Arctic. *Journal of Geophysical Research* 69: 4755–4766.

Uotila, P., Holland, P.R., Vihma, T., Marsland, S.J., Kimura, N. (2014) Is realistic Antarctic sea-ice extent in climate models the result of excessive ice drift. *Ocean Modelling* 79: 33–42.

Vyas, N.K., Dash, M.K., Bhandari, S.M., Khare, N., Mitra, A., Pandey, P.C. (2003) On the secular trends in sea ice extent over the antarctic region based on OCEANSAT-1 MSMR observations. *International Journal of Remote Sensing* 24(11): 2277–2287.

Vyas, N.K., Dash, M.K., Bhandari, S.M., Khare, N., Mitra, A., Pandey, P.C. (2001) Oceansat MSMR Imaging of the Antarctic and the Southern Polar Ocean. *Current Science* 80: 1391–1322.

Vyas, N.K., Dash, M.K. (2000) Oceansat – MSMR observes interesting features on the frozen continent and surrounding sea. *Journal of the Indian Society of Remote Sensing* 28(2&3): 67.

Wang G., Hendon, H.H., Arblaster, J.M., Lim, E.-P., Abhik, S., van Rensch, P. (2019) Compounding tropical and stratospheric forcing of the record low Antarctic sea-ice in 2016. *Nature Communications* 10: 13.

Williams, G., Maksym, T., Wilkinson, J., Kunz, C., Murphy, C., Kimball, P., Singh, H. (2015) Thick and deformed Antarctic sea ice mapped with autonomous underwater vehicles. *Nature Geoscience* 8: 61–67.

Worby, A. P., Geiger, C.A., Paget, M.J., Van Woert, M.L., Ackley, S.F., DeLiberty, T.L. (2008) Thickness distribution of Antarctic sea ice. *Journal of Geophysical Research* 113: C05S92.

Xie, H., Tekeli, A.E., Ackley, S.F., Yi, D., Zwally, H.J. (2013) Sea ice thickness estimations from ICESat Altimetry over the Bellingshausen and Amundsen Seas, 2003–2009. *Journal of Geophysical Research: Oceans* 118: 2438–2453.

Yu, L., Zhang, Z., Zhou, M., Zhong, S., Lenschow, D.H., Gao, Z., et al. (2011) Interpretation of recent trends in Antarctic sea ice concentration. *Journal of Applied Remote Sensing* 5: 053557.

Yuan, X. (2004) ENSO-related impacts on Antarctic sea ice: A synthesis of phenomenon and mechanisms. *Antarctic Science* 16(4): 415–425.

Yuan, X., Martinson, D.G. (2000) Antarctic sea ice extent variability and its global connectivity. *Journal of Climate* 13: 1697–1717.

Yuan, X., Li, C. (2008) Climate modes in southern high latitudes and their impacts on Antarctic sea ice. *Journal of Geophysical Research* 113: C06S91.

Zhang, J. (2014) Modeling the impact of wind intensification on Antarctic sea ice volume. *Journal of Climate* 27: 202–214.

Zhang, J. (2007) Increasing Antarctic sea ice under warming atmospheric and oceanic conditions. *Journal of Climate* 20(11): 2515–2529.

Zunz, V., Goosse, H., Massonnet, F. (2013) How does internal variability influence the ability of CMIP5 models to reproduce the recent trend in Southern Ocean sea ice extent? *Cryosphere* 7: 451–468.

Zwally, H.J., Comiso, J.C., Parkinson, C.L., Cavalieri, D.J., Gloersen, P. (2002) Variability of Antarctic sea ice 1979–1998. *Journal of Geophysical Research* 107(5): 3041.

8 Effect of Climate Change on the Multifractal Properties of the Sea-Ice Concentration around the Indian Antarctic Stations Maitri and Bharti

Suneet Dwivedi

8.1 INTRODUCTION

The sea ice is a critical component of the Earth's climate system due to its strong influence on the albedo, turbulent heat fluxes, wind drag, and upper-ocean stratification (Maruyama, 2020). The Southern Ocean (e.g. Antarctic) sea ice significantly affects the global climate (Kumar et al., 2017, 2018 a,b). Studies have shown that the extratropical and polar Southern Ocean sea ice cover is substantially changing as a result of global warming (Parkinson and Cavalieri, 2012; Comiso et al., 2017; Parkinson, 2019; Hao et al., 2021; Dwivedi and Pandey, 2022). Parkinson and Cavalieri (2012) and Parkinson (2019) suggest a positive trend in the Antarctic sea ice in recent decades. The various characteristics of the Antarctic sea ice cover and changes in it under different future projection scenarios have been analysed using a variety of models participating in the Coupled Model Intercomparison Project Phase 5 and 6 (CMIP5 and CMIP6) (see, e.g., Taylor et al., 2012; Turner et al., 2013; Shu et al., 2015; Roach et al., 2020; Shu et al., 2020). Dwivedi and Pandey (2022) evaluated the performance of CMIP6 sea ice simulations in the southern Indian Ocean region covering both the Indian Antarctic stations, Maitri located at 11.7°E; 70.7°S and Bharati located at 76.1°E; 69.4°S.

The deformation and fracturing processes of the sea ice cover are characterized by spatial heterogeneity and intermittency (Girard et al., 2009). Since Antarctic sea ice remains mostly seasonal and is not constrained by the land in the north, its fracturing and divergent deformation features differ from Arctic sea ice. It is important

                    DOI: 10.1201/9781003364115-8

to understand the structure and dimensionality (e.g. scaling properties) of Antarctic sea ice in the present and future climate. This information is also required for better parameterization and modelling of the sea ice by dynamical models. Similar to various other objects in nature, Antarctic sea ice also shows self-similarity, i.e. a fractal property (Weiss and Marsan, 2004; Dwivedi and Pandey, 2022). The fractal and multifractal techniques have been applied to a variety of datasets to estimate the dimensionality and predictability of a variable, including the Antarctic and Arctic sea ice concentrations (Chmel et al., 2005; Agarwal et al., 2012; Agarwal and Wettlaufer, 2018; Dwivedi, 2012; Yadav et al., 2012, 2013; Baranowski et al., 2015; Kumar et al., 2018b; Laib et al., 2018; Agbazo et al., 2019; Cadenas et al., 2019; Kimothi et al., 2019; Moon et al., 2019; Rampal et al., 2019; and references therein). Recently, Dwivedi and Pandey (2022) applied (mono)fractal analysis on the east Antarctic sea ice to quantify the changes in the predictability of the sea ice concentration (SIC) as a result of global warming.

Studies have suggested that sea ice in different regions of the world shows multifractal nature, instead of being monofractal (Rampal et al., 2019 and references therein; Maruyama, 2020). In other words, the spatio-temporal variability in the fractal nature of sea ice has been noted. As such, it will be worthwhile analysing the multifractal nature of southern Indian Ocean sea ice (i.e. east Antarctic sea ice) around the Indian Antarctic stations Maitri and Bharti. To the best of our knowledge, there is no such study which quantifies the multifractality of the east Antarctic SIC around the Maitri and Bharati regions. We, therefore, apply multifractal analysis to the long-term SIC time series of the Maitri and Bharti regions. The multifractal detrended fluctuation analysis (MFDFA) technique has been used for this purpose. The MFDFA technique, which is a generalization of the DFA technique (Kantelhardt et al., 2002, 2006) has been applied in different types of studies, such as the study of extreme events with nonlinear long-term memory (Bogachev & Bunde, 2011) and the influence of additive noise on long-term correlations, etc. (Ludescher et al., 2011). With the help of a multifractal spectrum, it is also possible to identify a wide range of different scale-invariant structures of the sea ice. In this chapter, we use the MFDFA technique to examine the multifractal properties of the historical and future projection dataset of SIC (derived from the CMIP6 models) over the Maitri and Bharati regions. We not only compare and contrast the multifractal properties of the two Indian Antarctic stations but also quantify the change in the multifractality as a result of increased greenhouse gas (GHG) concentration by analysing the future projection dataset of CMIP6 models. The chapter is organized as follows. Section 8.2 describes the data used and gives a brief description of the MFDFA technique. The results are described in Section 8.3, and the conclusions are given in Section 8.4.

8.2   DATA AND METHOD

An effort is made to analyse the multifractality of the SIC dataset derived from the CMIP6 model output. The historical data for this purpose are taken for the period 1900–2014 (115 years). To quantify the effect of climate change on the multifractal properties of the SIC, we use the SSP5-8.5 future projection data for the period 2015–2100 (86 years). The SIC of models is interpolated to a $0.5° \times 0.5°$ regular grid in

TABLE 8.1
The CMIP6 Models were Used in the Study.
These Models Were Chosen Based on the
Availability of the SIC Dataset and their
Performance in Realistically Simulating the
Observed SIC

S. no.	CMIP6 model
1	ACCESS-CM2
2	ACCESS-ESM1-5
3	CESM2-WACCM
4	CESM2
5	CIESM
6	CMCC-CM2-SR5
7	CanESM5
8	EC-Earth3
9	FGOALS-f3-L
10	FIO-ESM-2-0
11	GFDL-CM4
12	GFDL-ESM4
13	IPSL-CM6A-LR
14	MRI-ESM2-0
15	NESM3

Source: Dwivedi and Pandey, 2022.

the region (10E–100E; 55S–75S) which covers both of the Indian Antarctic stations, Maitri and Bharati. The historical dataset of a total of 33 CMIP6 models is analysed to select good and bad models (in terms of their ability to simulate the SIC over the region of interest), following the selection criterion described in Dwivedi and Pandey (2022). In this chapter, of these 33 models, we analyse the multifractal properties of SIC of those good 15 CMIP6 models for which both the historical and SSP5-8.5 data are available. The models are summarized in Table 8.1. We create the SIC time series of historical and SSP5-8.5 data for each of these 15 CMIP6 models for the Maitri region by area-averaging the SIC around the region (10E–14E; 65S–69S). Similarly, the SIC time series for the Bharati region is created by area-averaging the SIC around the region (65E–77E; 65S–69S).

The multifractal analysis of the SIC time series helps us to know about the structure and dimensionality of the sea ice. Specifically, it helps us to determine the spatio-temporal variations in the scale-invariant structure of the SIC time series. This is in contrast to monofractal analysis which assumes that scale invariance is independent of space and time (Ihlen, 2012). We carry out the MFDFA analysis on the SIC time series data of the Maitri and Bharati regions derived from the CMIP6 models. A brief description of the MFDFA technique is given as follows. For a series X_k of length N, the profile $Y(i)$ (or the cumulative sum of departure of X_k from its mean value) is constructed as $Y(i) = \sum_{k=1}^{i} (X_k - \overline{X_k})$ for $i=1,... .,N$. These profiles are divided into

N_s = int(N/s) nonoverlapping segments of equal length s. Here, N may not be an exact multiple of s, and as such some part of the profile may remain unaccounted for. To include this part, the same procedure is repeated from the end of the profile and returned to the beginning, thereby creating $2N_s$ segments altogether. We then compute the variance $F^2(s,v)$ or root mean square error $F(s,v)$ within each of these $v = 1,$ ... ., $2N_s$ segments as follows:

$$F^2(s,v) = \frac{1}{s} \sum_{i=1}^{s} \left[Y(\{v-1\}s+i) - y_v(i) \right]^2 \tag{8.1}$$

Here the variance $F^2(s,v)$ is computed to local least-squares polynomial fits $y_v(i)$ in segment v. The q-order fluctuation function $F_q(s)$, also known as generalized fluctuation function, is then computed as:

$$F_q(s) = \left(\frac{1}{2N_s} \sum_{v=1}^{2N_s} \left[F^2(s,v) \right]^{q/2} \right)^{1/q} \tag{8.2}$$

Equation (8.2) is obtained by taking the mean over the q-order variances $\left[F^2(s,v) \right]^{q/2}$. Here q can take any real value other than 0. In the MFDFA analysis, we examine the variation of $F_q(s)$ with scale s for different values of q. The $F_q(s)$ follows power-law behaviour with s as: $F_q(s) \propto s^{h(q)}$, where $h(q)$ is the generalized Hurst exponent. For a monofractal time series, $h(q)$ remains independent of q and is the same as the classical Hurst exponent H (Kantelhardt et al. 2002). In general, we obtain $h(q)$ using the log-log plot of $F_q(s)$ versus s for different q. In other words, the $h(q)$ is obtained as the slope of the linear portion of the plot between the log $F_q(s)$ and log s. We obtain q-order mass exponent $\tau(q)$ with the help of $h(q)$ as

$$\tau(q) = qh(q) - 1 \tag{8.3}$$

The singularity strength α, also known as the Hölder exponent is computed as $\alpha = \tau'(q)$, i.e. differentiation of $\tau(q)$ to q. The singularity spectrum $f(\alpha)$ is related to the mass exponent that is related to $\tau(q)$ via a Legendre transform (Kantelhardt et al., 2002):

$$f(\alpha) = q\alpha - \tau(q). \tag{8.4}$$

Using Eqs. (8.3) and (8.4), it is straightforward to note that

$$\alpha = h(q) + qh'(q) \text{ and } f(\alpha) = q\{\alpha - h(q)\} + 1 \qquad (8.5)$$

8.3 RESULTS AND DISCUSSION

The multifractal analysis (e.g. MFDFA) is carried out on the SIC time series of the Maitri and Bharati Indian Antarctic stations for all 15 CMIP6 models (mentioned in Table 8.1). The power law behaviour of the fluctuation function $F_q(s)$ as a function of scale (s) is shown on a (log-log scale) in Figure 8.1 for the Maitri region. The plots are obtained corresponding to q-order statistical moments ranging from $q = -5$ to 5. A similar plot for the Bharati region is shown in Figure 8.2. We also compute the $F_q(s)$ for the multimodel mean SIC time series of the Maitri and Bharati regions. The results are shown in Figure 8.3a,b corresponding to each q. We find that $F_q(s)$ for different q values (ranging from negative to positive) show a larger difference for small s. With the increase in s (i.e. for larger segment sizes), the difference becomes indistinguishable. The local periods with large and small fluctuations (e.g. positive and negative q) are clearly distinguished when the segment size s is small. On the other hand, since the large segments cross several local periods with both small and large fluctuations, therefore, their differences in magnitude get average out (Ihlen, 2012). For large segment sizes, the $F_q(s)$ values of multifractal time series become similar to monofractal time series.

The linear dependence of $F_q(s)$ on s in a log-log scale gives the value of q-order generalized Hurst exponent $h(q)$. We compute the slope of the linear portion of $\log_2 F_q(s)$ and $\log_2 s$ to obtain $h(q)$ corresponding to each q and for all the models. For a monofractal time series, the $h(q)$ remains constant with q (classical Hurst exponent H), whereas, for a multifractal time series, the $h(q)$ nonlinearly depends on q. The $h(q)$ as a function of q is shown in Figure 8.4(a,b) for Maitri and Bharati regions. We note that for all the CMIP6 models used in the study, the $h(q)$ is a nonlinear decreasing function of q. This fact indicates the multifractal nature of the SIC time series of the Maitri and Bharati regions. The figure also shows that $h(q)$ of various CMIP6 models are very different from each other and show a lot of variabilities, especially for $q < 0$. For example, the minimum and maximum values of $h(q)$ are obtained at $q = -5$. Moreover, the values of $h(q)$ for a particular model also change substantially between Maitri and Bharati regions. Thus, we not only notice a significant intermodal variability of $h(q)$ but a substantial change in values for a particular model is also seen between Maitri and Bharati regions. It is also observed from the figure that the $h(q)$ of the Bharati region is more variable as compared to the Maitri region.

We compute the q-order mass exponent $\tau(q)$ with the help of $h(q)$ and q using Eq. (8.3). The $\tau(q)$ corresponding to the SIC time series of all the CMIP5 models is shown as a function of q $(-5 < q < 5)$ in Figure 8.5 for the Maitri and Bharati regions. We note from the figure that the $\tau(q)$ varies nonlinearly with q. It increases with q (ranging from -5 to 5) in a nonlinear sense, with higher variability observed for

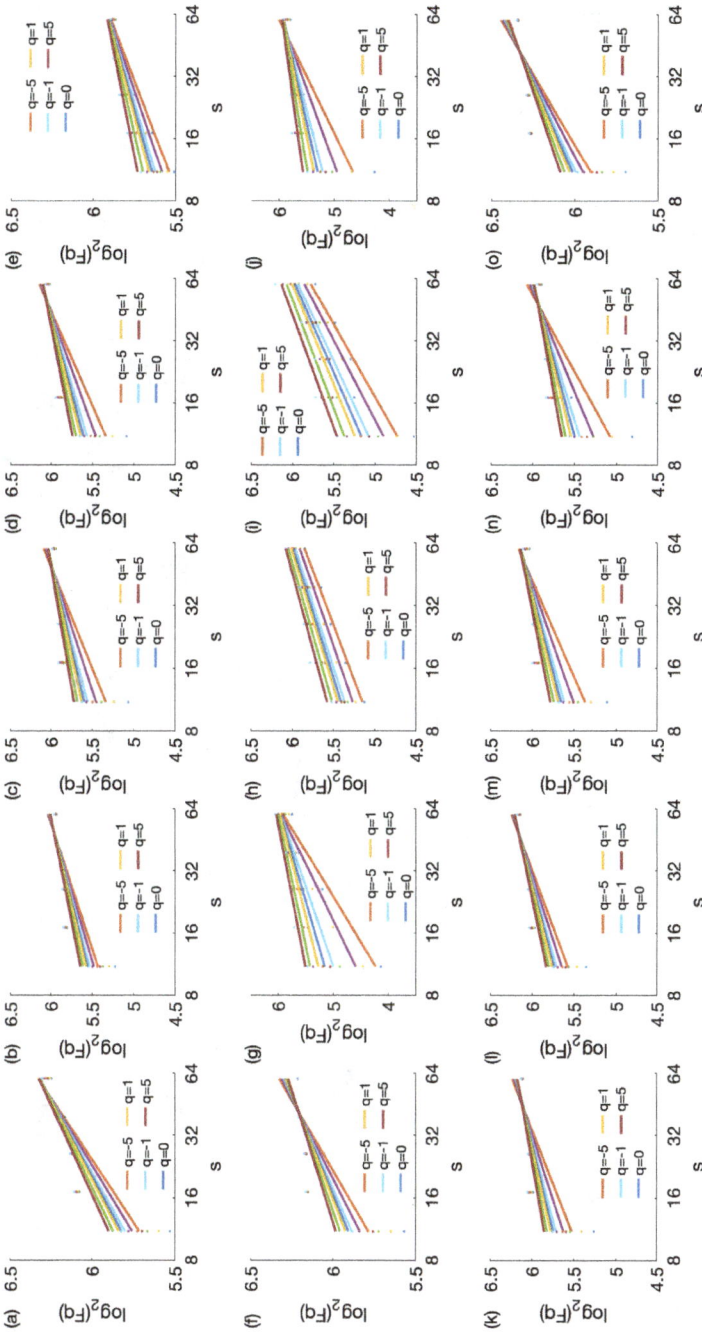

FIGURE 8.1 The fluctuation function $F_q(s)$ as a function of scale (s) on a log-log scale for the SIC time series of the Maitri region for q ranging between −5 to 5. The subplots from (a) to (o) corresponding to the CMIP6 models (1–15) mentioned in Table 8.1.

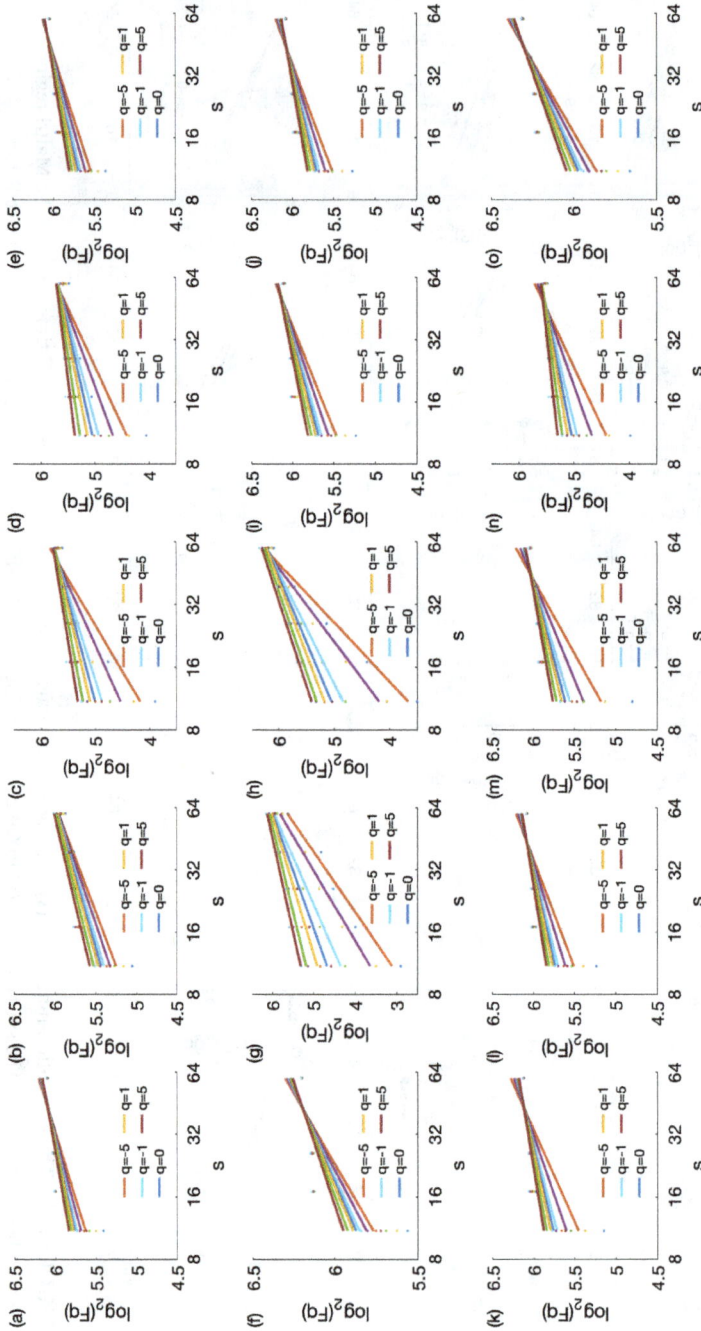

FIGURE 8.2   The fluctuation function $F_q(s)$ as a function of scale (s) on a log-log scale for the SIC time series of the Bharati region for q ranging between −5 to 5. The subplots from (a) to (o) corresponding to the CMIP6 models (1–15) mentioned in Table 8.1.

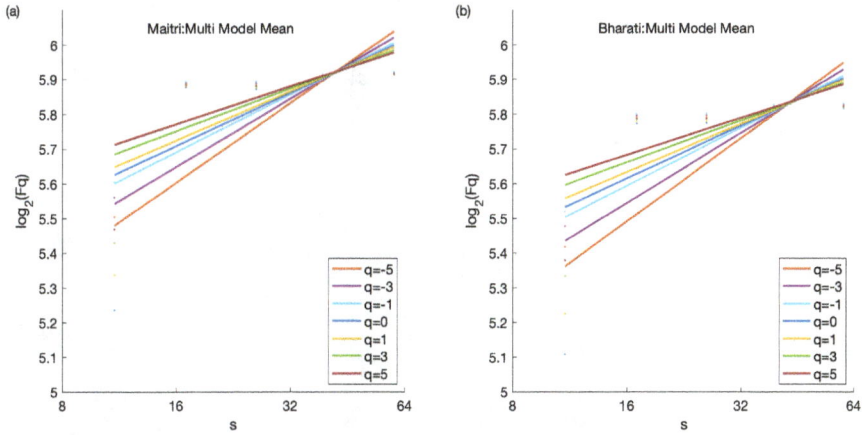

FIGURE 8.3 $F_q(s)$ as a function of s on a log-log scale using the multimodel mean SIC time series of (a) Maitri and (b) Bharati regions.

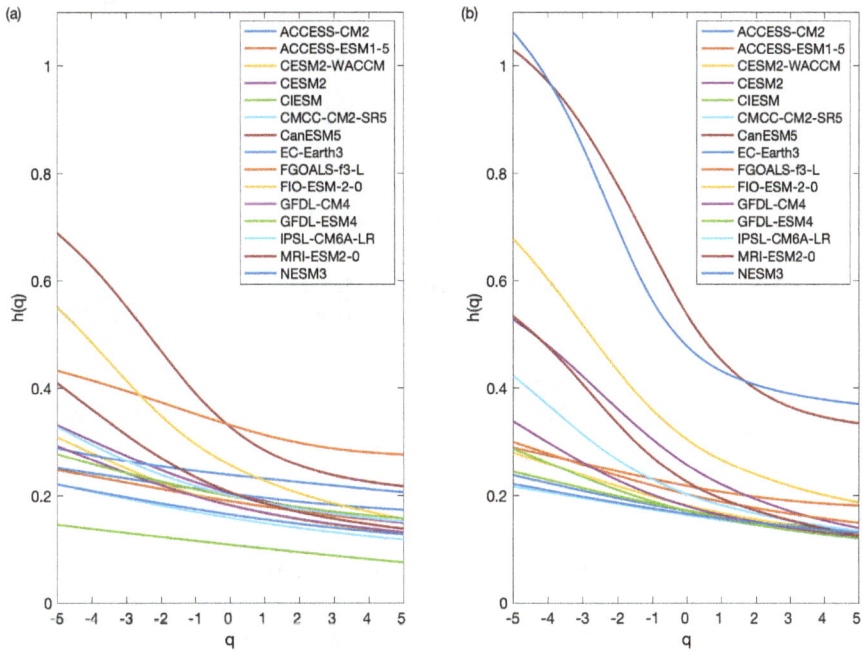

FIGURE 8.4   The q-order generalized Hurst exponent $h(q)$ as a function of q corresponding to the SIC time series of all the CMIP5 models for (a) Maitri region and (b) Bharati region.

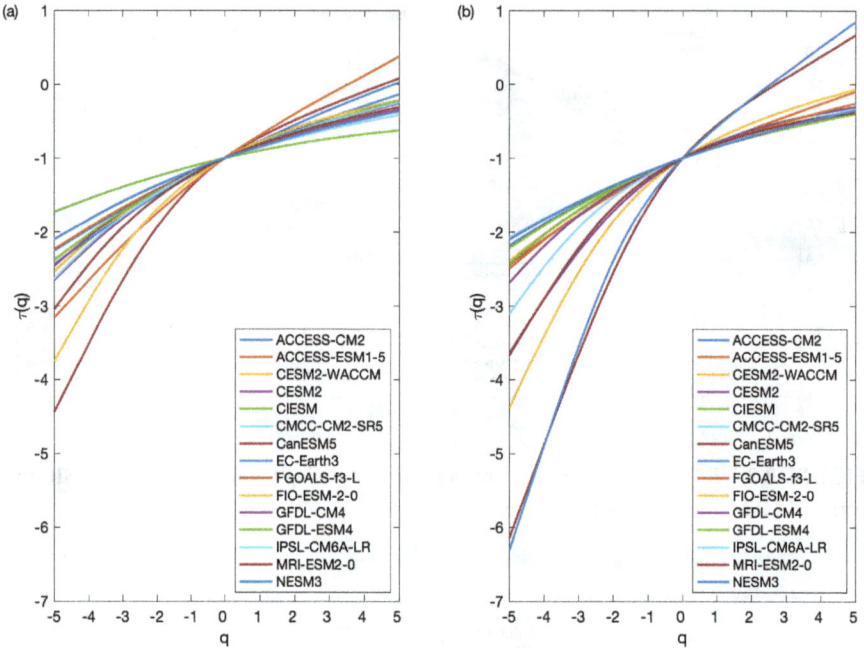

FIGURE 8.5 The q-order mass exponent $\tau(q)$ as a function of q for (a) Maitri region and (b) Bharati region.

negative q values of higher magnitude. This nature of the variability of mass exponent leads to a nonlinearly decreasing q-order singularity strength α curve (not shown). The highest intermodal variability of $\tau(q)$ (and hence multifractality of SIC time series) is seen in the case of the CanESM5 model over the Maitri region and the EC-Earth3 model over the Bharati region.

We also compute the $h(q)$ and $\tau(q)$ values for the multimodal mean time series of the SIC over the Maitri and Bharati regions. The results are shown in Figure 8.6(a,b). It is clear from Figure 8.6(a) that the $h(q)$ of multimodal mean, SIC time series decreases nonlinearly both for the Maitri and Bharati regions. The Bharati region shows greater variability (and higher multifractality) as compared to the Maitri region. For example, the $h(q)$ values of the Bharati region are higher (lower) than the corresponding values of the Maitri region for $q = -5$ ($q = 5$). Similarly, the mass exponent $\tau(q)$ vs. q curves shown in Figure 8.6(b) also suggest that the SIC time series of the Maitri and Bharati regions are multifractal. Moreover, the SIC of the Bharati region shows higher multifractality as compared to the Maitri region.

We compute the α and q-order singularity dimension $f(\alpha)$ using the method described in Section 8.2. The curve showing $f(\alpha)$ as a function of α is known as the multifractal spectrum. We show in Figure 8.7, the multifractal spectrum of the SIC time series of CMIP6 models over the Maitri and Bharati regions. It is found that the shape and width of the multifractal spectrum of the SIC time series are quite different for different CMIP6 models. It can be seen that the multifractal spectrum

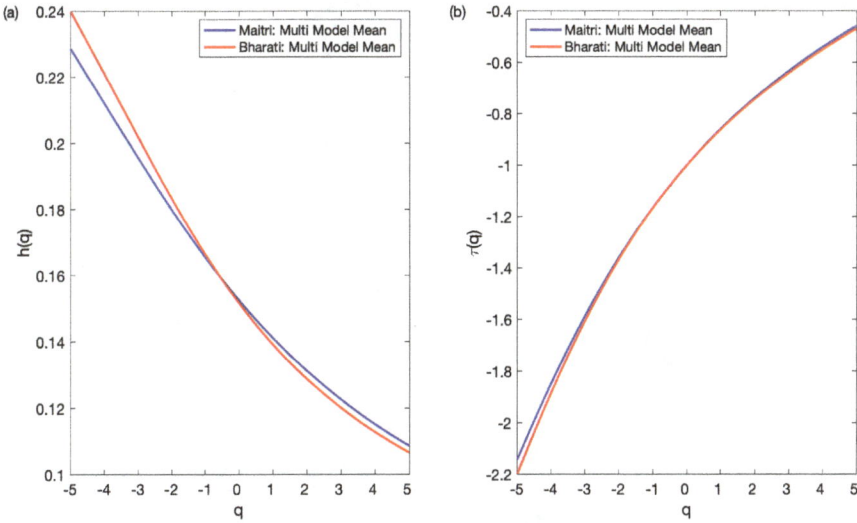

FIGURE 8.6 (a) $h(q)$ as a function of q and (b) $\tau(q)$ as a function of q using the multimodal mean time series of the SIC over the Maitri (blue) and Bharati (red) regions.

takes the shape of a symmetric/asymmetric arc around the maximum value of $f(\alpha)$ occurring at the particular α. For most of the CMIP6 models, the spectrum is either left-tailed or right-tailed (i.e. found with either left truncation or right truncation). The truncation of curves happens due to the levelling of $h(q)$ for negative/positive q's. When the SIC time series of a model represents a multifractal spectrum which becomes insensitive to the local fluctuations with small (large) magnitudes, then the spectrum will have a long left (right) tail. The minimum (α_{min}) and maximum (α_{max}) values of α are the singularity strengths corresponding to the least and most singular values obtained in the multifractal analysis of the SIC time series for a particular model, respectively. The difference between the maximum and minimum values of α in a multifractal spectrum is known as the multifractal spectrum width $\Delta\alpha$. The corresponding difference between maximum and minimum values of the singularity dimension (i.e. $f(\alpha)$ at $\alpha = \alpha_{min}$ and α_{max}, respectively) may be denoted as $\Delta f(\alpha)$. We quantify $\Delta\alpha$ and $\Delta f(\alpha)$ for expressing the strength of the multifractality of SIC time series data of different models used in the study. The values are summarized in Table 8.2 for the Maitri and Bharati regions for all the CMIP6 models. From Table 8.2 and Figure 8.7(a,b), we see that, in general, for the models in which the peak value of the $f(\alpha)$ occurs for higher α, the value of $\Delta\alpha$ is higher. In other words, with the shifting of the peak value of $f(\alpha)$ towards the right side in a multifractal spectrum, the width of the spectrum and hence multifractality increases. Table 8.2 suggests that the CanESM5 (CIESM) model has the highest (lowest) value of $\Delta\alpha$ around the Maitri station. On the other hand, the highest (lowest) value of $\Delta\alpha$ around the Bharati station is obtained for the EC-Earth3 (CMCC-CM2-SR5) model.

We also carry out a multifractal analysis on the multimodel mean time series of the SIC around the Maitri and Bharati regions using the historical dataset of

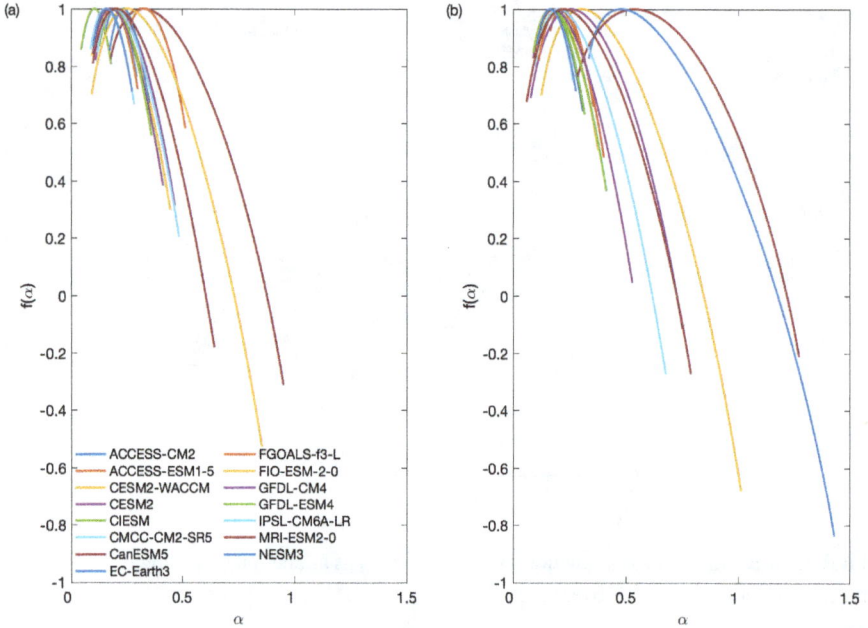

FIGURE 8.7 The multifractal spectrum ($f(\alpha)$ vs. α) of the SIC time series of CMIP6 models over the (a) Maitri and (b) Bharati regions.

the CMIP6 models. The multifractal spectrum is shown in Figure 8.8. We find that the multifractal spectrum of the Bharati region is wider as compared to the Maitri region. For example, $\Delta\alpha$ values for the Maitri and Bharati regions are 0.2305 and 0.2513, respectively. Similarly, $\Delta f(\alpha)$ values are 0.2536 and 0.3068, respectively (Table 8.3). These values suggest that the strength of multifractality around the Bharati region is greater as compared to the Maitri region. In other words, the sea ice structure of the Bharati region is more variable (in terms of the fractal nature and hence dimensionality) as compared to the Maitri region.

It will be worthwhile also analysing how global warming is affecting the multifractal properties of the SIC around the Maitri and Bharati Indian Antarctic stations. For this, we construct the multimodel mean time series of the SIC around the Maitri and Bharati using the SSP5-8.5 data of all 15 CMIP6 models used in the study (Table 8.1). We then compute the multifractal spectrum of these time series using the techniques described above. The $h(q)$ versus q as well as $f(\alpha)$ versus α curves for this purpose are shown in Figure 8.9(a,b) for the Maitri and Bharati regions. The output from historical data is also shown as dashed curves in these figures for comparison. We see from Figure 8.9(a,b) that due to the effect of global warming (as a result of the high GHG concentration scenario of SSP5-8.5 data), a complete reversal of multifractal properties over the Maitri and Bharati region is obtained. For example, Figure 8.9(a) shows that the maximum value of $h(q)$ occurs for $q = -5$ for the Maitri region in the case of SSP5-8.5 data, whereas, for historical data, the maximum value was obtained for the SIC data of Bharati region. The difference in the maximum value of $h(q)$ at

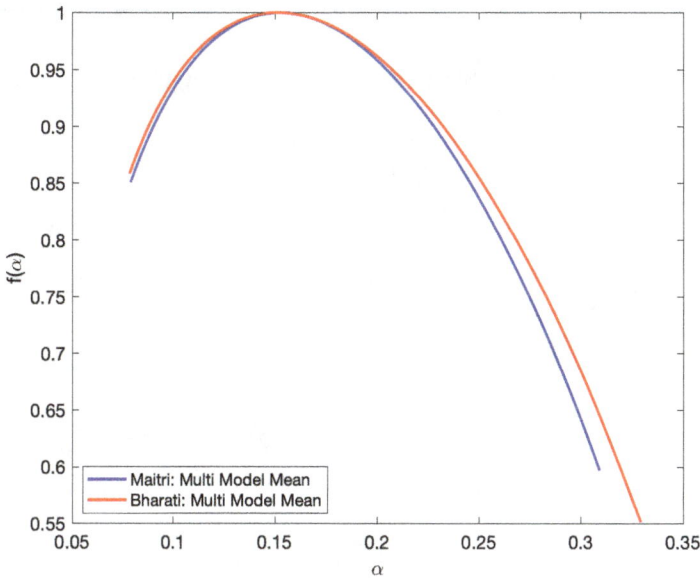

FIGURE 8.8 The multifractal spectrum obtained using the multimodel mean SIC time series of the historical data around the Maitri (blue) and Bharati (red) regions.

TABLE 8.2
Strength of Multifractality (i.e., the Width of Singularity Spectrum $\Delta\alpha$ and Corresponding Difference of Singularity Dimension $\Delta f(\alpha)$) of the SIC Time Series of CMIP6 Models around the Maitri and Bharati Indian Antarctic Stations for the Historical Period of 1900–2014

Models	Maitri station		Bharati station	
	$\Delta\alpha$	$\Delta f(\alpha)$	$\Delta\alpha$	$\Delta f(\alpha)$
ACCESS-CM2	0.1432	0.1440	0.1988	0.2367
ACCESS-ESM1-5	0.1822	0.1422	0.1911	0.2671
CESM2-WACCM	0.3515	0.5358	0.8872	1.3776
CESM2	0.3554	0.5054	0.6735	0.8037
CIESM	0.1375	0.0497	0.2239	0.2060
CMCC-CM2-SR5	0.1984	0.1914	0.1697	0.1073
CanESM5	0.7685	1.1393	0.9834	0.9797
EC-Earth3	0.1774	0.1833	1.0949	-1.6634
FGOALS-f3-L	0.2572	0.3312	0.2898	0.3391
FIO-ESM-2-0	0.7601	1.2271	0.2824	0.3261
GFDL-CM4	0.3151	0.4598	0.4402	0.7839
GFDL-ESM4	0.2345	0.3118	0.3241	0.4820
IPSL-CM6A-LR	0.3664	0.6439	0.5835	1.0734
MRI-ESM2-0	0.5454	0.9890	0.7290	0.9492
NESM3	0.1773	0.1593	0.1727	0.1646

TABLE 8.3
Comparison of Multifractal Strength (i.e. Width of Singularity Spectrum $\Delta\alpha$ and Difference of Singularity Dimension $\Delta f(\alpha)$) Between the Multimodel SIC Mean Time Series of Maitri and Bharati Indian Antarctic Stations for the Historical and SSP5-8.5 Future Projection Dataset

Station	Historical data		SSP5-8.5 data	
	$\Delta\alpha$	$\Delta f(\alpha)$	$\Delta\alpha$	$\Delta f(\alpha)$
Maitri	0.2305	0.2536	0.2202	0.2504
Bharati	0.2513	0.3068	0.1811	0.1384

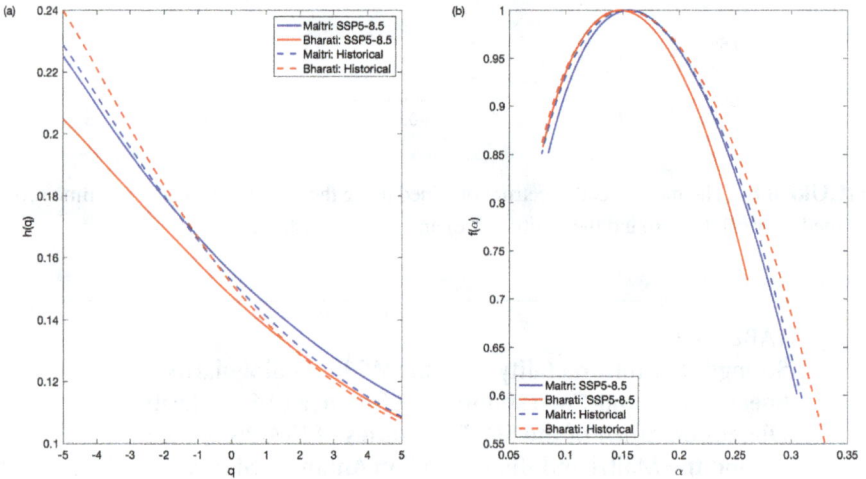

FIGURE 8.9 (a) $h(q)$ as function of q, and (b) $f(\alpha)$ as a function of α obtained using the multimodel mean SIC time series of the SSP5-8.5 data for the Maitri (solid blue) and Bharati (solid red) regions. The dashed blue and red curves are for the historical data of Maitri and Bharati regions, respectively.

$q = -5$ between the Maitri and Bharati regions also becomes higher with an increase in global warming. We also find that for historical data of SIC, the nonlinear decrease of $h(q)$ with q is higher in the case of the Bharati region, whereas, for SSP5-8.5 data, $h(q)$ decreases nonlinearly with q at a much higher rate for the Maitri region. The result suggests that with an increase in GHG concentration, the multifractality of the Maitri region will become more pronounced as compared to the Bharati region.

The multifractal spectrum curves shown in Figure 8.9(b) are used to compute the values of $\Delta\alpha$ and $\Delta f(\alpha)$ over the Maitri and Bharati regions. These values are summarized in Table 8.3. It is clear from Table 8.3 and Figure 8.9(b) that with an increase in global warming, the multifractal properties over the Maitri and Bharati regions are getting reversed. For the historical data, the singularity (multifractal) spectrum of the Bharati region is wider as compared to the Maitri region. On the other hand, for the SSP5-8.5 data, we find that the singularity spectrum of the Maitri region becomes wider than for the Bharati region. For example, the $\Delta\alpha$ values over the Maitri and Bharati regions which were obtained as 0.2305, and 0.2513, respectively, for the historical data of SIC, become 0.2202 and 0.1811, respectively, for the SSP5-8.5 data over the same region. The finding is further substantiated by the corresponding change in $\Delta f(\alpha)$ values (Table 8.3). This analysis (Table 8.3 and Figure 8.9) suggests that with an increase in GHG concentration, the strength of multifractality of the SIC time series becomes weaker. In other words, the $\Delta\alpha$ and $\Delta f(\alpha)$ values of SSP5-8.5 data become smaller as compared to the historical data. In addition to this, the multifractal nature of the SIC around the Maitri and Bharati regions also gets reversed in a climate change scenario. The $\Delta\alpha$ and $\Delta f(\alpha)$ values which were higher over the Bharati region over the historical period of 1900–2014 are likely to become higher over the Maitri region during the period 2015–2100 for the SSP5-8.5 data.

8.4 CONCLUSIONS

In this chapter, an effort is made to investigate the effect of global warming on the multifractal nature of the SIC time series of Indian Antarctic stations Maitri and Bharati. The historical and SSP5-8.5 SIC dataset of a total of 15 CMIP6 models are analysed for this purpose using the MFDFA technique. The multimodel mean time series of SIC is also analysed. It was found that the generalized Hurst exponent $h(q)$ of the SIC time series of Maitri and Bharati regions is a nonlinear decreasing function of q for all the CMIP6 models (and their multimodel mean) used in the study. This result confirms that the SIC of the Maitri and Bharati regions is multifractal. It is also found that $h(q)$ and mass exponent $\tau(q)$ of the Bharati the region is more variable (and hence exhibits higher multifractality) as compared to the Maitri region. The shape and width of the multifractal spectrum of the SIC time series of different CMIP6 models are quite different from each other, with the spectrum being either left- or right-tailed. The multifractal spectrum of the Bharati region is found to be wider (with higher $\Delta\alpha$ and $\Delta f(\alpha)$) as compared to the Maitri region, which conforms with the fact that the strength of multifractality is stronger around the Bharati region as compared to the Maitri region. A complete reversal of multifractal properties over the Maitri and Bharati regions is seen as a result of global warming. With the increase in GHG concentration (SSP5-8.5 data), the multifractality of the Maitri region will become greater as compared to the Bharati region (i.e. singularity spectrum of the Maitri region will become wider than the Bharati region). It is also found that the overall strength of multifractality of the SIC time series will become weaker in a global warming scenario. In future, efforts will be made to check the robustness of the results obtained in the study by applying other multifractal techniques such as wavelet transformation-based analysis and multifractal detrended moving average (MFDMA) methods. In addition to this, the SIC time series data of the other future projection scenarios, such as SSP1-2.6 and SSP3-7.0, shall also be analysed to understand the effect of global warming on the multifractal strength of the SIC under varying GHG concentrations.

ACKNOWLEDGEMENTS

SD is thankful for infrastructural support provided to the KBCAOS under the DST-FIST scheme of DST, GoI. The CMIP6 data were obtained from https://esgf-node.llnl.gov/search/cmip6/.

REFERENCES

Agarwal, S., Moon, W., Wettlaufer, J.S. (2012) Trends, noise and re-entrant long-term persistence in Arctic sea ice. *Proceedings of the Royal Society* A 468: 2416–2432.

Agarwal, S., Wettlaufer, J.S. (2018) Fluctuations in Arctic sea-ice extent: comparing observations and climate models. *Philosophical Transactions of the Royal Society A*, 376: 20170332.

Agbazo, M., Koto N'gobi, G., Alamou, E., Kounouhewa, B., Afouda, A., Kounkonnou, N. (2019) Multifractal behaviors of daily temperature time series observed over Benin synoptic stations (West Africa). *Earth Sciences Research Journal* 23(4): 365–370.

Baranowski, P., Krzyszczak, J., Slawinski, C., Hoffmann, H., Kozyra, J., Nieróbca, A., Siwek, K., Gluza, A. (2015) Multifractal analysis of meteorological time series to assess climate impacts. *Climate Research* 65: 39–52.

Bogachev, M.I., Bunde, A. (2011) On the predictability of extreme events in records with linear and nonlinear long-range memory: efficiency and noise robustness. *Physica A* 390: 2240–2250.

Cadenas, E., Campos-Amezcua, R., Rivera, W., et al. (2019) Wind speed variability study based on the Hurst coefficient and fractal dimensional analysis. *Energy Science & Engineering* 7: 361–378.

Chmel, A., Smirnov, V.N., Astakhov, A.P. (2005) The Arctic sea-ice cover: fractal space-time domain. *Physica A* 357: 556–564.

Comiso, J.C., Gersten, R., Stock, L.V., Turner, J., Perez, G.J., Cho, K. (2017) Positive trend in the Antarctic Sea ice cover and associated changes in surface temperature. *Journal of Climate* 30(6): 2251–2267.

Dwivedi, S. (2012) Quantifying predictability of Indian summer monsoon intraseasonal oscillations using nonlinear time series analysis. *Meteorologische Zeitschrift* 21(4): 413–419.

Dwivedi, S., Pandey, L.K. (2022) Quantifying the predictability of Southern Indian Ocean sea ice concentration in a changing climate scenario, Chapter 9. In Khare, N. (Ed.) *Climate Variability of Southern High Latitude Regions: Sea, Ice, and Atmosphere Interactions* (1st ed.). CRC Press.

Girard, L., Weiss, J., Molines, J. M., Barnier, B., Bouillon, S. (2009) Evaluation of high-resolution sea ice models on the basis of statistical and scaling properties of Arctic sea ice drift and deformation. *Journal of Geophysical Research* 114: C08015.

Hao, G., Shen, H., Sun, Y. et al. (2021) Rapid decrease in Antarctic sea ice in recent years. *Acta Oceanologica Sinica* 40: 119–128.

Ihlen, E.A. (2012) Introduction to multifractal detrended fluctuation analysis in MATLAB. *Frontiers in Physiology* 3: 141.

Kantelhardt, J.W., Zschiegner, S.A., Koscielny-Bunde, E., Havlin, S., Bunde, A., Stanley, H.E. (2002) Multifractal detrended fluctuation analysis of nonstationary time series. *Physica A* 316: 87–114.

Kantelhardt, J.W., Koscielny-Bunde, E., Rybski, D., Braun, P., Bunde, A., Havlin, S. (2006) Long-term persistence and multifractality of river runoff and precipitation records. *Journal of Geophysical Research: Atmospheres* 111: D01106.

Kimothi, S., Kumar, A., Thapliyal, A., Ojha, N., Soni, V.K., Singh, N. (2019) Climate predictability in the Himalayan foothills using fractals. *MAUSAM* 70: 357–362.

Kumar, A., Dwivedi, S., Rajak, D.R. (2017) Ocean sea-ice modelling in the Southern Ocean around Indian Antarctic stations. *Journal of Earth System Science* 126(5): 70.

Kumar, A., Dwivedi, S., Rajak, D.R., Pandey, A.C. (2018a) Impact of air-sea forcings on the Southern Ocean sea ice variability around the Indian Antarctic stations. *Political Science* 18: 197–212.

Kumar, A., Dwivedi, S., Pandey, A.C. (2018b) Quantifying predictability of sea ice around the Indian Antarctic stations using coupled ocean sea ice model with shelf ice. *Political Science* 18: 83–93.

Laib, M., Golay, J., Telesca, L., Kanevski, M. (2018) Multifractal analysis of the time series of daily means of wind speed in complex regions. *Chaos, Solitons and Fractals* 109: 118–127.

Ludescher, J., Bogachev, M.I., Kantelhardt, J.W., Schumann, A.Y., Bunde, A. (2011) On spurious and corrupted multifractality: the effects of additive noise, short-term memory and periodic trends. *Physica A* 390: 2480–2490.

Maruyama, F. (2020) Relationship between the Northern Hemisphere sea ice area and global temperature by multifractal analysis. *Journal of Applied Mathematics and Physics* 8: 896–909.

Moon, W., Nandan, V., Scharien, R.K., Wilkinson, J., Yackel, J.J., Barrett, A., Lawrence, I., Segal, R.A., Stroeve, J., Mahmud, M., Duke, P.J. (2019) Physical length scales of wind-blown snow redistribution and accumulation on relatively smooth Arctic first-year sea ice. *Environmental Research Letters* 14: 104003.

Parkinson, C.L., Cavalieri, D.J. (2012) Antarctic sea ice variability and trends, 1979–2010. *Cryosphere* 6: 871–880.

Parkinson, C.L. (2019) A 40-y record reveals gradual Antarctic sea ice increases followed by decreases at rates far exceeding the rates seen in the Arctic. *Proceedings of the National Academy of Sciences of the United States of America* 116(29): 14414–14423.

Rampal, P., Dansereau, V., Olason, E., Bouillon, S., Williams, T., Korosov, A., Samaké, A. (2019) On the multi-fractal scaling properties of sea ice deformation. *The Cryosphere* 13: 2457–2474.

Roach, L.A., Dörr, J., Holmes, C.R., Massonnet, F., Blockley, E.W., Notz, D., et al. (2020) Antarctic sea ice area in CMIP6. *Geophysical Research Letters* 47: e2019GL086729.

Shu, Q., Song, Z., Qiao, F. (2015) Assessment of sea ice simulations in the CMIP5 models. *The Cryosphere* 9(1): 399–409.

Shu, Q., Wang, Q., Song, Z., Qiao, F., Zhao, J., Chu, M., Li, X. (2020) Assessment of sea ice extent in CMIP6 with comparison to observations and CMIP5. *Geophysical Research Letters* 47: e2020GL087965.

Taylor, K.E., Stouffer, R.J., Meehl, G.A. (2012) An overview of CMIP5 and the experiment design. *Bulletin of the American Meteorological Society* 93: 485–498.

Turner, J., Bracegirdle, T.J., Phillips, T., Marshall, G.J., Hosking, J.S. (2013) An initial assessment of Antarctic sea ice extent in the CMIP5 models. *Journal of Climate* 26: 1473–1484.

Weiss, J., Marsan, D. (2004) Scale properties of sea ice deformation and fracturing. *Comptes Rendus Physique* 5: 735–751.
Yadav, R.P., Dwivedi, S., Mittal, A.K., Kumar, M., Pandey, A.C. (2012) Fractal and multifractal analysis of LiF thin film surface. *Applied Surface Science* 261: 547–553.
Yadav, R.P., Pandey, R.K., Mittal, A.K., Dwivedi, S., Pandey, A.C. (2013) Multifractal analysis of sputtered CaF2 thin films. *Surface and Interface Analysis* 45: 1775–1780.

9 Spatial and Temporal Variations in Surface Energy Balance of Antarctic Snow and Blue-Ice

P.K. Srivastava, H.S. Gusain, P. Datt, K.K. Singh, Paramvir Singh, V. Bharti, M. Kumar, and V.D. Mishra

9.1 INTRODUCTION

Absorption, reflection, sublimation, melting, condensation, deposition, and other physical phenomena at the surface drives the interaction between Antarctic ice-sheet and the atmosphere. These physical processes exchange energy between the snow-ice surface and the atmosphere above. The components of the surface energy balance, i.e., net shortwave radiation flux, net longwave radiation flux, sensible heat flux, latent heat flux, and sub-surface conductive heat flux, depend greatly on the prevailing meteorological, topographical, and snow-ice surface conditions. Large seasonal fluctuations in surface climate at these high latitudes create substantial changes in the energy exchange processes at snow-ice surfaces. Long-term fluctuations in the surface energy exchange process of the Antarctic ice sheet are of considerable interest because of significant contribution to the mass balance and its fluctuations resulting from greenhouse-induced heating, melting, and sublimation. Surface albedo, which controls the amount of reflected solar radiation, constitutes a positive feedback component of the Earth's climate system and is a key factor controlling the global energy balance and pattern of atmospheric circulations. The incoming solar energy hardly balances the outgoing surface energy losses in Antarctica because of the high reflectivity and extreme thermal emissivity of snow-ice surfaces. In order to understand the long-term effects of changing climate on the Antarctic snow-ice fields, it is important to investigate the variability of surface reflectance and surface energy fluxes at multiple temporal and spatial scales. The problem thus involves quantifying the competing effects of energy balance components versus net energy gain or loss over the surface. Large-scale modelling of Antarctic ice-sheet dynamics and response to changing climate conditions also requires a thorough understanding of the atmospheric boundary layer and the energy and mass exchange processes across the polar ice sheet.

Shortwave and longwave radiation fluxes are components of radiative energy fluxes and are generally measured directly using pyranometers or pyrgeometers. Several parametrization schemes (Moritz, 1978; Dozier, 1980; Prata, 1996; Dilley and O'Brien, 1998) are also proposed to estimate different radiative energy fluxes using in situ meteorological observations. The sensible and latent heat fluxes, which are components of turbulent energy fluxes, can be either measured directly using an eddy covariance system or estimated using a bulk transfer approach (Oke, 1970; Moore, 1983; Srivastava, 2002). Bulk transfer equations use meteorological variables over the surface and at screen level height above the surface as input to estimate turbulent energy fluxes. The contribution of sub-surface heat flux in net surface energy balance is generally very low compared to radiative and turbulent energy fluxes and can be estimated using temperature measurements at different vertical levels inside the snowpack or ice. Sub-surface heat flux can also be measured directly using heat flux sensors placed inside snow/ice (Datt et al., 2015).

The majority of the continent of Antarctica consists of a snow-covered interior ice sheet. In coastal areas, the snow and ice drain through ice streams that, further downstream, form floating ice shelves along the periphery of the continent. While coastal regions of Antarctica are characterized by mild temperatures and high katabatic winds, the interior Antarctic ice sheet experiences colder temperatures and low wind speeds. Approximately 2% of the area of the Antarctic continent, also known as oases, is ice-free. The Schirmacher Oasis, close to the Indian Antarctic station Maitri, is one of the smallest oases in Antarctica. Blue-ice areas of the continental ice-sheet in Antarctica are of special interest as their surface mass balance is negative and can be considered as climate-change indicators (Orheim and Lucchitta, 1990). A comprehensive review of the glaciological, meteorological, and climatological significance of blue-ice areas can be found in Bintanja (1999). Several studies on the annual surface energy balance of different parts of the Antarctic ice-sheet have been reported previously (Bintanja and Van den Broeke, 1995; Bintanja et al., 1997; Srivastava, 2002; Van den Broeke et al., 2004; 2005). While the radiative and turbulent energy fluxes are dependent on the location, surface characteristics, and local weather conditions, sublimation dominates the surface energy balance in the ablation zone of glaciers. A major difficulty lies in the fact that due to the sparse observatory network over the Antarctic continent, area-averaged values for the surface energy balance components in different snow-ice media are still not precisely known.

Defence Geoinformatics Research Establishment (erstwhile SASE) initiated scientific studies on the reflectance and surface energy balance of different snow-ice media during the 15th Indian Scientific Expedition to Antarctica (ISEA). Initial research efforts during 1996–1998 were focused on the measurements of surface albedo and associated snow-meteorological parameters at several locations on the continental-shelf and continental-ice in Dronning Maud Land, East Antarctica. Mishra (1999) analysed the dependence of albedo on various snow-meteorological parameters (e.g. solar elevation angle, cloud amount, snow density and grain-size, snow wetness, etc.) and utilized a simple energy balance model to estimate the surface energy balance components for a few days during which the measurements were available. Subsequently, two automatic weather stations (AWSs) were installed on the snow covered continental-shelf (unofficial name: DG) and blue-ice surface of continental ice-sheet (unofficial name: Sankalp) during summer of 1999 to collect high-quality continuous measurements of snow-meteorological parameters (Figure 9.1). DGRE further established a third AWS at the edge of the east Antarctic ice-sheet (unofficial name-Dozer) close to Schirmacher Oasis during the summer of 2007. Gusain et al. (2014) analysed the 4-year record (March 2007–February 2011) of meteorological parameters and radiative energy fluxes at the Dozer AWS location and reported that net radiative flux was the main heat source to the ice-sheet during summer (46.8 Wm^{-2}) and heat-sink during winter (–42.2 Wm^{-2}) months. Further, sensible heat flux (annual mean 32 Wm^{-2}) was the heat-source and latent heat flux (annual mean –61 Wm^{-2}) was the heat sink to the glacier surface, averaged over all seasons. The surface energy balance at the edge of the Antarctic ice-sheet close to oases (i.e. ice-free land regions) is crucial in the case of retreating ice sheet and growing oasis areas.

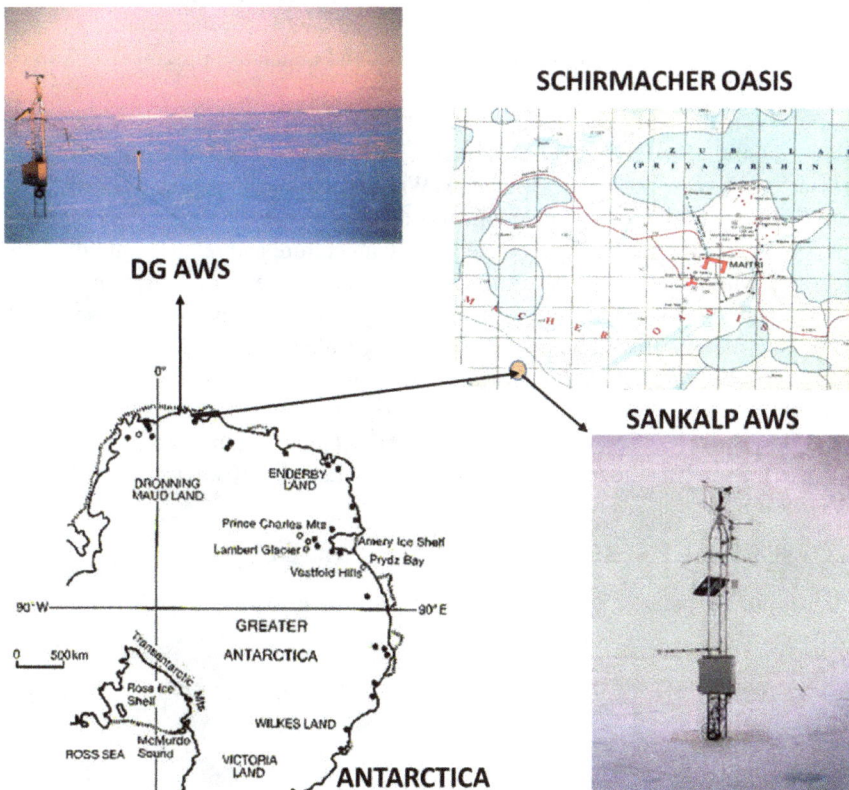

FIGURE 9.1   Location of DG AWS on the continental-shelf and Sankalp AWS on continental-ice (blue-ice surface) in Dronning Maud land East Antarctica.

In the present study, a comparative analysis of the meteorological conditions, reflectance characteristics, and surface energy balance of snow-covered continental-shelf and blue-ice areas of continental ice-sheet during 1999 is discussed first. Subsequently, we present a detailed analysis of the seasonal and diurnal variations of the snow-meteorological parameters, surface albedo, and radiative and turbulent energy fluxes over the blue-ice surface derived from the 6-year data (Mar 2006–Jul 2012) recorded by the Sankalp AWS.

9.2   STUDY AREAS AND INSTRUMENTATION

The DG AWS (Figure 9.1) was installed at about 18 km from the ocean on the continental-shelf ($70^\circ05'$ S, 12°E). The surface over the continental-shelf is always covered with a thick layer (~1.5 meter) of high-density snow (450 kg m^{-3}). The Sankalp AWS (Figure 9.1) was installed at about 10 km from the Schirmachar Oasis on the blue-ice surface ($70^\circ45'$ 52" S, $11^\circ44'$ 3" E) of continental-ice sheet. The blue-ice site, which is almost always free of snow, is of particular interest because though Antarctic blue-ice areas make up only a small part of the Antarctic continent, it is one of the few areas on the continent where ablation exceeds accumulation (Bintanja and Van den Broeke, 1994). As such, this blue-ice area is similar to other Antarctic blue-ice areas located in coastal regions and around nunataks (Bintanja and Van den Broeke, 1995).

The automatic weather stations recorded hourly observations of air temperature, relative humidity, wind speed and direction, incoming and reflected solar radiation, pressure, and snow surface temperature. Wind speed and direction were measured with RM Young 05103 3-cup anemometers and wind vane with sensor accuracies of $\pm$ 0.3 m s^{-1} and $\pm$ 3 degrees, respectively. Air temperature and relative humidity were measured with a Campbell Scientific 41372V-90 AT/Rh sensor having accuracies of $\pm$ 0.3°C and $\pm$ 3-4%, respectively. Incoming solar radiation and reflected solar radiation were measured with a Kipp & Zonen CM3 albedometer with an accuracy of $\pm$ 10%. Air pressure and glacier surface temperature were measured with an Anika PTB210B2A1B and Everest 4000, with sensor accuracies of 0.5 mbar and $\pm$ 0.1–0.3°C, respectively. Three-hourly observations of cloud amount and type were also recorded manually at the Indian research station "Maitri", ~ 10 km from the AWS site.

9.3   SURFACE ENERGY BALANCE (SEB) MODEL

The net energy balance or the residual energy flux at the ice surface, ΔQ, can be expressed as:

$$\Delta Q = S_{net} + L_{net} + H + LE + G = R_{net} + H + LE + G \qquad (9.1)$$

where

$$S_{net} = S_{in} - S_{out}$$

$$L_{net} = L_{in} - L_{out}$$

S_{in} = incoming shortwave radiation flux
S_{out} = reflected shortwave radiation flux
L_{in} = incoming longwave radiation flux
L_{out} = emitted longwave radiation flux
S_{net} = net shortwave radiation flux
L_{net} = net longwave radiation flux
R_{net} = net radiative flux
H = sensible heat flux
LE = sensible heat flux
G = sub-surface ice heat flux

All the surface energy balance components are defined as positive when directed towards the surface. The net energy balance ΔQ indicates the net change in energy storage of the glacier and is defined by net radiative flux, turbulent fluxes, and sub-surface ice heat flux. The contribution due to conduction into the ice or snow from the ground is generally neglected.

9.3.1   NET RADIATIVE FLUX

Net radiative flux is the sum of the down-welling global solar radiation, the reflected solar radiation, emitted terrestrial longwave radiation, and down-welling longwave radiation emitted by clouds and atmospheric gases. The proportion of the down-welling solar radiation, which is absorbed at the surface is called effective shortwave radiation and is given by:

$$S_{net} = S_{in} - S_{out} = S_{in}(1-\alpha) \tag{9.2}$$

where α is the albedo of the surface. Incoming (S_{in}) and outgoing (S_{out}) shortwave radiation fluxes were observed directly from the albedometer.

The net longwave radiation (L_{net}) at the glacier surface is the sum of the upwards terrestrial component L_{out}(-ve) and the downwards L_{in}(+ve) radiation emitted by the clouds and gases. The upward component is usually greater than the down-welling flux, so net longwave flux is generally negative and expresses the loss of energy from the glacier surface in the longwave part of the electromagnetic spectrum. The down-welling longwave radiation flux is estimated using a model developed by Prata (1996) which computes atmospheric emissivity (ε_m) as a function of precipitable water content (w) and is given by:

$$L_{in} = \varepsilon_m \sigma T_a^4 \tag{9.3}$$

where $\varepsilon_m = 1 - (1+w)\exp\left\{-(1.2+3\,w)^{1/2}\right\}$, and $w = 46.5(e_a/T_a)$, e_a is vapour pressure (Pa), and T_a the absolute temperature measured at 2.0 m above the surface. The emitted longwave radiation (L_{out}) is computed as:

$$L_{out} = \varepsilon_s \sigma T_s^4 \tag{9.4}$$

where σ is the Stephan-Boltzmann constant (5.67×10^{-8} W m^{-2} K^{-4}), T_s the radiative temperature of the surface in Kelvin, and ε_s the surface emissivity. The surface emissivity of the snow/ice surface is assumed to be unity (Bintanja & Van den Broeke, 1994). In the absence of clouds, the net longwave radiation absorbed by the glacier surface is estimated as:

$$L_{net} = L_{in} - L_{out} = \varepsilon_m \, \sigma T_a^4 - \varepsilon_s \, \sigma T_s^4 \tag{9.5}$$

In overcast sky conditions, the net longwave radiation is given by:

$$L_{net} = \varepsilon_m \, \sigma T_a^4 - \varepsilon_s \, \sigma T_s^4 \left(1 - kN\right)$$

where the coefficient k depends on type and height of clouds ($k = 0.76, 0.52$, and 0.26 for low, medium, and high clouds, respectively), and N is the amount of cloudiness in terms of fraction of sky covered (Upadhyay, 1999). The net radiative flux in the model is then computed as:

$$R_{net} = S_{in} \left(1 - \alpha\right) + \varepsilon_m \, \sigma T_a^4 - \varepsilon_s \, \sigma T_s^4 \left(1 - kN\right) \tag{9.6}$$

9.3.2 TURBULENT FLUXES

Sensible heat flux and latent heat flux are the turbulent energy fluxes. The glacier surface is in general colder than the air above the surface and the heat transfers from warm air to the colder surface by the process of sensible heating and so the glacier surface gains energy. In the case of temperature inversion, the heat transfer takes place in the opposite direction and the glacier surface loses energy. The vertical turbulent sensible heat flux H is expressed as (Ambach and Kirchlechner, 1986; Paterson, 1994):

$$H = \left(C_p \, \rho_0 / P_0\right) K_n \, P u \left(T_a - T_s\right) \tag{9.7}$$

$$K_n = k^2 / \left[\log\left(z_a / z_0\right)\right]^2$$

where ρ_0 is the density of air (1.29 kg m^{-3}) at the standard atmospheric pressure P_0 (1.013×10^5 Pa), C_p is the specific heat of air (1005 J kg^{-1} K^{-1}), K_n is a dimensionless transfer coefficient, P is the mean atmospheric pressure at the measuring site, and u is the measured wind speed at a height of 2 m above the glacier surface. k is von Karman's constant (0.41), z_a is sensor height above ground (2 m), and z_0 is aerodynamic roughness length and a value of 0.001m is adopted for an ice surface (van de Wal and Russell, 1994; Bintanja and Van den Broeke, 1995; Konzelmann and Braithwaite, 1995).

Analogous to sensible heat flux, the latent heat flux is estimated as:

$$LE = L_v \left(0.623 \rho_0 / P_0\right) K_n \, u \left(e_a - e_s\right) \tag{9.8}$$

where L_v is the latent heat of vaporization, e_a is the vapour pressure at height z above glacier surface and e_s is the saturation vapour pressure at the glacier surface. e_s is a function of the surface temperature and is 611 Pa for a melting surface (Paterson, 1994). When $\left(e_a - e_s\right)$ is positive, and $T_s = 0°C$, water vapour condenses as liquid water on the melting glacier surface with $L_v = 2.514$ MJ kg^{-1}; when $\left(e_a - e_s\right)$ is negative, there is sublimation with $L_v = 2.849$ MJ kg^{-1}. Also, when $\left(e_a - e_s\right)$ is positive and $T_s < 0°C$, there is deposition from vapour to solid ice with $L_v = 2.849$ MJ kg^{-1}.

The equations of sensible and latent heat fluxes are applicable for neutral atmospheric conditions. Stability corrections were applied using the transfer coefficient K_n in terms of bulk Richardson number (R_i). For unstable conditions ($R_i < 0$) the effective transfer coefficient is given by $K_n\left(1 - 10R_i\right)$; for neutral condition ($R_i = 0$), it is given by K_n; and for stable conditions ($R_i > 0$), it is given by $K_n/\left(1 + 10R_i\right)$ (Price and Dunne, 1976).

9.4   RESULTS AND DISCUSSION

9.4.1   SURFACE ENERGY BALANCE STUDIES OVER SNOW AND BLUE-ICE (JAN–JUL 1999)

9.4.1.1   Meteorological Conditions

Figure 9.2a shows the daily variation of ambient temperature over the snow-covered continental-shelf (DG AWS) and blue-ice areas of continental-ice (Sankalp AWS) during Jan–Feb 1999. Over blue-ice, measured average ambient temperature was –5.6°C, while over the continental-shelf average T_a was –8.3°C. Thus, average T_a over the continental-shelf was 2.6°C lower than over the blue-ice, which indicates the relatively large radiative heating of the blue-ice surface. The variation pattern shows that a sharp rise in average air temperature corresponds to fair weather, followed by a drop in air temperature corresponding to bad weather days. One interesting feature to be noted is that variation in air temperature over both surfaces followed each other, although there seems to be a time delay in the appearance of short-term temperature maxima at the two locations. The time delay patterns can be classified into two groups. In one, the temperature maximum appeared earlier at the blue-ice location than at the continental-shelf (Figure 9.2b, events P1, P2, and P3). In the other, the temperature maximum at the continental-shelf location preceded that at the blue-ice site (Figure 9.2c, events P4 and P5). In both cases the maximum propagated within one or two days. The nature of the delay pattern depends whether the centre of the low-pressure disturbance is closest to the shelf or blue-ice site.

Figure 9.2d shows the variation of monthly mean ambient temperature, surface temperature, relative humidity, and wind speed from February 1999 to July 1999 over blue-ice. A remarkable feature was the temperature rise in June, otherwise the air temperature continued to drop with the onset of austral winter. The phenomenon of the winter temperature having no distinct minimum is known as a "coreless" winter for the Antarctic. Average surface temperature was lower than average air temperature for all months except February 1999. This was because of a longer period of

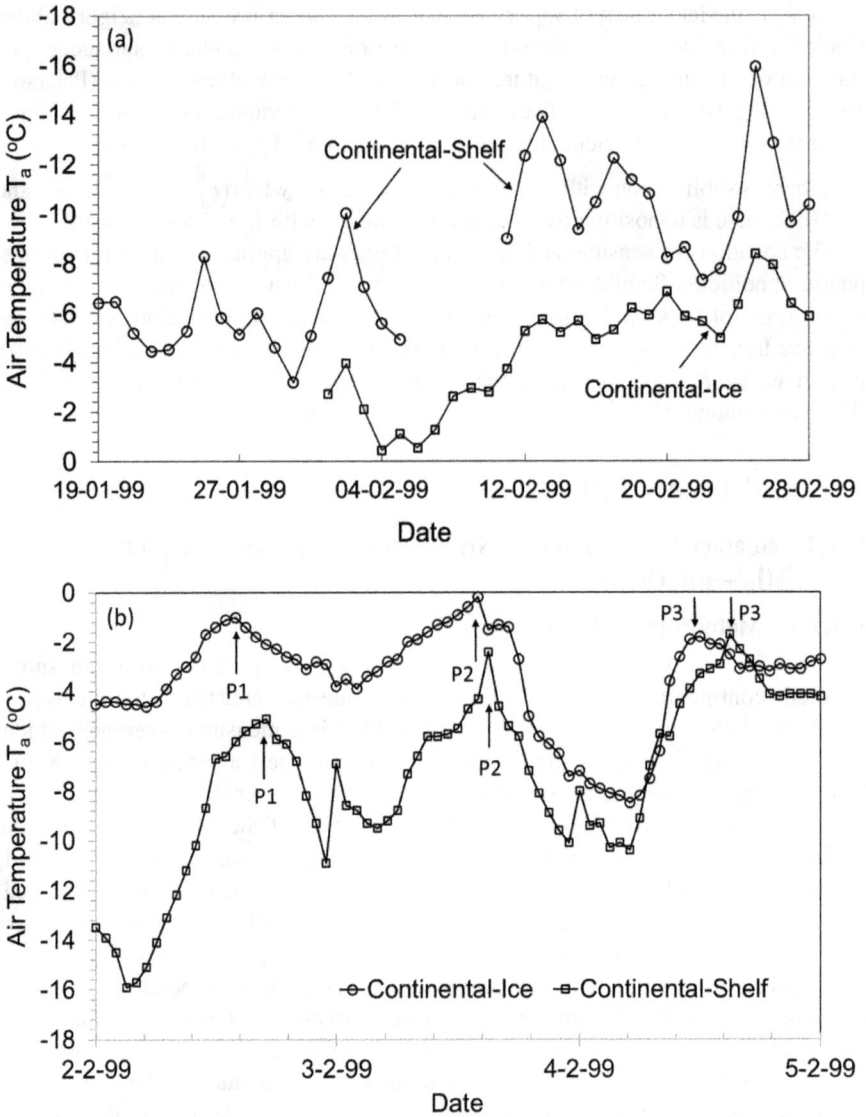

FIGURE 9.2 (a) Comparison of daily mean air temperature (T_a) over continental-shelf and continental-ice and (b) Time delay pattern in short-term air temperature maximum (P1, P2 and P3). (c) Time delay pattern in short-term air temperature maximum (P4 and P5) and (d) Monthly variation of air temperature (T_a), surface temperature (T_s), relative humidity (RH) and wind speed (u).

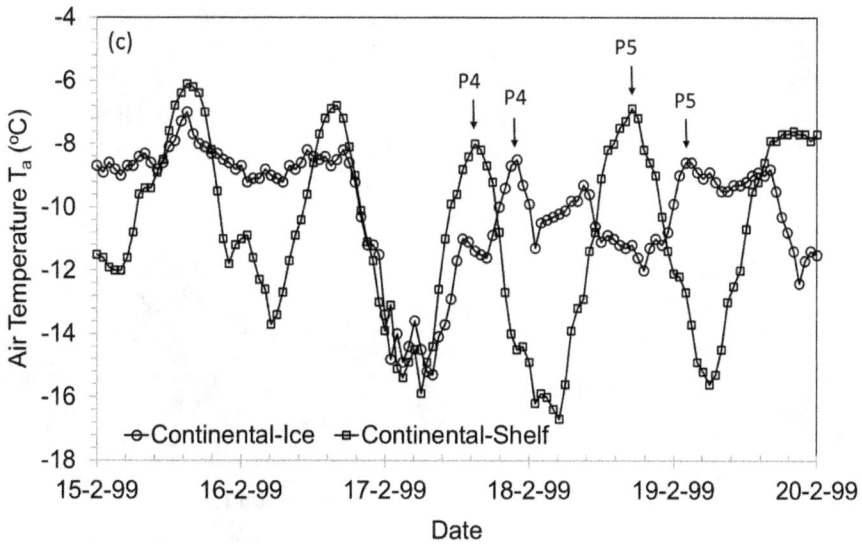

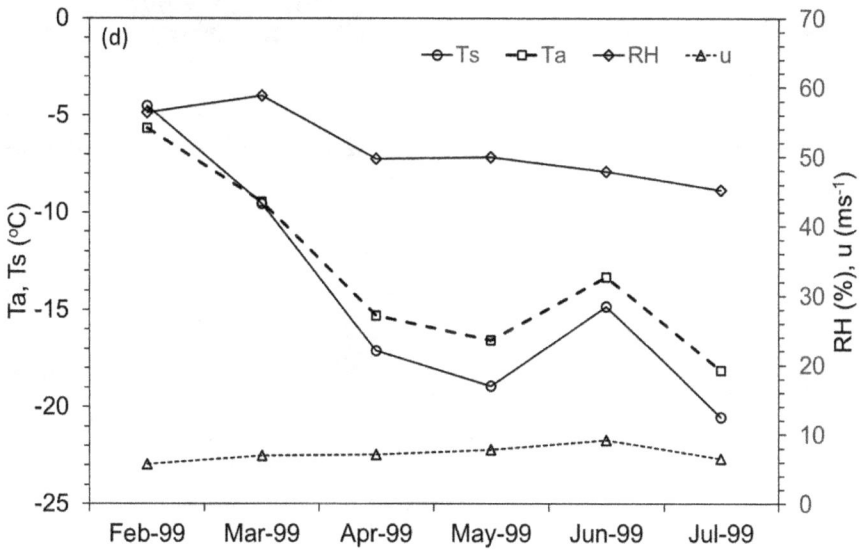

FIGURE 9.2   (Continued)

unstable stratification caused by strong radiative heating of the ice surface during February. Wind speed was highest in June with an average value of 9.24 m s^{-1}.

9.4.1.2   ALBEDO OVER CONTINENTAL-SHELF AND BLUE-ICE

We define daily albedo as the ratio of daily amount of reflected solar radiation to the daily insolation. Note that this differs from the daily mean albedo, which is the mean of all instantaneously determined albedo values during daylight. However, when the

solar elevation angle is very low, this quantity may be influenced greatly by instrumental errors. In view of this we will concentrate on daily albedos for analysing
measurements.

Figure 9.3a shows the daily variation of albedo over both surfaces during January
and February 1999. The plot shows that albedo over the continental-shelf is much

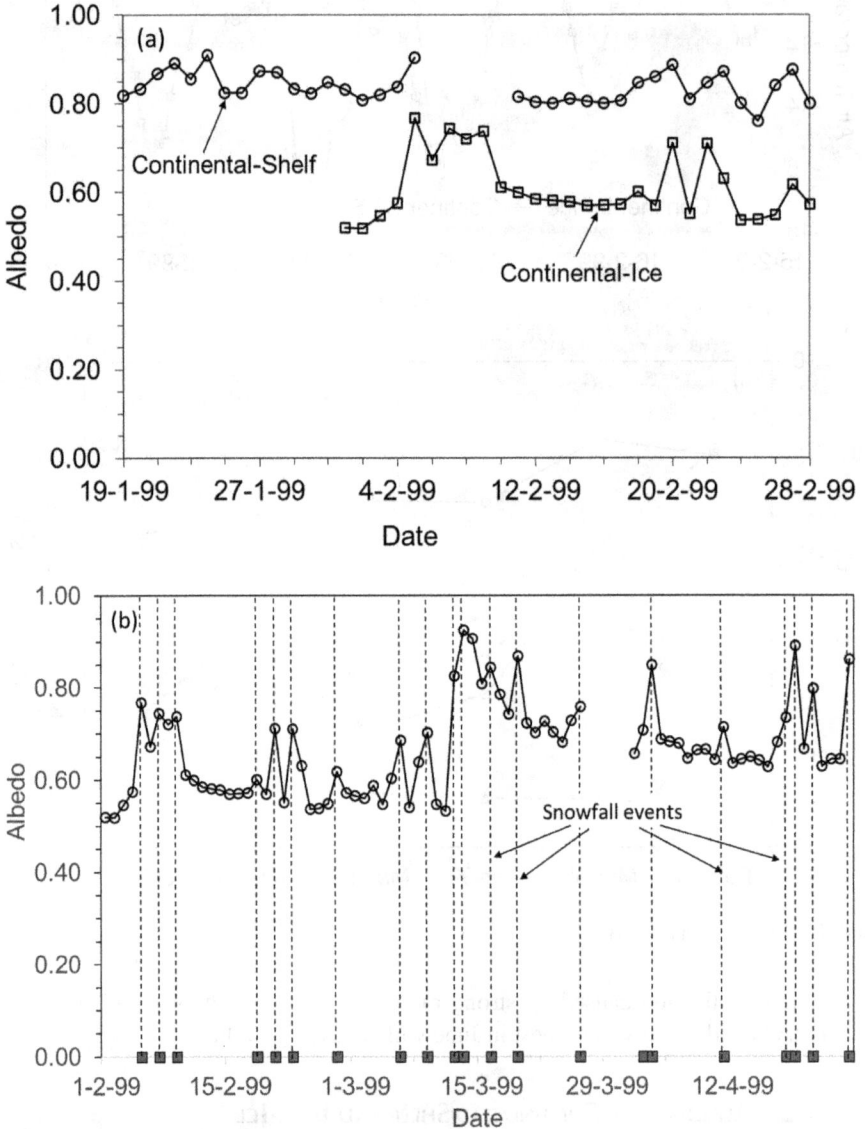

FIGURE 9.3 (a) Mean diurnal variation of global and reflected radiation over Sea Ice for the
observation period and (b) The surface albedo as a function of sea ice concentration. Values are
averaged as a function of ice concentration.

higher than over blue-ice as the albedo over the shelf is due to snow while over blue-ice it is due to ice only. The average albedo over the continental-shelf is 82.4%, while that over blue-ice is 60.9% for the corresponding period. These values are in general agreement with those reported by various investigators. Figure 9.3b shows daily albedos from February to April 1999 over blue-ice. Also shown in the Figure 9.3b are the days with snowfall events. As expected, there is a clear relation between albedo and snowfall events.

9.4.1.3 ALBEDO MEASUREMENTS OVER SEA-ICE

The albedo of sea-ice was investigated during the sea voyage to Antarctica, through pack ice in the Southern Ocean. The time of observation was close to mid-summer. During 8-13 January 1999, continuous radiation and meteorological measurements were carried out from the ship. On 9 January 1999, the first sea-ice was seen. The mean diurnal variation of the incident (S_{in}) and reflected (S_{out}) shortwave radiation for the observation period is presented in Figure 9.4a. Daily courses for individual days varied widely, depending mostly on the amount and type of cloud and to a lesser extent on the atmospheric turbidity. A maximum mean daily value of 225 W m^{-2} was observed on 12 January 1999, a partly cloudy day, the minimum (140 W m^{-2}) was observed on 11 January 1999, with complete overcast. The mean daily incident radiation was measured as 205 W m^{-2} and was strongly dependent on the amount of cloudiness.

For open water, albedo values varied between 7% and 16%. The exact value depended on the solar elevation, the absence or presence of clouds, and the sea-state. The sea-ice we encountered was normally covered with snow. Values of up to 80% were recorded as mean hourly values for 10/10 snow-covered sea ice. In Figure 9.4b the ice concentration, which was observed hourly, is plotted against the albedo. It can be seen that the albedo increases with increasing ice concentration. Ice concentration and albedo were averaged for 2-10 ice concentration classes. For open water, a mean value of about 12% was observed. This relatively high value may be because of long time periods with low solar elevation during which the surface reflectivity is increased. For 10/10 sea-ice concentrations, the mean albedo value was 61%. The mean value for the albedo for the whole observational period was 43%.

9.4.1.4 SURFACE ENERGY BALANCE OVER CONTINENTAL-SHELF AND BLUE-ICE

Mean surface energy fluxes for the period 19 January-28 February 1999 over snow and blue-ice are shown in Figure 9.5a. Due to the differences in albedo, the blue-ice surface has greater net shortwave radiation (S_{net}) at the surface. Averaged over the analysis period, surface temperature is slightly higher over blue ice than snow, leading to slightly greater longwave loses (L_{net}) over the blue-ice surface. For the snow surface, net longwave radiation is comparable with net shortwave radiation, thereby making the net radiation (R_{net}) very small. However, R_{net} over blue-ice is quite high and positive, making it a significant heat source term. Sensible heat flux over snow is on an average positive (the air warms the surface) in contrast to that over the blue-ice surface. We calculated a mean sensible heat flux of -13.4 W m^{-2} for the

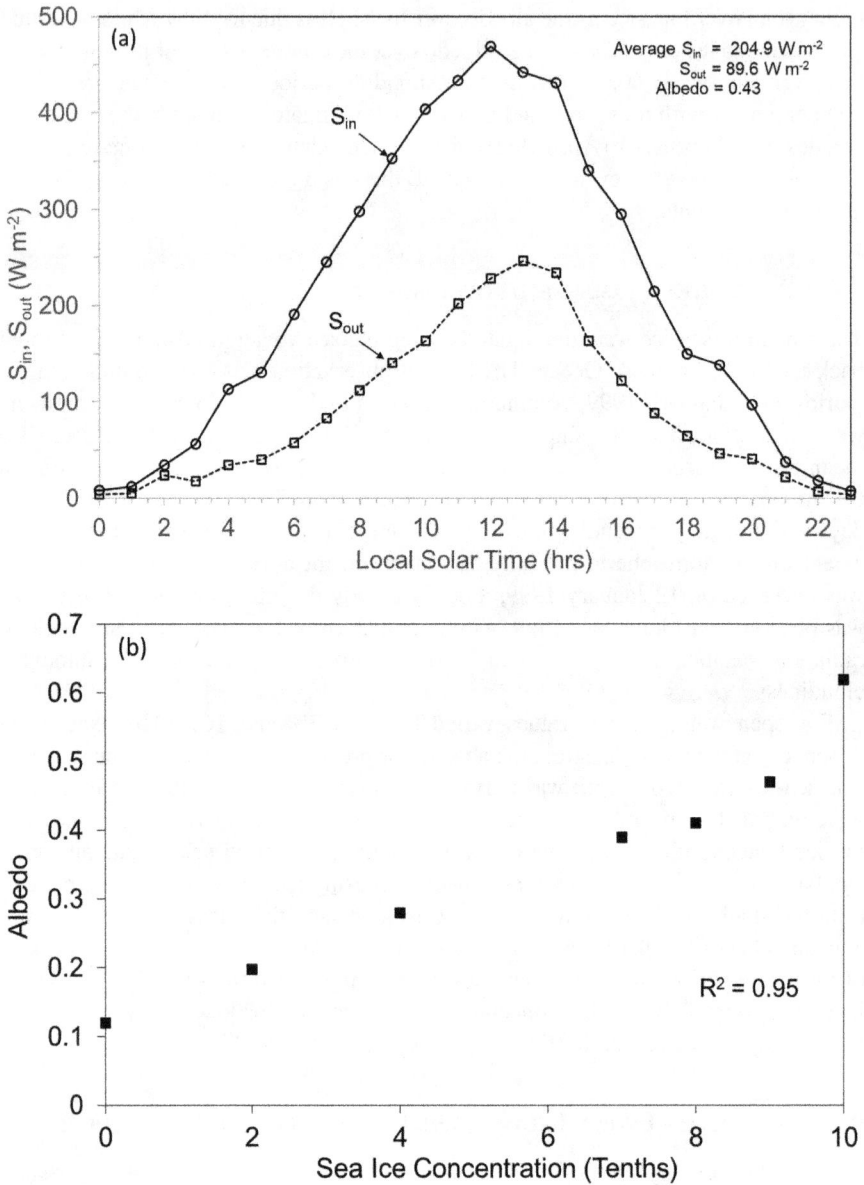

FIGURE 9.4 (a) The mean diurnal variation of the incident (S_{in}) and reflected (S_{out}) shortwave radiation over sea ice for the observation period and (b) The surface albedo as a function of sea ice concentration. Values are averaged as a function of ice concentration.

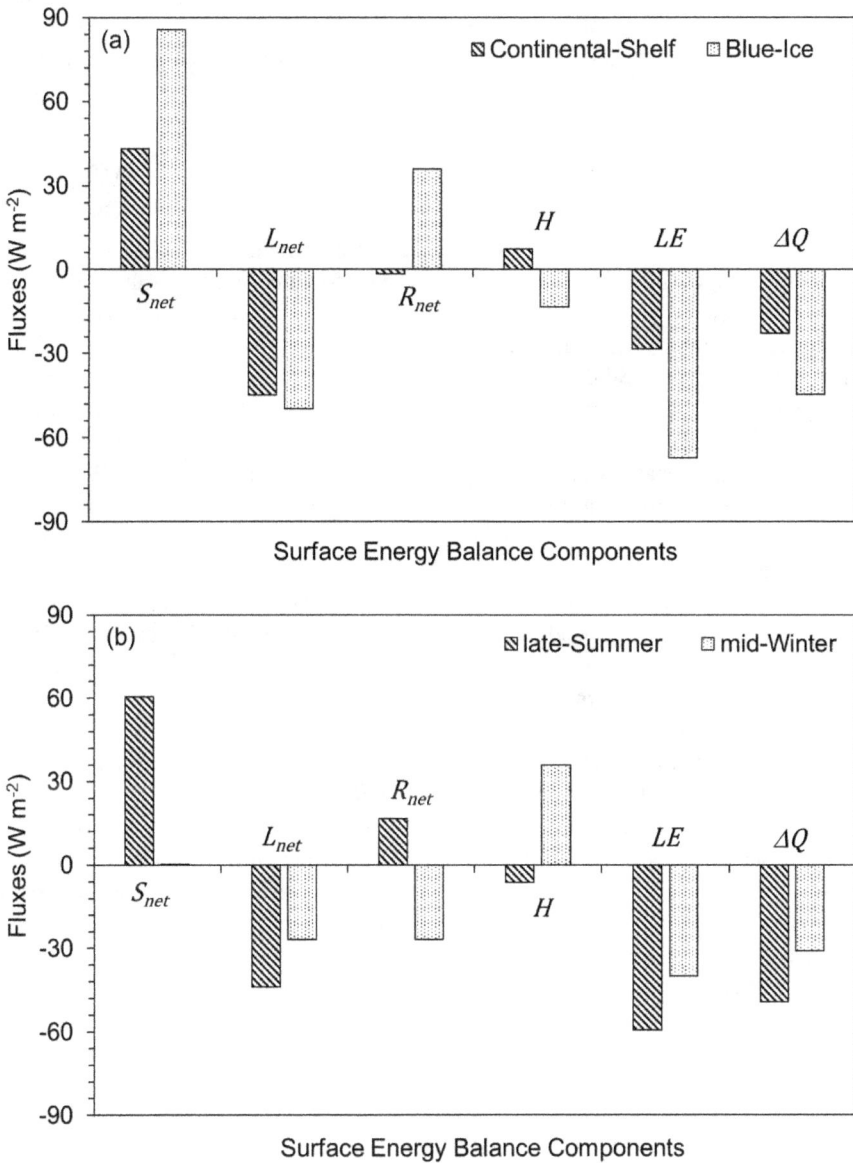

FIGURE 9.5   (a) Mean surface energy fluxes over blue-ice during 19 January–28 February 1999 and (b) Mean surface energy fluxes over blue-ice during late-summer (Jan–Feb 1999) and mid-winter (June–July 1999) periods.

blue-ice surface compared to the value of -7.2 W m^{-2} which Bintanja and Van den Broeke (1995) obtained for sensible heat for the blue-ice zone in Antarctica. The latent heat flux is on an average negative over both surfaces, indicating that evaporation is more prevalent than condensation. The latent heat flux over snow is much less than that over blue-ice. The net surface energy balance (ΔQ), which indicates the change in energy storage of the surface, might be used for phase change (melting or sublimation) or storage change (change of temperature of ice sheet). The values of ΔQ were found to be negative over both snow and blue-ice surfaces.

9.4.1.5   LATE-SUMMER AND MID-WINTER SURFACE ENERGY FLUXES OVER BLUE-ICE

Figure 9.5b show the components of surface energy balance over blue-ice for late-summer (average values for Jan–Feb 1999) and mid-winter (average values for June–July 1999) periods. The average value of net shortwave flux was found to be 60.4 W m^{-2} for late-summer, which decreases to almost zero during the mid-winter period when the sun was below the horizon. The longwave losses also decreased from an average value of –43.9 W m^{-2} for the late-summer to –26.8 W m^{-2} for the mid-winter period. The longwave losses are found to be strongly affected by the fractional cloud cover. While the mean value for complete overcast conditions was –26 W m^{-2}, the longwave losses more than doubled for clear-sky conditions with an average value of –60 W m^{-2}. The net radiation flux was positive during the late summer period and changes sign between seasons. The average value of R_{net} for February 1999 was 35.9 W m^{-2}, which reduced to –2.9 W m^{-2} in March 1999, with an average of 16.5 W m^{-2} for the late-summer period. During mid-winter the net radiation consists of only longwave fluxes and was a strong sink term. The average value of R_{net} during the mid-winter period was 26.8 W m^{-2}.

The sensible heat-flux also changes sign and shows an increasing trend between late-summer and mid-winter. The average value of H for February 1999 was –13.4 W m^{-2}, which increased to 0.8 W m^{-2} in March 1999 with an average of –6.3 W m^{-2} for the late-summer period. During February 1999, the chief turbulent production mechanism is buoyancy of the surface layer due to strong heating of the ice. Warming of ice is very effective through a large penetration depth for ice and low backscattering coefficient for the large grained blue-ice surface. Larger surface heating causes a longer period of unstable stratification, which is indicated by a negative sensible heat flux in February 1999. The average value of H during the mid-winter period (35.9 W m^{-2}) was quite high compared to the summer period. Thus, during the late-summer period sensible heat flux acts as a weak sink term, while in winter it is strongly positive and acts as a source term. The blue-ice surface gains latent heat when atmospheric water vapour condenses on it and loses heat when moisture evaporates from it. Figure 9.5b shows that latent heat flux is one of the dominant components of the energy balance. Latent heat flux was found to be negative in both summer and winter periods, though the mid-winter LE loses (–40.1 W m^{-2}) are less than those in the late-summer (–59.3 W m^{-2}) period.

9.4.1.6   DIURNAL VARIATIONS IN SURFACE ENERGY FLUXES OVER BLUE-ICE (FEBRUARY 1999)

The mean daily variation in each of the energy balance components is large and can vary considerably. The average diurnal cycle of shortwave and longwave radiation, sensible heat flux, latent heat flux, and surface energy budget for the month

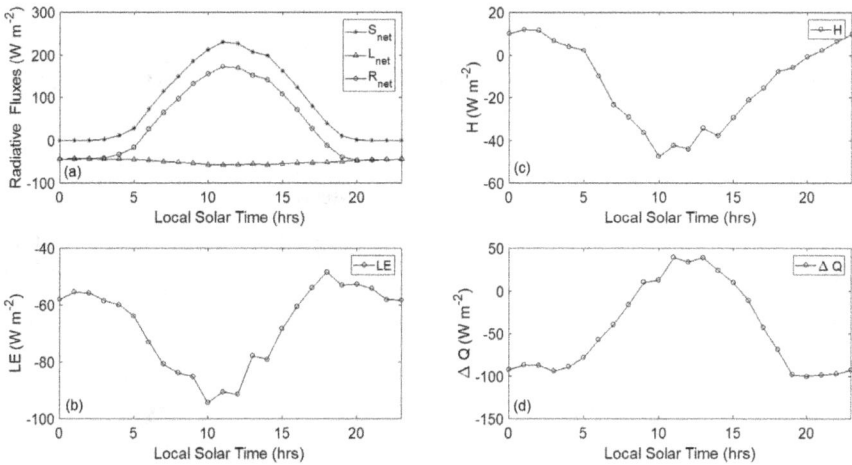

FIGURE 9.6   Mean diurnal cycle of surface energy balance components over blue-ice during Feb 1999: (a) net shortwave, net longwave and net radiation fluxes, (b) sensible heat flux, (c) latent heat flux, and (d) net energy balance.

of February at the blue-ice site is depicted in Figure 9.6(a–d). Shortwave radiation peaks at near 230 Wm^{-2} at 11:00 hrs (approximate local solar noon) and then drops to a minimum at about 20:00 hrs. This is the driving force for the latent and sensible heat fluxes, which show similar diurnal cycles. Latent heat varies from –55 to –90 W m^{-2}, with the greatest flux at solar noon when air temperature is highest and therefore drives the greatest vapour exchange. The large surface heating causes a longer period of unstable stratification indicated by negative sensible heat flux during daytime. Since the radiation fluxes are higher than the turbulent heat fluxes in February, it is obvious that the overall surface energy budget will also follow the diurnal pattern of net radiation. The net surface energy budget was positive between 09:00 hrs and 15:00 hrs UTC, i.e. the surface was gaining energy between this period. For the remaining part of the day the surface was continuously losing energy. The surface energy budget was found to fluctuate between 40 to –100 W m^{-2}.

9.4.2   SIX-YEAR (2006–2012) RECORD OF METEOROLOGICAL CONDITIONS AND SEB OVER BLUE-ICE

This section presents the temporal, seasonal, and diurnal variations in the meteorological parameters and surface energy balance components over blue-ice surface near Sankalp AWS. The energy fluxes are analysed for the summer season (November, December, January, February), winter season (May, June, July, August), transition period from summer to winter (March, April; hereafter referred to as T1), and transition period from winter to summer (September, October; hereafter referred to as T2). Table 9.1 shows the seasonal and annual mean of meteorological parameters and surface energy fluxes over blue-ice for 6 years. Figures 9.7 and 9.8 show the time-series of meteorological variables and surface energy fluxes from 1 March 2006 to

TABLE 9.1

Seasonal and Annual Mean of Meteorological Variables and Surface Energy Fluxes over Blue-ice During 2006–2012

Parameter	Transition 1 Mean	Winter Mean	Transition 2 Mean	Summer Mean	Annual Mean
T_a (°C)	−11.8	−17.3	−16.6	−4.4	−12.4
T_s (°C)	−12.4	−17.9	−17.2	−4.8	−12.9
RH (%)	50	49	45	58	51
u (m s^{-1})	11.7	11.9	9.4	8.9	10.5
Cloud amount (okta)	4.2	4.7	3.9	4	4.2
Albedo*	–	–	–	0.69	–
S_{in} (W m^{-2})	68	6	137	275	118
S_{out} (W m^{-2})	43	3	84	188	78
S_{net} (W m^{-2})	25	3	53	87	40
L_{net} (W m^{-2})	−51	−44	−52	−55	−50
R_{net} (W m^{-2})	−26	−41	1	32	−10
H (W m^{-2})	19	19	14	9	15
LE (W m^{-2})	−83	−60	−55	−89	−70
ΔQ (W m^{-2})	−90	−82	−40	−48	−65

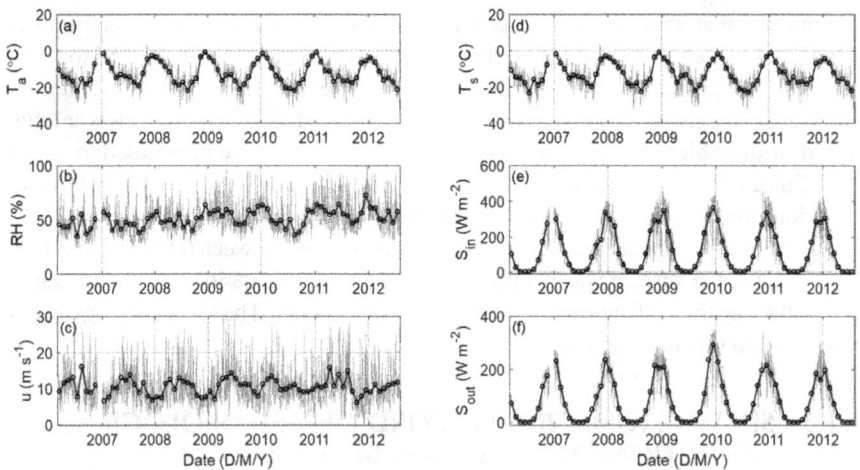

FIGURE 9.7 Daily averages of meteorological variables: (a) air temperature, (b) relative humidity, (c) wind speed, (d) surface temperature, (e) incident shortwave radiation, and (f) reflected shortwave radiation recorded over blue ice surface at Sankalp AWS location during Mar 2006 to July 2012. The dark black line shows the 30-day running means.

31 July 2012, respectively. Figures 9.9 and 9.10 depict the mean diurnal variations in meteorological variables and surface energy balance components for the summer, winter, T1, and T2 periods. The overall seasonal variations (averaged over 6 years during 2006–2012) in meteorological variables and surface energy fluxes are shown in Figure 9.11.

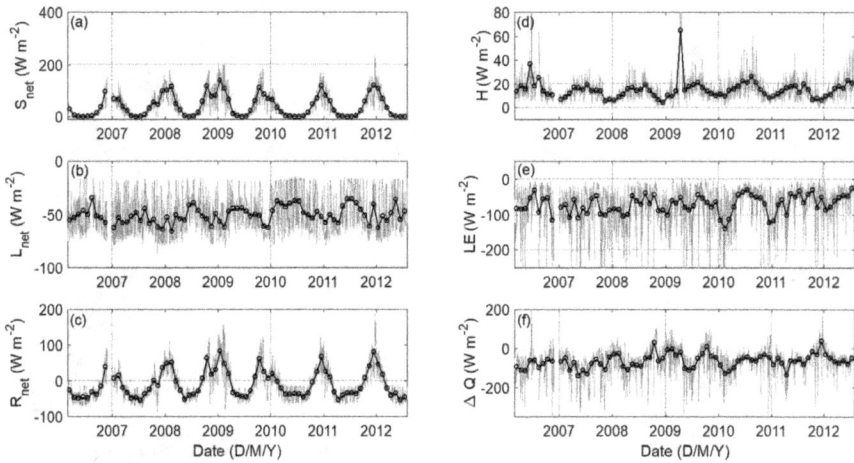

FIGURE 9.8   Daily averages of surface energy balance components: (a) net shortwave flux, (b) net longwave flux, (c) net radiation flux, (d) sensible heat flux, (e) latent heat flux, and (f) residual energy balance over blue ice surface at Sankalp AWS location during Mar 2006 to July 2012. The dark black line shows the 30-day running means.

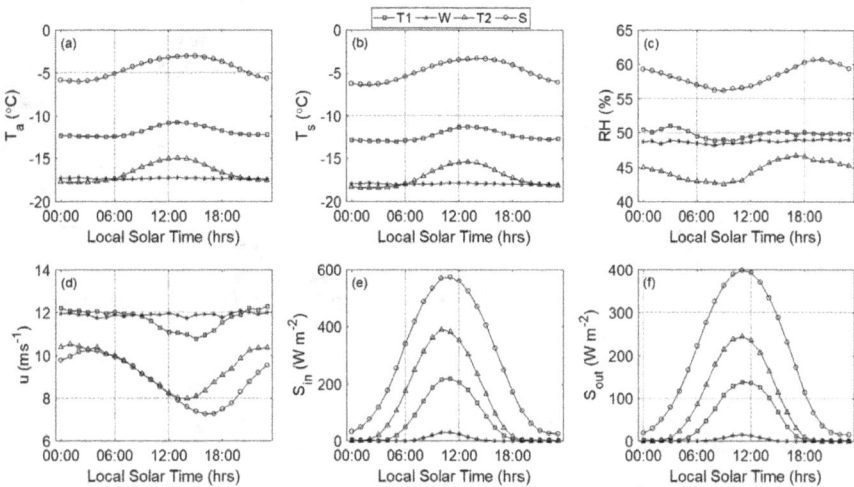

FIGURE 9.9   Diurnal variations in meteorological variables: (a) air temperature, (b) surface temperature, (c) relative humidity, (d) wind speed, (e) incident shortwave radiation, and (f) reflected shortwave radiation recorded over blue ice surface at Sankalp AWS location during Mar 2006 to July 2012.

9.4.2.1   Meteorological Parameters over Blue-Ice (2006–2012)

Daily averages of T_a and T_s at Sankalp AWS from 1 March 2006 to 31 July 2012 are shown in Figures 9.7a and 9.7d. A strong annual cycle with significant short-term variability was observed in the T_a and T_s for all 6 years. The maximum and minimum values of daily averaged T_a during the study period were found to be 5.0°C

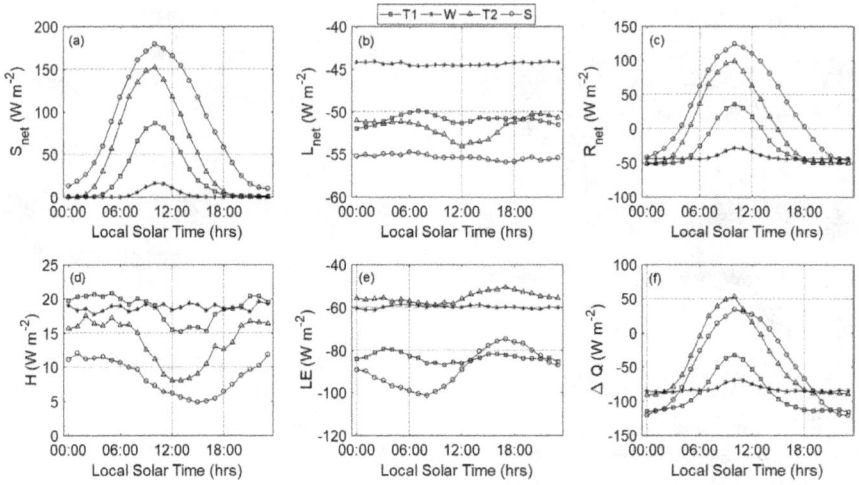

FIGURE 9.10 Diurnal variations in surface energy balance components: (a) net shortwave flux, (b) net longwave flux, (c) net radiation flux, (d) sensible heat flux, (e) latent heat flux, and (f) residual energy balance over blue ice surface at Sankalp AWS location during Mar 2006 to July 2012.

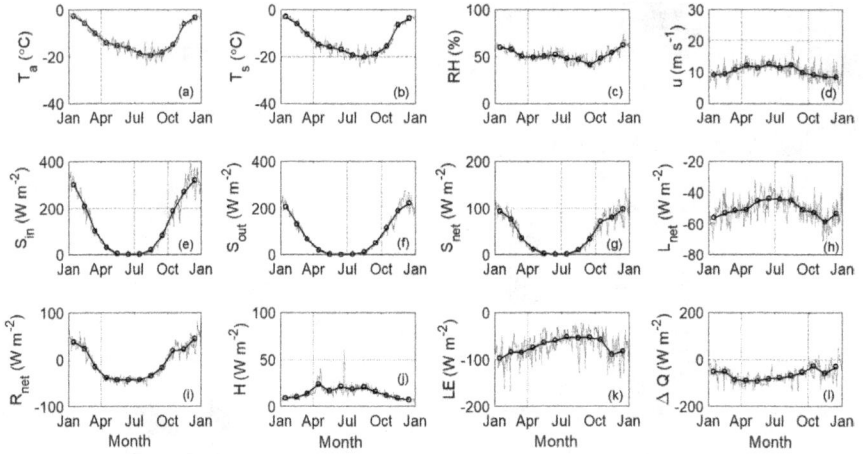

FIGURE 9.11 Overall seasonal variation in (a) air temperature, (b) surface temperature, (c) relative humidity, (d) wind speed, (e) incident shortwave radiation, (f) reflected shortwave radiation, (g) net shortwave radiation, (h) net longwave radiation, (i) net radiation flux, (j) sensible heat flux, (k) latent heat flux, and (l) residual energy flux (Q) over blue ice during 2006–2012.

and –30°C, respectively. Over the entire period, averages of T_a and T_s were –12.4°C ± 7.3°C and –12.9 °C ± 7.5°C, respectively. Mean diurnal variations in the T_a and T_s for different seasons are shown in Figures 9.9a and 9.9b. The amplitudes of diurnal variation of T_a and T_s were highest during the summer (2.98°C and 3.10°C) followed by

T2 (2.82°C and 2.99°C) and T1 (1.65°C and 1.76 °C) periods. The lowest amplitude was observed during the winter period (0.17°C and 0.21°C). The high amplitudes of diurnal variations in T_a and T_s during the summer and transitions periods were due to high diurnal variation in solar heating (Figure 9.9e). The maximum temperatures occurred in the diurnal cycle around local solar noon. The average values of T_a were higher compared to the average values of T_s in the diurnal cycle for all the seasons at the observation location, which contributed to the downward sensible heat flux. Overall seasonal variations in T_a and T_s indicate August and September as the coldest months during all 6 years (Figures 9.11a and 9.11b). This is because the energy loss by net longwave radiation (L_{net}) dominates over energy gained by net shortwave radiation (S_{net}) till the month of September (Figures 9.11g and 9.11h). From October onwards, net radiation becomes a heat source to the ice surface as energy gained by net shortwave radiation flux dominates energy loss due to net longwave radiation flux.

Daily mean relative humidity (*RH*) also exhibits significant short-term variability and seasonal cycle (Figure 9.7b). Daily mean *RH* values varied between 24–100% and monthly averaged *RH* varied between 35–73% during the study period. In general, the atmosphere over the blue-ice field is dry, with a 6-year mean *RH* of 51%. Relative humidity was higher during summer than in other seasons (Table 9.1). Figure 9.9c shows the mean diurnal variation in *RH* during different seasons. The amplitudes of the diurnal cycle for the summer, winter, T1, and T2 seasons were 4.5%, 0.9%, 2.3%, and 4.2%, respectively. High diurnal variation in RH during the summer and T2 seasons corresponds to high diurnal variation in incident radiation and lower katabatic winds (Figures 9.9e and 9.9d). Low diurnal variation during winter and T1 may be due to higher katabatic winds (Figure 9.9d) causing air moisture to be well mixed in the boundary layer. Relative humidity was high during summer months (Figure 9.11c) when there was warm air advection at the observation location which corresponds to frequent cyclonic events in the coastal regions.

Strong katabatic winds were observed at the Sankalp AWS location due to its topographic setting down a steep slope from the Antarctic Plateau. Variation of daily mean wind speed (*u*) is shown in Figure 9.7c. The wind is characterized by a short period of calm and light wind followed by a period of high wind. Generally, high-wind events correspond to the passage of low-pressure systems at the observation site. The maximum value of daily averaged *u* during the 6-year period was 30.5 m s^{-1} at the study site. Monthly averaged *u* varied from 6.1 m s^{-1} to 16.1 m s^{-1} with a 6-year average of 10.5 m s^{-1}. Figure 9.9d shows the diurnal variations in the wind speed during the seasons. The amplitude of the diurnal cycle was 2.96 m s^{-1} for summer, 0.33 m s^{-1} for winter, 1.49 m s^{-1} for T1, and 2.53 m s^{-1} for T2 seasons. High diurnal variation during summer and transition periods corresponds to high diurnal variation in insolation (Figure 9.9e). Insolation heats the blue-ice surface and destroys the surface temperature inversion, which results in weakening of katabatic forcing. Low wind speed in the afternoon during summer and transition periods corresponds to weakening of katabatic forcing, which may be due to break up of surface inversion. A strong seasonal cycle was observed in the wind speed and during the winter season the wind was high compared to other seasons (Table 9.1, Figure 9.11d).

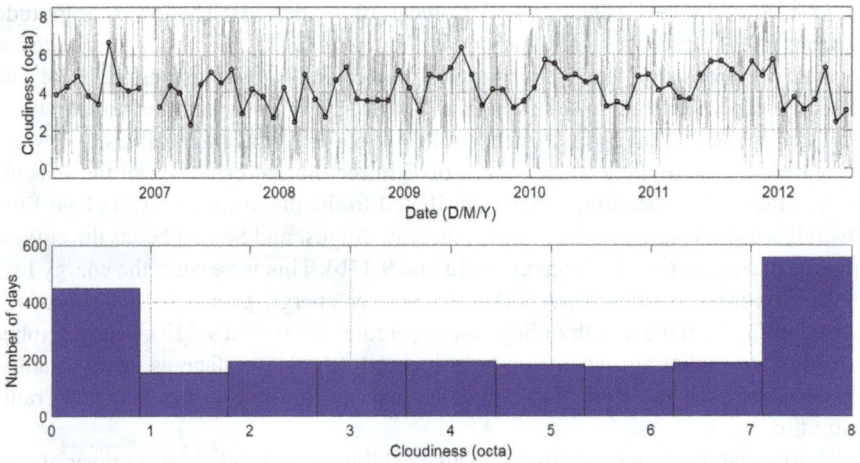

FIGURE 9.12 Daily variation of cloudiness over blue-ice at Sankalp AWS location during Mar 2006 to July 2012 and histogram of cloud amount. The dark black line shows the 30-day running means.

Daily variation of cloudiness at Sankalp AWS location during March 2006 to July 2012 along with a histogram of cloud amount is shown in Figure 9.12. Most of the days were observed to be partly cloudy to cloudy at the observation location and only 450 days were observed as cloud-free in the 6 years. Mean seasonal cloud amount was 4.2 octa, 4.7 octa, 3.9, octa and 4.0 octa during the T1, winter, T2, and summer seasons, respectively (Table 9.1).

Significant short-term variability and strong seasonality were observed in the incident and reflected shortwave radiation fluxes (S_{in} and S_{out}) at the study site (Figures 9.7e and 9.7f). While hourly (not shown) and daily averages of S_{in} varied up to 1078 W m^{-2} and 485 W m^{-2}, respectively, the corresponding hourly and daily averages of S_{out} varied up to 820 W m^{-2} and 349 W m^{-2}, respectively. The 6-yearly average values of S_{in} and S_{out} are calculated as 118 W m^{-2} and 78 Wm^{-2}, respectively. As expected, both S_{in} and S_{out} were highest during the summer season, followed by the T2 and T1 seasons, and were close to zero for the winter months (Table 9.1, Figure 9.11e, 9.11f). Figures 9.9e and 9.9f show the diurnal variation of S_{in} and S_{out} respectively during different seasons. The amplitudes of S_{in} diurnal cycle during the summer, winter, T1 and T2 seasons were 548 W m^{-2}, 30 W m^{-2}, 217 W m^{-2}, and 390 W m^{-2}, respectively. On the other hand, the amplitudes of S_{out} diurnal cycle during corresponding seasons were found to be 384 W m^{-2}, 15 W m^{-2}, 137 W m^{-2}, and 244 W m^{-2}, respectively.

The effective transmissivity of the atmosphere (τ_{eff}) is defined as the ratio of incident shortwave radiation S_{in} and extraterrestrial radiation I on a horizontal surface. The result of a number of atmospheric processes, i.e. scattering and absorption of solar radiation by air, aerosol, and clouds is included in τ_{eff}. Daily mean extraterrestrial radiation (I), was calculated at the Sankalp AWS location using mean Earth–Sun distance, solar constant, and solar zenith angle values during the study period. Figure 9.13a shows the variation of I along with daily mean incident shortwave

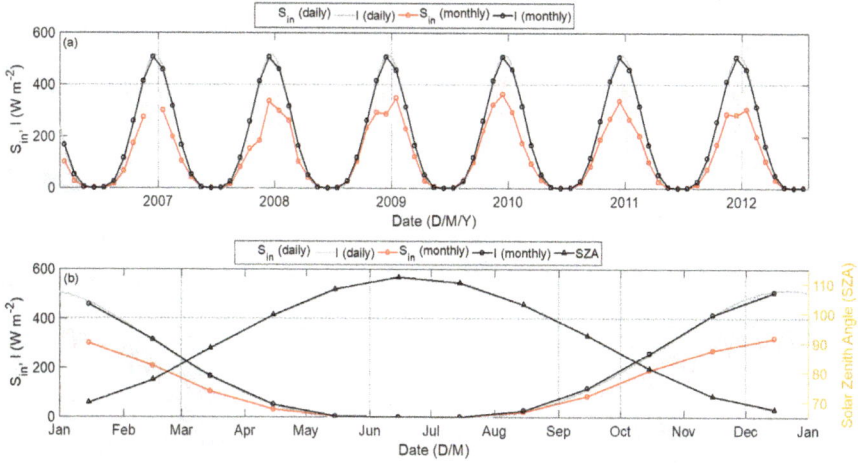

FIGURE 9.13   (a) Daily and monthly mean of incident shortwave radiation (S_{in}) and extra-terrestrial radiation (I) over blue ice at Sankalp location during Mar 2006 to July 2012. (b) Mean annual cycle (seasonal variation averaged over 2006–2012) of S_{in} and I along with solar zenith angle (SZA).

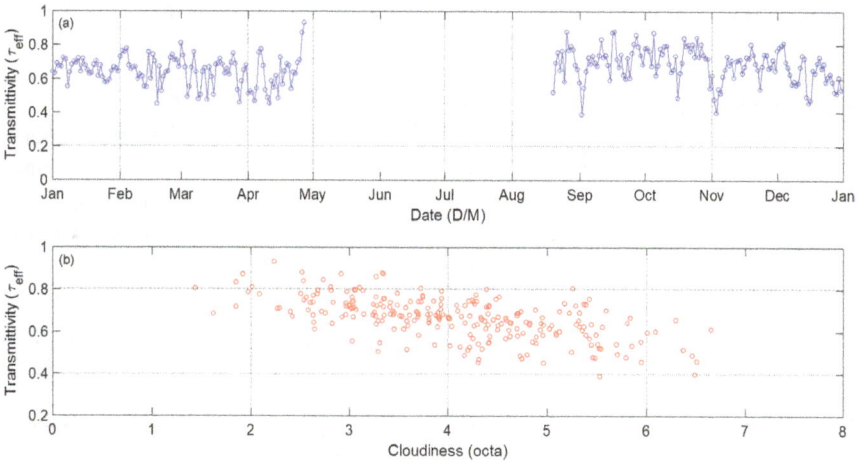

FIGURE 9.14   (a) Mean annual (averaged over 2006–2012) atmospheric transmissivity τ_{eff} over blue ice at Sankalp location. (b) Variation of τ_{eff} with cloudiness.

radiation S_{in} during March 2006 to July 2012. The mean annual cycle (averaged over 2006–2012) of S_{in} and I is strongly dependent on the solar zenith angle (SZA) and is depicted in Figure 9.13b. The overall seasonal variation of τ_{eff}, averaged over the study period, is in the range of 0.39–0.93 (Figure 9.14a). The variations in τ_{eff} are mainly due to differences in cloudiness, as shown in Figure 9.14b. The mean annual atmospheric transmissivity at the Sankalp AWS location was observed as 0.66, which

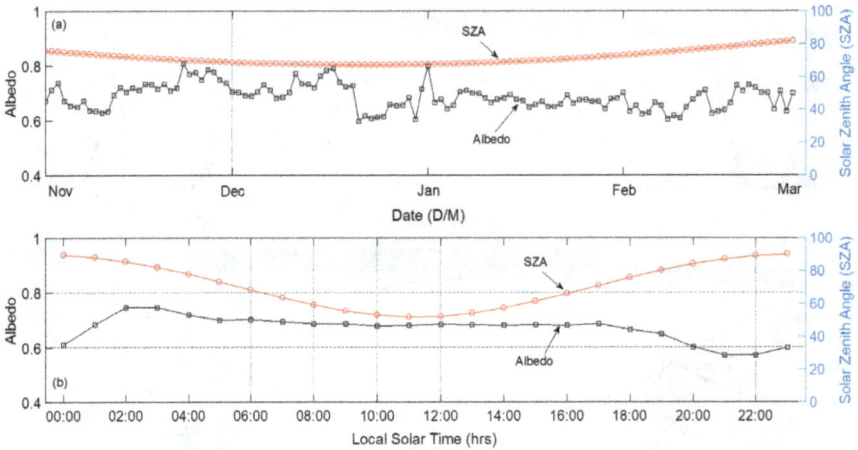

FIGURE 9.15 (a) Overall seasonal (daily mean from 2006–2012) variation of albedo over blue-ice at Sankalp AWS location along with solar zenith angle (SZA) for the period (November to March) when mean SZA < 80°. (b) Overall mean diurnal cycle of snow albedo along with SZA (average of November to March diurnal cycles during 2006–2012).

is comparable with the values reported for coastal regions in Antarctica by Van den Broeke et al. (2004).

Reflected shortwave radiation flux (S_{out}) mainly depends on S_{in} and surface albedo characteristics. Figure 9.15a show the overall daily mean albedo (averaged over 2006–2012) over blue-ice at Sankalp location for the period during November to February when SZA is < 80°. The albedo varied from 0.60 to 0.81 with a mean annual albedo of 0.69 over the 6 years. Figure 9.15b shows the overall mean diurnal cycle (average of November to February diurnal cycles during 2006–2012) of snow albedo along with SZA.

9.4.2.2 RADIATIVE AND TURBULENT ENERGY FLUXES OVER BLUE-ICE (2006–2012)

Figures 9.8a, 9.8b, and 9.8c show the temporal variation of net shortwave radiation flux (S_{net}), net longwave radiation flux (L_{net}), and net radiation flux (R_{net}) during the study period, respectively. Hourly and daily mean S_{net} varied up to 719 W m^{-2} and 236 W m^{-2}, respectively, with a 6-year average of 40 W m^{-2}. Mean seasonal S_{net} was computed as 87 W m^{-2}, 3 W m^{-2}, 25 W m^{-2}, and 53 W m^{-2} for the summer, winter, T1, and T2 periods, respectively (Table 9.1). Hourly L_{net} varied from –8 W m^{-2} to –89 W m^{-2} and the daily mean L_{net} varied from –13.6 W m^{-2} to –86 W m^{-2} with a 6-year average of –50 W m^{-2}.

Figures 9.10a, 9.10b, and 9.10c show diurnal variations of S_{net}, L_{net} and R_{net} respectively during different seasons. The amplitude of the S_{net} diurnal cycle was 170 W m^{-2}, 16 W m^{-2}, 86 W m^{-2}, and 152 W m^{-2} for the summer, winter, T1, and T2 periods, respectively. The amplitude of the L_{net} diurnal cycle was 1.2 W m^{-2}, 0.6 W m^{-2},

2.1 W m^{-2}, and 3.8 W m^{-2} for the summer, winter, T1, and T2 seasons, respectively. High diurnal variations in L_{net} during summer, T1, and T2 were due to high diurnal variations in the T_a and T_s during this period (Figures 9.9a and 9.9b). Strong diurnal variations were also observed in R_{net} with amplitudes of 170 W m^{-2}, 16 W m^{-2}, 87 W m^{-2}, and 150 W m^{-2} for the summer, winter, T1, and T2 seasons, respectively, and R_{net} was positive for about 14 hours, 6 hours, and 10 hours during the summer, T1, and T2 seasons, respectively, in the diurnal cycle (Figure 9.10c). High diurnal variations in R_{net} during the summer and transition periods were due to high diurnal variations in S_{net} during these periods.

Figures 9.11g and 9.11h show strong seasonal variability in S_{net} and L_{net}. L_{net} was high during the summer compared to other seasons, due to high surface temperature of the blue-ice. A strong seasonal cycle in R_{net} was also observed and during October, November, December, January, and February R_{net} was positive at the study location (Figure 9.11i). In fact, R_{net} changes from positive to negative twice a year over all 6 years (Figures 9.8c and 9.11i). At these points R_{net} is zero and S_{net} balances the L_{net}. Above these points S_{net} dominates L_{net} and blue-ice surface absorbs radiation. Absorbed radiation increases the blue-ice surface temperature and melting can start when the surface temperature reaches 0°C. Below these points L_{net} dominates the S_{net} and the blue-ice surface loses energy, tending to decrease the surface temperature. Table 9.1 indicates that R_{net} was the main heat source to the blue-ice during summer (32 W m^{-2}) and heat sink during winter (–41 W m^{-2}) and T1 (–26 W m^{-2}) periods (Table 9.1). Further, S_{net} almost balances L_{net} during the T2 period.

High temporal variability was observed in the sensible (H) and latent heat (LE) fluxes (Figures 9.8d, 9.8e). A strong seasonal cycle was observed in H during all 6 years. Sensible heat flux was lowest during the summer period compared to the winter and transition periods (Table 9.1, Figure 9.11j). The monthly average value of sensible heat flux varied from 4 W m^{-2} to 65 W m^{-2} with a 6-year average of 15 W m^{-2}. Figure 9.10d shows the diurnal variation of H during different seasons and the amplitude during the summer, winter, T1, and T2 periods was 7.2 W m^{-2}, 1.9 W m^{-2}, 5.6 W m^{-2}, and 9.5 W m^{-2}, respectively.

Monthly average LE varied from –25.6 W m^{-2} to –138.4 W m^{-2} with a 6-yearly mean of –70 Wm^{-2}. The highest LE was estimated for the summer period (seasonal mean –89 W m^{-2}) followed by T1 (seasonal mean –83 W m^{-2}), winter (seasonal mean –60 W m^{-2}), and T2 (seasonal mean –55 W m^{-2}) periods (Table 9.1). Comparatively low values of LE during the winter and T2 periods was due to low values of T_s during these periods. Figure 9.10e shows diurnal variation of LE during different periods and the amplitude of the diurnal cycle was 26.5 W m^{-2}, 2.6 W m^{-2}, 7.4 W m^{-2}, and 8.6 W m^{-2} for the summer, winter, T1, and T2 periods, respectively. High diurnal variation in LE during summer may be a result of high diurnal variation in surface temperature and wind speed in the corresponding season.

Net energy balance (ΔQ) at the blue-ice surface is the sum of R_{net}, H, LE, and sub-surface heat flux (G). During 19 January 2012–1 February 2012, direct measurement of the net sub-surface ice heat flux at Sankalp AWS location was carried out to estimate the magnitude of G. Two heat-flux plates (make Hukseflux) were placed horizontally at a depth of 10 cm below the blue-ice surface. The heat-flux plate

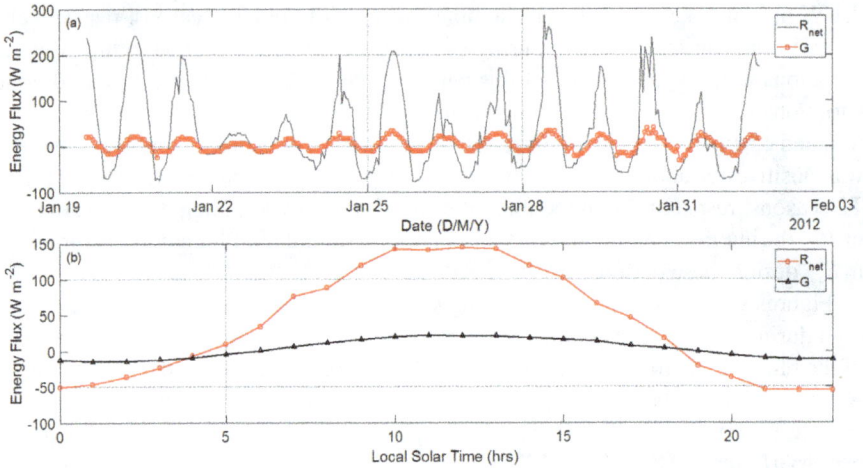

FIGURE 9.16 (a) Hourly mean net sub-surface ice heat flux (G) and mean net radiative flux (R_{net}) over blue-ice at Sankalp AWS location during 19 Jan–01 Feb 2012. (b) Mean diurnal variations in G and R_{net} over blue-ice during the measurement period.

measures the combined effect of net conductive and radiative energy flux across a cross-section in a horizontal plane. The temporal variation of hourly mean sub-surface ice heat flux and hourly mean net radiative flux is shown in Figure 9.16a. Hourly mean G varied from -32.1 W m^{-2} to 39.8 W m^{-2}. Mean diurnal cycle for both R_{net} and G are shown in Figure 9.16b, which indicates that net sub-surface ice heat flux is positive between 06:00 hrs and 18:00 hrs. The overall mean value of G for the measurement period was computed as -3.2 W m^{-2} which is very small and can be neglected for the calculation of ΔQ. Positive ΔQ is the energy available for melt if blue-ice surface is at 0°C. Figure 9.8f shows the strong temporal variability in ΔQ during the observation period. Daily mean ΔQ varied from -358 W m^{-2} to 226 W m^{-2} and monthly averaged ΔQ varied from -140 W m^{-2} to 39 W m^{-2} with a 6-year average of -65 W m^{-2}. The highest ΔQ was computed for the T1 (seasonal mean -90 W m^{-2}) followed by the winter (seasonal mean -82 W m^{-2}), summer (seasonal mean -48 W m^{-2}), and T2 (seasonal mean -40 W m^{-2}) periods. Figure 9.10f shows the diurnal variation of ΔQ during different seasons and amplitude during the summer, winter, T1, and T2 periods was 155 W m^{-2}, 17 W m^{-2}, 84 W m^{-2}, and 144 W m^{-2}, respectively.

9.5 CONCLUSIONS

Measurements of meteorological parameters over snow-covered continental-shelf, sea-ice, and blue-ice regions on the continental-ice in the general area of Dronning Maud Land, East Antarctica, helped us to describe the reflectance characteristics and surface energy balance of different snow-ice media in Antarctica. Albedo measurements during a sea voyage (8–13 January 1999) indicated a mean value of 12% for open water and 61% for 10/10 sea-ice concentrations. The mean albedo of continental-shelf (82.4%) was found to be much higher than that of blue-ice sur-face (60.9%) during January–February 1999. Energy balance analysis during 19 January–28 February 1999 indicated that: (i) net shortwave radiation flux (S_{net}) over blue-ice was signifiicantly higher compared to snow surface, (ii) greater (positive) net radiation flux (R_{net}) over blue-ice as compared to small negative values over snow surface, (iii) mean sensible heat flux (H) was positive over snow and negative over blue-ice, and (iv) latent heat flux (LE) was negative for both snow and blue-ice surfaces, but the magnitude of LE for snow surface was much smaller than that over blue-ice.

In order to more accurately characterize the seasonal and diurnal variations of radiative and turbulent energy fluxes over blue-ice surface, we further analysed the meteorological data recorded at Sankalp AWS over the 6-year measurement period (Mar 2006–Jul 2012). The surface energy fluxes were analysed for the summer season, winter season, and transition periods. The meteorological conditions over the blue-ice site during the 6-year analysis period are characterized by mild air tem-perature (annual mean –12.4°C), low relative humidity (annual mean 51%), and high katabatic winds (annual mean 10.5 m s⁻¹). Incident shortwave radiation flux during the T2 period (seasonal mean 137 W m⁻²) was high compared to the T1 period (68 W m⁻²). Net shortwave radiation flux was maximum during the summer (seasonal mean 87 W m⁻²) followed by T2 (seasonal mean 53 W m⁻²) and T1 (seasonal mean 25 W m⁻²) periods. The mean annual atmospheric transmissivity was 0.66 for the 6-year analysis period. Mean seasonal net longwave radiation flux varied from –44 W m⁻² (winter) to –55 W m⁻² (summer). Net radiative flux was positive for the summer (seasonal mean 32 W m⁻²) and negative for the winter (seasonal mean –41 W m⁻²) and T1 (seasonal mean –26 W m⁻²) periods. During the T2 period, net shortwave radiation flux almost balances the net longwave radiation flux. Sensible heat flux was positive throughout the year, with the lowest values during summer. Latent heat flux was observed to be highest during the summer (–89 W m⁻²) and lowest during the T2 (–55 W m⁻²) period.

Our analysis suggests that over blue-ice surface, net radiation flux is the dominant energy source and latent heat flux is the dominant energy sink in the summer season. On the other hand, sensible heat flux is the major energy source, while latent heat flux and net radiation are the major energy sinks in the winter and T1 periods. During the T2 period, average net radiation is close to zero, sensible heat flux is the major heat source, and latent heat flux is the major heat sink. In general, the sensible heat flux provides the net energy for the sublimation during winter and both transition periods. Meanwhile, in summer, the excess energy from the net radiation is used primarily for sublimation. The net energy balance (ΔQ) over blue-ice was found to be negative for all the four seasons.

ACKNOWLEDGEMENT

The authors are thankful to Dr P.K. Satyawali, Outstanding Scientist and Director Defence Geoinformatics Research Establishment (DGRE) for encouragement during the work. We great fully acknowledge Defence Research and Development Organisation (DRDO) for funding the project and NCAOR (National Centre for

Antarctic and Ocean Research) for logistic support. We would also like to thank all the erstwhile SASE members of Indian Scientific Expedition to Antarctica for sincere efforts in data collection and maintenance of AWS in Antarctica.

REFERENCES

Ambach, W., Kirchlechner, P. (1986) Nomographs for the determination of meltwater from ice and snow surfaces by sensible and latent heat. *Wetter Leben* 38: 181–189.

Bintanja, R. (1999) On the glaciological, meteorological and climatological significance of Antarctic blue ice areas. *Reviews of Geophysics* 37(3): 337–359.

Bintanja, R., Jonsson, S., Knap, W.H. (1997) The annual cycle of the surface energy balance of Antarctic blue ice. *Journal of Geophysical Research* 102(D2): 1867–1881.

Bintanja, R., Van den Broeke, M.R. (1994) Local climate, circulation and surface-energy balance of an Antarctic blue-ice area. *Annals of Glaciology* 20: 160–168.

Bintanja, R., Van den Broeke, M.R. (1995) The surface energy balance of Antarctic snow and blue ice. *Journal of Applied Meteorology* 34(4): 902–926.

Datt, P., Gusain, H.S., Das, R.K. (2015) Measurements of net subsurface heat flux in snow and ice media in Dronning Maud Land. Antarctica. *Journal of the Geological Society of India* 86: 613–619.

Dilley, A.C., O'Brien, D.M. (1998) Estimating downward clear sky long-wave irradiance at the surface from screen temperature and precipitable water. *Quarterly Journal of Royal Meteorological Society* 124: 1391–1401.

Dozier, J. (1980) A clear sky spectral solar radiation model for snow-covered mountainous terrain. *Water Resource Research* 16: 709–718.

Gusain, H.S., Mishra, V.D., Arora, M.K. (2014) A four-year record of the meteorological parameters, radiative and turbulent energy fluxes at the edge of the East Antarctic Ice Sheet, close to Schirmacher Oasis. *Antarctic Science* 26(1): 93–103.

Konzelmann, T., Braithwaite, R.J. (1995) Variations of ablation, albedo and energy balance at the margin of the Greenland ice sheet, Kronprins Christian Land, Eastern North Greenland. *Journal of Glaciology* 41: 174–182.

Mishra, V.D. (1999) Albedo variations and surface energy balance in different snow-ice media in Antarctica. *Defence Science Journal* 49(5): 347–362.

Moore, R.D. (1983) On the use of bulk aerodynamic formulae over melting snow Nordic. *Hydrology* 14: 193–206.

Moritz, R.E. (1978) *A model for estimating global solar radiation, Energy budget studies in relation to fast-ice breakup processes in Davis Strait.* Occasional Paper of University of Colorado, Institute of Arctic and Alpine Research 26: 121–142.

Oke, T.R. (1970) Turbulent transport near the ground in stable conditions. *Journal of Applied Meteorology* 9: 778–786.

Orheim, O., Lucchitta, B. (1990) Investigating climate change by digital analysis of blue ice extent on satellite images of Antarctica. *Annals of Glaciology* 14: 211–215.

Paterson, W.S.B. (1994) *The Physics of Glaciers*, 3rd ed., 480 pp. Oxford: Elsevier.

Prata, A.J. (1996) A new longwave formula for estimating downward clear-sky radiation at the surface. *Quarterly Journal of the Royal Meteorological Society* 122: 1127–1151.

Price, A.G., Dunne, T. (1976) Energy balance computation of snowmelt in a sub-Arctic area. *Journal of Resource* 12: 686–694.

Srivastava, P.K. (2002) A comparative study of glacio-meteorological parameters, reflectance and surface energy exchange over different snow-ice media in Dronning Maud Land in

East Antarctica. *Eighteenth Indian Expedition to Antarctica, Scientific Report*, Technical Publication Number 16: 153–190.

Upadhyay, D.S. (1999) *Cold Climate Hydrometeorology*, 218 pp. New Delhi: New AGE International.

Van den Broeke, M.R., Reijmer, C.H., Van As, D., Van de Wal, R.S.W., Oerlemans, J. (2005) Seasonal cycles of Antarctic surface energy balance from automatic weather stations. *Annals of Glaciology* 41: 131–139.

Van den Broeke, M.R., Reijmer, C.H., Van de Wal, R.S.W., Van As, D. (2004) A study of the Antarctic surface energy and mass balance using automatic weather stations. *Journal of Geophysical Research* 6: 07593.

Van de Wal, R.S.W., Russell, A.J. (1994) A comparison of energy balance calculations, measured ablation and melt water runoff near Sondre Stromfjord, West Greenland. *Global and Planetary Change* 91: 29–38.

10 Spatio and Temporal Variability of Physical Parameters in the Prydz Bay for Climate Change

S.M. Pednekar

10.1 INTRODUCTION

Antarctica is the coldest, windiest, and driest continent on Earth. The Antarctic weather condition is highly variable and can change dramatically in a short period of time. The temperature in summer can exceed +10°C near the coast and fall to below −40°C in winter. At times the resulting wind speeds can exceed 100 km/h for days. Antarctica has just two seasons, namely six months of daylight in its summer and six months of darkness in its winter. The seasons are caused by the tilt of the Earth's axis with the sun. The average annual temperature on the Antarctic coast ranges from −10°C to −60°C in the interior region. The inland temperature rises to about −30°C in summer but falls below −80°C in winter. The lowest temperature recorded was −89.2°C at Vostok station on 21 July 1983. The polar night occurs from late May to late July, and the polar day from late November to early February with the lowest monthly mean air temperature in July (−18.6°C) and the highest in January (0.4°C) (Ruibo, 2010).

Climate change refers to variations in physical parameters at considerably longer time and places. According to the IPCC (2007) report, climate change is defined as any change in climate over time; whether due to natural variability or as a result of human activities. The IPCC (2007) pointed out that, in the past 50 years the western coastline of the Antarctic Peninsula has undergone some of the severest warming globally. The warming over the eastern coastline of the Antarctic Peninsula is fairly slow, with the faster warming seasons being summer and autumn. The southern Indian Ocean is described in the IIOE (International Indian Ocean Expedition) Atlas by Wyrtki (1971) and the Soviet Antarctic Atlas (Tolstikov, 1966). The thermohaline circulation taking place on the continental shelf of Antarctica is on a large scale and the most active region of the world.

Prydz Bay is one of the regions undergoing the revolution of climate change and is an important area because of its V-shape and unique nature in the polar region. It is the third-largest embayment in the Antarctic continent (Figure 10.1) and lies in

DOI: 10.1201/9781003364115-10

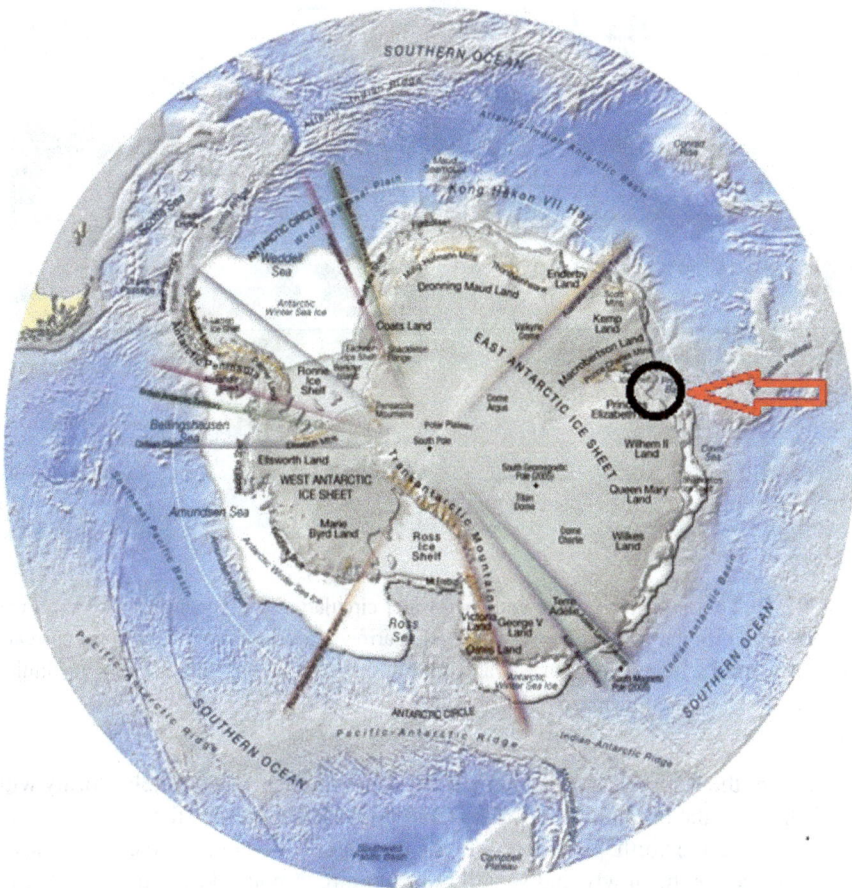

FIGURE 10.1 Antarctica is a landmass surrounded by the Southern Ocean. The red arrow indicates the Prydz Bay region within the circle. Magnification of the Prydz Bay in the circle is shown in Figure 10.4. The ocean seafloor topography is plotted.

the Indian Ocean sector. The southeastern coast of Prydz Bay is associated with the groups of islands known as the Larsemann Hills, and the Amery Ice Shelf, Princess Elizabeth Land, and East Antarctica (Chauhan et al., 2015). The annual mean temperature is increasing by 0.28°C/10 years and the winter mean temperature by nearly 0.3°C/10 years. The lowest monthly mean air temperature occurred in August and the highest in January. In summer, the surface radiation budget directly affects the change in physical parameters of the surface water. The open surface can absorb more energy from downward short-wave radiation as compared to that gained by the ice-covered surface. The hydrological balance is maintained by the variability of ice conditions temporally and spatially, particularly regarding the surface structure (Kupetskll, 1959; Denisov and Myznhikova, 1978). Prydz Bay continental shelf water is warmer and colder in deeper parts (Zverev, 1959, 1963; Izvekov, 1959).

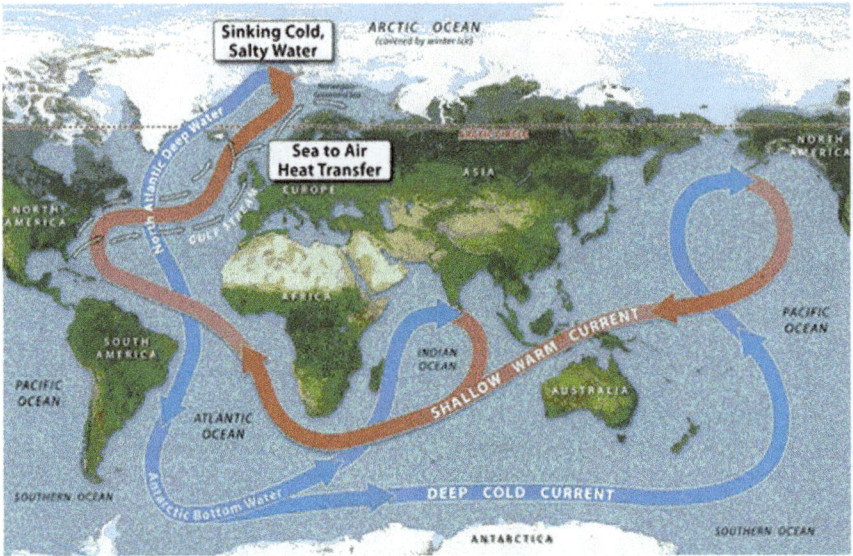

FIGURE 10.2 The major surface and deep water circulation components of the ocean that combined to form the global conveyor belt. Blue arrows represent deep-cold water currents, while red arrows represent surface warm currents. Credit to the Smithsonian Institution websites for providing the image.

North of the Antarctic continental slope the environment is extremely stormy with wind stress on the ocean toward the east and thus the transport in the surface Ekman layer is toward the north. To conserve mass, the water column below the Ekman layer moves to the south, slowly and steadily driving circumpolar deep water (CDW) up the continental slope that has been in the deep ocean for a long time (~1000 years), which is relatively salty, warm, low in oxygen, and rich in nutrients as compared to the Antarctic coastal waters.

Surface ocean currents are the main sources of deep water masses and the deepwater masses also feed into one another. The surface and deep ocean currents are an integrated system known as the global conveyor belt (Figure 10.2). This explains heat transport, bottom water ageing, and nutrient supply in the oceans. It also redistributes heat taken up from the Pacific through thermohaline circulation. Through thermohaline circulation ocean heat is transported over large geographic distances and it takes approximately 1000 years to complete a round trip.

In the last decade, the major challenges among researchers have been to understand change in the climate system which is one of the major physical processes that contributes to the world's oceans. Little is known about Prydz Bay as compared to the Weddell Sea and Ross Sea in terms of spatial and temporal variability due to a lack of data. In this chapter, spatial and temporal variability of the physical parameters in Prydz Bay are explained using long-term in situ climatology data collected by seals since 2004. The following sections include:

10.2 BACKGROUND

The Prydz Bay topographical features described are located in the Indian sector of the Southern Ocean occupying about 80,000 km^2 between 65°E and 80°E (Taylor and McMinn, 2002). It has a direct connection with the open water in the north adjacent to the Indian Ocean sector of the Southern Ocean. Towards the south, the borders are on the ice shelf and the Antarctic continent to its east and west. The water stretches more inland (southward) as compared to the Weddell Sea and Ross Sea. Prydz Bay is narrow in the southwest and wide in the northeast, with the furthest eastern end at 70°S, 76°E near Four Ladies Bank, and the furthest western end at 68°S, 69°E near Fram Bank. The bank's performs as a barrier to water exchange between the bay and the deep ocean (Smith and Treguer, 1994). The water on the continental shelf of the bay is shallow, with a depth from 400 m to 600 m (Alberts, 1995).

The majority the bay area is covered by ice about 2 m thick in winter. The Lambert Glaciers extend from the inland to the bay (Christie et al., 1990), and connect with Amery Ice Shelf. The ice becomes broken and partly molten, but the ice amount is still great and the floe-distributed area is changeable in summer (Dong et al., 1984). The water becomes sharply deeper in the north of the bay and has a flat bottom further north on the continental shelf which extends to the slope break at about 67°S. The depth of the water column is as deep as 3000 m or deeper, and is distributed to the north of the continental slope in the Southern Ocean. The Antarctic Coastal Current (CoC) and Antarctic Slope Current (ASC) flow westward around Prydz Bay and the Antarctic Circumpolar Current flows eastward. The cyclonic Gyre in Prydz Bay and the Antarctic Divergence Zone (ADZ) in the vicinity rotate anti-clockwise (Smith et al., 1984; Taylor and McMinn, 2002; Vaz and Lennon, 1996; Williams et al., 2016; Yabuki et al., 2006). Seasonal sea ice formed in the Prydz Bay grows up to ~58°S in austral winter and withdraws back to the continental shelf in austral summer, although some of the previous year's ice may remain in coastal areas (Smith and Treguer, 1994; Worby et al., 1998).

The geographical condition in the Prydz Bay region has special features that enable the shelf water of the bay to retain the memory of the extremely cold events which occur in the continent in winter for a longer time. Such memory can be evidenced by the bottom water with low temperature and high salinity on the shelf of the bay. The formation of the bottom waters is explained as follows. In late autumn or winter, the cold air cools the surface water and the strong wind above the sea surface causes deep convection in the shelf water so that the cooled and denser water near the sea surface subsides downward to the bottom and at the same time takes part in the coastal circulation which includes the wind drive current by the polar easterly. After the sea surface has become frozen, the ice-covered water can still be further cooled by surface ice rather than wind mixing. Furthermore, the salt-release process starts to take effect and make the salinity and density of the ice-covered water increase when surface water becomes frozen. Therefore, the extremely cold, salty, and dense water will go down to the bottom of the bay to form the shelf water which holds the memory of the extreme events in winter. During summer the density of the upper layer of water decreases through the effect of seasonal warming and the freshwater flux coming from the molten ice in the seasonal ice zone around Prydz Bay. The seasonal thermoclines are formed, and the vertical stability of the water column is increased. However, these surface changes cannot penetrate through the thermoclines and reach the lower layer because the wind force seasonally decreases and wind mixing becomes weaker. Therefore, the winter water or bottom water still bears the memory of the coldest events in the previous winter.

10.3   DATA AND METHODOLOGY

This chapter explains the spatial and temporal variability of the physical parameter using long-term data, and was possible only because of the data collected by the new techniques of instrumented elephant seals. Hydrographic profiles obtained with instrumented elephant seals analyze spatial and temporal variability of the physical parameter in Prydz Bay. Seal-derived data are making a growing contribution to climatologists in building upon existing oceanographic databases (Fabien et al., 2014), such as the World Ocean Database (Figure 10.3). The climatic factor is particularly responsible for the distribution of temperature and salinity near the ocean surface as a result of complicated air–sea–ice interactions. The hydrographic temperature and salinity parameters obtained from instrumented elephant seals were used to illustrate the variability in Prydz Bay (Figure 10.4). The calibration processing procedure for the seal CTD data is described in Roquet et al. (2014), with the temperature accuracy estimated to be within ±0.03°C and salinity within ±0.05. CTDSRDL (CTD as Conductivity, Temperature, and Depth; SRDL as Satellite, Relayed, Data, and Loggers) for the instrumental southern elephant seals measures vertical profiles of temperature and salinity during their foraging trips on the continental slope and shelf regions of Antarctica. The data processing steps involved calibration using delayed mode techniques and cross-comparison with the existing in situ profiles to establish similar protocols within the Argo community. All stations located in the region bounded 66°–70°S, 65°–85°E were utilized since 2004. The data format is provided following the Argo netCDF format so that it can be easily processed.

10.4   CLASSIFICATIONS OF WATER MASSES

The spatial distribution of various water masses identified based on the horizontal distribution of potential temperature (T) and salinity (S) diagram is shown in Figure 10.5 as a scatter plot during the austral summer and austral winter. Instrumented seal climatological profiles obtained during austral summer (November to April, Figure 10.5a) and austral winter (May to October, Figure 10.5b) seasons are used since 2004.

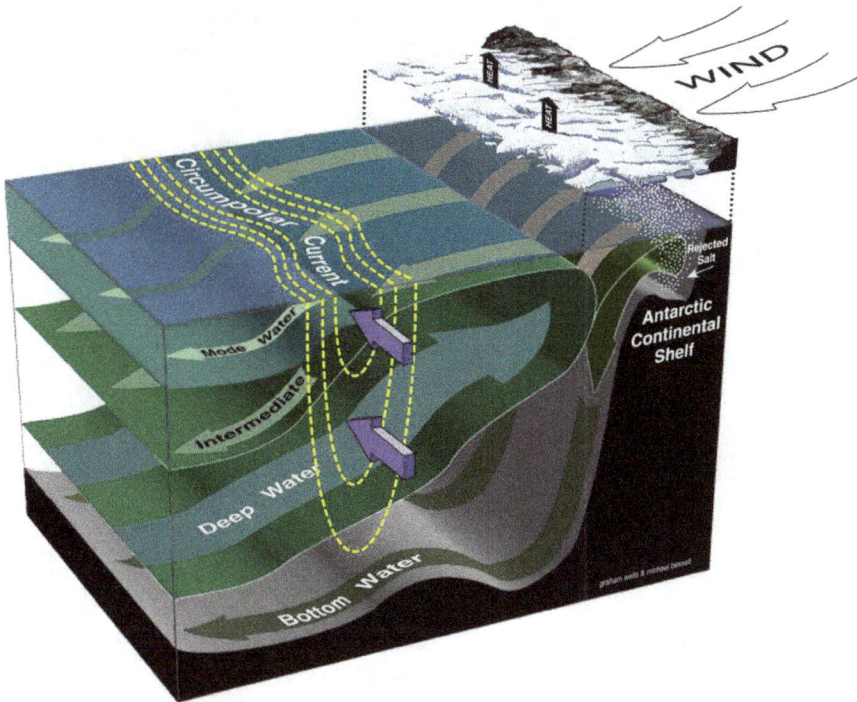

FIGURE 10.3 Simplified view of the Antarctic polar system showing subglacial hydrology along the Antarctic continental shelf and Southern Ocean circulation. Antarctic Bottom Water (AABW); Antarctic Circumpolar Current (ACC); Antarctic Slope Current (ASC) and Circumpolar Deep Water (CDW). (Figure adopted from CSIRO.)

Oceanic water mass south of the Polar Frontal Zone comprises three major source masses, namely Antarctic Surface Water (AASW), Circumpolar Deep Water (CDW), and Antarctic Bottom Water (AABW). Based on these primary sources, water masses identified in Prydz Bay and the nearby offshore region are AASW subdivided as Antarctic Summer Surface Water (AASSW) and Winter Water (WW), Shelf Water (SW) subdivided as High Salinity Shelf Water (HSSW) and Low Salinity Shelf Water (LSSW), modified SW (mSW), Ice Shelf Water (ISW), CDW, modified CDW (mCDW), and AABW. The TS characteristics of the different water masses have been explained in various earlier studies (Orsi and Wiederwohl, 2009; Smith et al., 1984; Whitworth et al., 1998; Wong et al., 1998). In this section, all the water mass TS characteristics are explained briefly.

10.4.1 AASW

AASW is identified above the thermocline with warmer surface water in summer and colder WW in winter with a temperature nearer to the freezing point. The WW formed during the previous winter is found below the seasonal thermocline. WW

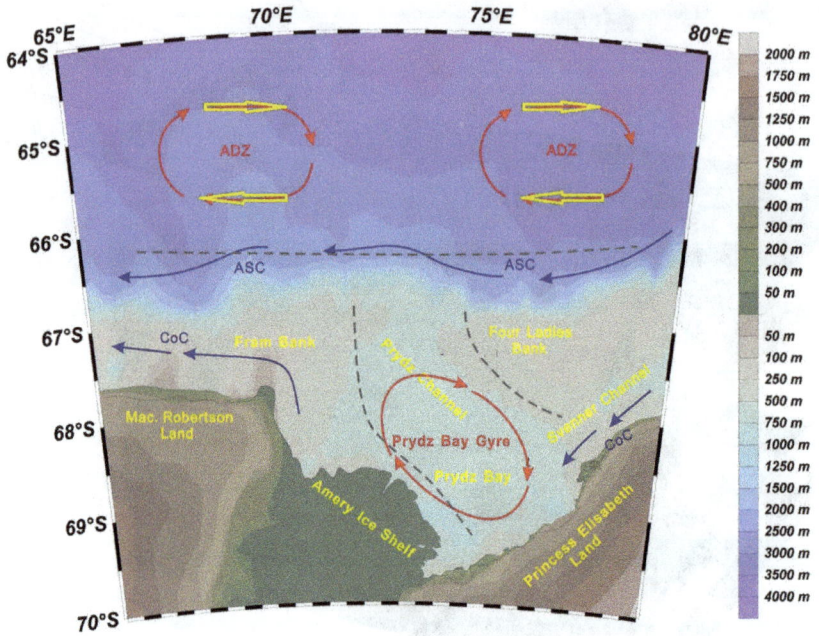

FIGURE 10.4 Map boundaries are 65°E to 80°E, 64°S to 70°S, and the circulation process involved in Prydz Bay, East Antarctica. The blue lines represent the Antarctic Coastal Current (CoC) and Antarctic Slope Current (ASC). The Prydz Bay Gyre and Antarctic Divergence Zone (ADZ) are denoted as red circles. The currents are traced following Cooper and O'Brien (2004) and Wu et al. (2017). The light gray dotted lines indicate the inner bay, banks, and deep ocean (Wang et al., 2015). CD: Cape Darnley; FLB: Four Ladies Bank. Locations of the Mackenzie Polynya (MP) and Davis Polynya (DP) are shown as cyan lines following Williams et al. (2016).

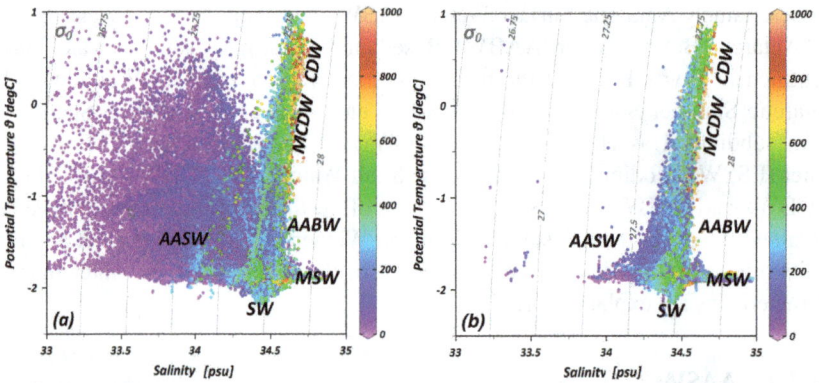

FIGURE 10.5 Climatology of potential temperature and salinity scatterplot with color bar indicating depth: (a) left panel for summer and (b) right panel for winter. Major water masses are labeled as Antarctic Surface Water (AASW), Modified Circumpolar Deep Water (mCDW/ CDW), Modified Shelf Water (mSW/SW), and Antarctic Bottom Water (AABW).

mass is not well defined close to the continental shelf. The AASW (Guo et al., 2019) is subdivided into AASSW ($-1.8°C < T < 2.1°C$, $30.6 < S < 34.2$ psu) and WW.

104.2   SW, mSW, and ISW

Mosby (1934) segregated continental SW as cold continental SW having low salinity and high salinity and near-freezing ISW. SW close to near-freezing temperature but varying salinity in the water mass in the shelf domain is described as LSSW and HSSW. Dense SW is referred to as HSSW, having a temperature range from $-1.95°C < T < -1.85°C$ and salinity $S \geq 34.5$. Whitworth et al. (1998) redefined the LSSW in Prydz Bay. LSSW and HSSW are usually separated at a salinity of 34.6 psu (Smith et al., 1984). Smith and Treguer (1994) draw the boundary between these two types at a salinity of 34.5 psu. Nunes Vaz and Lennon (1996) investigated the HSSW with a salinity of more than 34.6 psu. When the salinity of the SW is over 34.6 psu, the dense bottom water which is formed due to the mixing process can descend the slope of the continental shelf and is termed Prydz Bay Bottom Water (Middleton and Humphries, 1989).

The water in the bottom layer with a temperature above $-1.85°C$ has revealed the mixing of SW with the warmer water above. SW with near-freezing temperature – $1.89°C$, salinity 34.4 psu, and density equal to zero is constantly mixing vertically with the relatively warm inflow of mCDW above, producing transitional dense waters known as mSW. In the Weddell Sea, mSW produces AABW through a classic mechanism (Gill, 1973; Foster and Carmack, 1976). At the mouth of the Drygalski Trough, mSW was produced during the strongest spring tidal currents (Whitworth and Orsi, 2006; Muench et al., 2009; Padman et al., 2008).

SW colder than $-1.95°C$ interacts with the ice shelf from below and adopts this temperature as an upper limit to ISW (Sverdrup, 1940; Lusquinos, 1963). The signature of ISW at the bottom of the troughs is traced toward the sills as an intermediate temperature minimum core (Orsi et al., 2009). The spatial distribution of ISW was studied by Zheng et al. (2011) in the west near the Amery Ice Shelf and suggested its origin near Amery Ice Shelf as dense water. The distribution of ISW is restricted within the western Amery Basin. ISW in the eastern part of the bay is probably related to the West Ice Shelf (Yabuki et al., 2006). The ISW close to the Amery Ice Shelf is characterized as salty and cool due to the loss of heat and rejection of salt below the ice shelf. Prydz Bay topography depression together with Fram Bank and Four Ladies Bank control the exchange of water with the ocean, due to which they increased the effect of the deep waters within the bay. The deeper depth next to the Amery Ice Shelf leads to freezing below the shelf and the presence of cold and salty water (Morgan, 1972; Budd et al., 1982). ISW was found to be less in volume (about 2.4% of the total volume) and appears to have only a marginal influence on large-scale oceanography in terms of thermodynamic properties (Carmack, 1977).

10.4.3   CDW and mCDW

CDW is the most abundant water mass in the Southern Ocean which contributes 55% of the water from the Indian Ocean (Carmacks, 1977). According to Challahan (1972) the North Atlantic Deep Water is converted into CDW through a complicated process along isopycnal surfaces. Water columns are divided into three neutral density layers according to Jackett and McDougall (1997) and Orsi and Wiederwohl (2009). Whitworth et al. (1998) considered a density of 28.00 kg/m^3 as the lower limit of AASW since it is present near the warmest subsurface temperature maximum of CDW and, according to Orsi et al. (1999), the density of 28.27 kg/m^3 separates CDW from the AABW. Therefore the middle-density layer is occupied by CDW and mCDW in the north and south of the shelf break (Figures 10.5a,b). The characteristics of the CDW in Prydz Bay, according to Callahan (1972), are the relatively warm temperature (>0.5°C) and salinity (>34.65 psu) in deepwater and modified CDW (mCDW) as cold shelf water in the shelf region (Nunes Vaz and Lennon, 1996).

CDW was identified at a depth of ~400 m, having the warmest layer temperature of > 1.2°C (Whitworth et al., 1998). Separated from mCDW by a marked alongshore front (Jacobs, 1991), the Antarctic slope front, SW, is cooled almost down to the surface freezing point on the continental shelf. The mixing processes between CDW, SW, and AASW result in the production of a transition water mass, mCDW. The mixing of CDW, SW, and AASW produces mCDW with temperature ranges from $-1.85°C < T < -0.5°C$ and potential density between 27.72 kg/m^3 and 27.85 kg/m^3 (Herraiz-Borreguero et al., 2015, 2016; Wong et al., 1998).

10.4.4   AABW

AABW is generally recognized by temperatures below 0°C and salinity in the range of 34.60–34.72 psu (Gordon, 1971; Carmack, 1977). The key element in the formation of AABW is the possibility of CDW rising and extending onto the shelf (Gill, 1973; Foster and Carmack, 1976) and it may be responsible for breaking pack ice (Jacobs et al., 1970). AABW is formed due to the mixing of mCDW with salty shelf water (Foster and Carmack, 1976) and spread over the abyssal layer of the world ocean. Jacobs and Georgi (1977) identified cold and highly oxygenated water near the bottom of the continental slope at 60°E which is the source of AABW coming from Enderby Land/Prydz Bay. Orsi et al. (1999) inferred AABW formation based on the large-scale flow pattern and chlorofluorocarbon (CFC) distribution and inferred its formation not only in the Weddell Sea and Ross Sea but also in the Prydz Bay region. Jacobs et al. (1970) discovered regions having temperatures near the freezing point at the base of the Ross Ice Shelf and suggested the formation of AABW in summer due to the mixing of Ross Sea ISW and CDW having low salinity. Mosby (1934) showed that if continental shelf waters became sufficiently salty due to sea-ice formation, it will sink the continental slope due to greater density than bottom water (Deacon, 1937). The mixing processes are complex and have been undertaken in the Weddell Sea, the principal source region for Antarctic Bottom Water (e.g. Foster et al., 1987).

10.5   SPATIAL DISTRIBUTION OF MCDW

This section explains about spatial and temporal variability of the mCDW in the Prydz Bay region. Spatial distribution of mCDW is perceived using potential temperature and salinity diagram. The austral summer (November to April) and the austral winter (May to October) seasons are demonstrated in Figures 10.5a and 10.5b, respectively. Foster and Carmack (1976) considered −1.7°C isotherms as the lower limit of temperature for mCDW. The range of temperature varies from −1.7°C to 0.2°C and the neutral density is between 28.00 kg/m³ and 28.27 kg/m³. The spatial display of mCDW on average neutral density $\gamma N= 28.135$ kg/m³ is the intermediate value. Therefore, potential temperature and salinity are chosen on the average neutral density isopycnal surface $\gamma N= 28.135$ kg/m³ to analyze horizontal distribution during austral summer and austral winter. The distribution of potential temperature and salinity for austral summer is shown in Figure 10.6 (left panel) and for austral winter is shown in Figure 10.6 (right panel). The distribution of data in the Prydz Bay region in austral winter is less as compared to austral summer. In both seasons (summer and winter), mCDW was identified north of 67°S across a zonal section near the continental slope of Prydz Bay. The mCDW water mass is warmer and saltier when it enters the southeastern inner shelf region, as demonstrated by Lin et al. (2016) between 72°E to 75°E adjacent to 67.25°S.

The vertical sections of potential temperature for two zonal transects are shown in Figure 10.7 along 66.3°S (Figure 10.7a) and 67°S (Figure 10.7b) across Prydz Bay between 65°–85°E. The two transects are randomly selected occupying the slope and shelf region of Prydz Bay. The variability in the vertical structure of potential temperature was noticed in both transects. Along the 66.3°S zonal transect, the slope of Prydz Bay occupies CDW below 200 m between 65°E to 75°E below 28.00 kg/m³ isopycnals. The shelf region of Prydz Bay is occupied by mCDW along 67°S (Figure 10.7b). The presence of mCDW flows into Prydz Bay between 28.00 kg/m³ and 28.27 kg/m³ isopycnal surface between 66°E–70°E and 72°E–79°E, with the

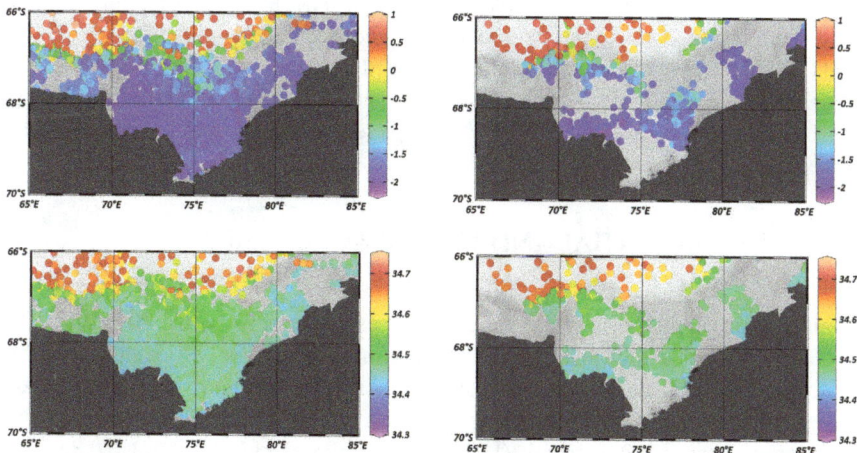

FIGURE 10.6 Climatological distribution of potential temperature and salinity on horizontal neutral density $\gamma N= 28.135$ kg/m³ isopycnal surface. The left panel is for summer and the right panel is for winter. Top row for potential temperature, and bottom row for salinity. A temperature of −1.7°C is considered the lower temperature limit of mCDW according to Foster and Carmack (1976).

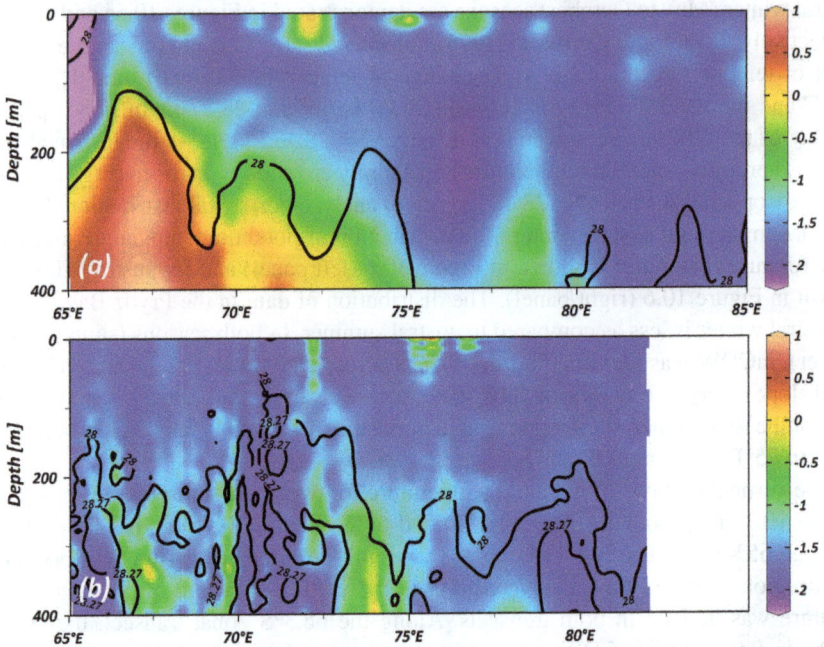

FIGURE 10.7 Vertical sections of potential temperature for two zonal transects: (a) 66.3°S and (b) 67°S during summer (November to April). The solid contour lines represent the $\gamma N = 28.00$ kg/m^3 and $\gamma N = 28.27$ kg/m^3 isotherms defining the boundary of mCDW.

slightly modified core of warm water below 200 m. The mCDW occurs further south below 100 m in pockets in the transect of 67°S between 72°E–78.5°E and 74°E (Lin et al., 2016). The warmer core identified near the surface at 75°E could be due to ocean ice–atmosphere interaction and the floe concentration at the sea surface (Pu et al., 2010). Thee intrusion of CDW is depicted in the vertical column of zonal transects.

10.6 ANNUAL SPATIAL AND TEMPORAL VARIABILITY OF MCDW

In this section, annual time scale variability is established using the 2011 and 2012 profiles obtained using instrumental seal climatology. The horizontal distribution of the potential temperature and salinity is shown in Figure 10.8 for 2011 and in Figure 10.9 for 2012 consecutive periods on selected isopycnal surface of $\gamma N = 28.00$ kg/m^3 (Figures 10.8a,b and 10.9a,b) and $\gamma N = 28.27$ kg/m^3 (Figures 10.8c,d and 10.9c,d). The isotherm -1.7°C is superimposed over the spatial distribution on potential temperature and salinity to highlight the extent of the mCDW signal to the interior of the bay as depicted in Figures 10.8 and 10.9, and there exists variability in the extension of mCDW each year in both isopycnal surfaces. In 2011, warm and saline

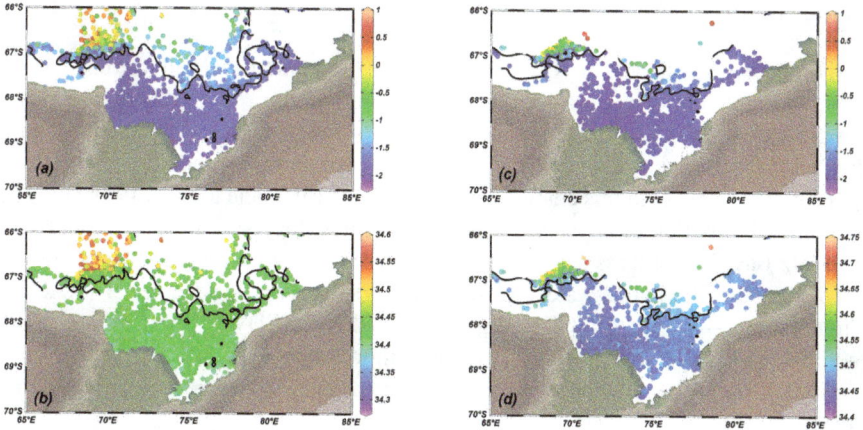

FIGURE 10.8 Horizontal distribution of the potential temperature (a, c) and salinity (b, d) on the neutral density isopycnal surface γN= 28.00 kg/m^3 and γN= 28.27 kg/m^3 for 2011. The thick dashed isotherm of -1.7°C is superimposed.

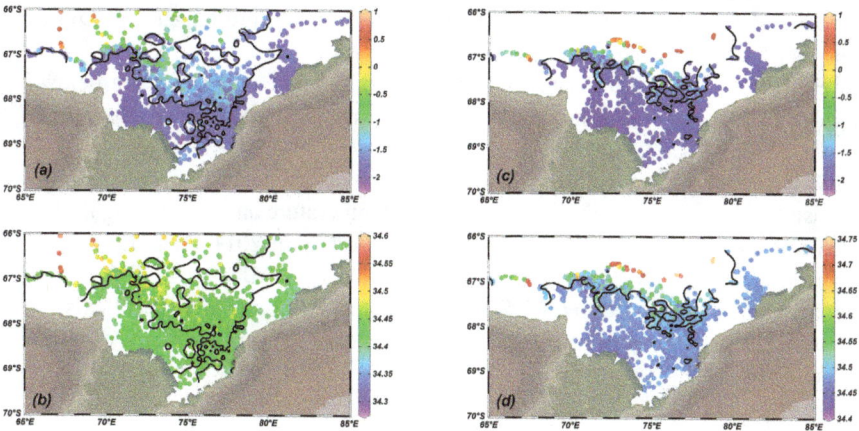

FIGURE 10.9 Horizontal distribution of the potential temperature (a, c) and salinity (b, d) on the neutral density isopycnal surface γN= 28.00 kg/m^3 and γN= 28.27 kg/m^3 for 2012. The thick dashed isotherm of -1.7°C is superimposed.

water is identified north of -1.7°C isotherm that is north of 67°S. The mCDW resides in the slope section between 67°E and 72°E. The mCDW occupies the shelf region between 73°E and 78°E north of 68°S. In 2012, warm and saline water occupied more of the interior region of Prydz Bay as -1.7°C isotherms extended more southward of 68.5°S with small pockets up to 69°S whereas, in Figures 10.9c,d, mCDW extended up to 68°S, which is less as compared to Figures 10.9a,b approaching toward the shelves and disappearing faster.

10.7 SUMMARY

CTD data were collected in the region using instrumented seals south of the polar front zone covered by the sea ice during the austral winter season. Instrumental seal data supported the scientific community to understand better the climate change process and fill important data gaps in the polar region, in studying the role of Antarctica in the global climate. In this chapter, an attempt has been made to show the spatio and temporal distribution of physical parameters and how they vary with space and time in Prydz Bay. The process seal data were analyzed to reveal the presence of various water masses. The major water masses, namely Antarctic Surface Water (AASW), Circumpolar Deep Water (CDW), modified CDW (mCDW), Shelf Water (SW), modified SW (mSW), Ice SW, and Antarctic Bottom Water (AABW), were identified during the summer and winter periods based on a potential temperature and salinity diagram.

Further, the spatial distribution of mCDW is perceived based on the horizontal distribution of the potential and salinity diagram. The analyses have shown the entrance of CDW into Prydz Bay near the shelf break during summer with warmer and saltier water characteristics being identified. In the vertical sections of potential temperature for two zonal transects, along the 66.3°S zonal transect, the slope of Prydz Bay occupies CDW below 200 m between 65°E and 75°E. The presence of mCDW flows onto Prydz Bay between 28.00 kg/m³ and 28.27 kg/m³ isopycnal surface between 66°E–70°E and 72°E–79°E and the slightly modified core of warm water below 200 m. The mCDW occurs further south below 100 m in pockets in the transect of 67°S between 72°E–78.5°E and 74°E

Over a year, scale variability was established using the 2011 and 2012 profiles. The isotherm –1.7°C distribution on potential temperature and salinity highlights the extent of the mCDW signal to the interior of the bay in 2011. In 2012, warm and saline water occupied the more interior region of Prydz Bay as –1.7°C isotherms extended more southward to 68.5°S with small pockets up to 69°S. There exists variability in the extension of mCDW each year in both isopycnal surfaces on a spatial and temporal scale.

ACKNOWLEDGEMENTS

The author is thankful to the Secretary, Ministry of Earth Sciences, and Director, National Centre for Polar and Ocean Research for their continuous support. The author expresses thanks also to Schlitzer, Reiner, and Ocean Data View, 2020 for providing a package for the interactive exploration, analysis, and visualization of oceanographic data. The marine mammal data were collected by the International MEOP Consortium and the national programs that contribute to it are gratefully acknowledged.

REFERENCES

Alberts, F.G. (1995) *Geographic Names of the Antarctic*, Second Edition. United States Board on Geographic Names.

Budd, W.F., Corry, M.J., Jacka, T.H. (1982) Results from the Amery Ice Shelf Project. *Annals of Glaciology* 3; 36–41.

Callahan, L.E. (1972) The structure and circulation of deep water in the Antarctic. *Deep-Sea Research* 19: 563–575.

Carmack, E.C. (1977) Water characteristics of the southern ocean south of the polar front. In: Angel, M. (Ed.) *A Voyage of Discovery*, Deacon 70th Anniversary volume, a supplement to Deep-Sea Research, pp. 15–61. Oxford: Pergamon Press.

CPCB (2002) *Climate Change: Special Issue*. Central Pollution Control Board, Delhi.

Christie, J., Bartholomew, J.C., Jones, R., Lewis, H.A.G., Lippard, S., Rothery, D., Whitehouse, D. (1990) *The Concise Atlas of the World*. London: Times Books Press.

Cooper, A.K., O'Brien, P.E. (2004) Leg-188 synthesis: Transitions in the glacial history of the Prydz Bay region, East Antarctica, from ODP drilling. In: A.K. Cooper, P.E. O'Brien, C. Richter (Eds) *Proceedings of the Ocean Drilling Program, Scientific Results*. Available from Ocean Drilling Program, Texas A&M University, College Station, TX, Vol. 188 [CD-ROM], 1–42.

de Boer, B., et al. (2015) Simulating the Antarctic ice sheet in the late-Pliocene warm period: PLISMIP-ANT, an ice-sheet model intercomparison project. *Cryosphere* 9: 881–903.

Deacon, G.E.R. (1937) The hydrography of the Southern Ocean. *Discovery Report* 15: 124.

Denisov A.S., Myznhikova, M.N. (1978) Osobennosti gidrologicheskogo rezhima v zalive Priuds (fevral' 1973 g.) (Features of the oceanographic regime in Prydz Bay, February 1973) *Trudy Sovetskoi Antarkticheskoi Ekspeditsii* 68: 100–105.

Dong, Z.Q., Smith, N.R., Kerry, K.R., Wright, S. (1984) *Water masses and circulation in the region of Prydz Bay, Antarctica. A collection of Antarctica Scientific Exploration*, No. 2. Beijing: Ocean Press, 1–24 (in Chinese).

Dutton, A. et al. (2015) Sea-level rise due to polar ice-sheet mass loss during past warm periods. *Science* 349: aaa4019.

Fabien, R., et al. (2014) *A southern Indian Ocean database of hydrographic profiles obtained with instrumental elephant seal. Scientific data 2014 Sep 2*, 1:140028. doi: 0.1038/sdata.2014.28. eCollection 2014

Foster, T.D., Foldwk, A., Mdleton, J.H. (1987) Mixing and bottom water formation in the shelf break region of the southern Weddell Sea. *Deep-Sea Research* 34: 1771–1794.

Foster, T.D., Carmack, E.C. (1976) Frontal zone mixing and Antarctic Bottom Water formation in the southern Weddell Sea. *Deep-Sea Research* 23: 301–317.

Gill, A.E. (1973) Circulation and bottom water production in the Weddell Sea. *Deep-Sea Research* 20: 111–140.

Gordon, A.L. (1971a) Recent physical oceanography studies of Antarctic waters. In: *Research in the Antarctic*, L. Quam, (Ed.), pp. 609–629. Washington D.C.: American Association for the Advancement of Science.

Gordon, A.L. (1971b) Oceanography of Antarctic waters. In: *Antarctic Oceanography I: Antarctic Research Series*, Vol. 15, pp. 169–203. J.L. Reid (Ed.). American Geophysical Union.

Gordon, A.L., Tchernia, P. (1972) Waters of the continental margin off Adelie Coast, Antarctica. In: *Antarctic Oceanography 11: The Australian-New Zealand Sector*. Antarctic Research Series, 9, pp. 59–69. D.E. Hayes (Ed.). American Geophysical Union.

Guijun Guo, J.S., Gao, L., Tamura, T., Williams, G.D. (2019) Reduced sea ice production due to upwelled oceanic heat flux in Prydz Bay, East Antarctica. *Geophysical Research Letters* 46(9): 4782–4789.

Haozhuang, W., Chen, Z., Wang, K., Liu, H., Tang, Z., Huang, Y. (2015) Characteristics of heavy minerals and grain size of surface sediments on the continental shelf of Prydz Bay: implications for sediment provenance. *Antarctic Science* 28: 103–114.

Herraiz-Borreguero, L., Coleman, R., Allison, I., Rintoul, S.R., Craven, M., Williams, G.D. (2015) Circulation of modified Circumpolar Deep Water and basal melt beneath the Amery Ice Shelf, East Antarctica. *Journal of Geophysical Research: Oceans* 120: 3098–3112.

IPCC (2007) *Climate Change 2007 The Physical Science Basis. Contribution of Working Group I to the Fourth Assessment Report of the IPCC.* Solomon, S., Qin, D., Manning, M., Chen, Z., Marquis, M., Averyt, K.B., Tignor, M. Miller, H.L. (Eds.). Cambridge University Press, Cambridge, UK.

Izvegov, M.V. (1959): Results of observations on currents in the region of the West ice Shelf. *Soviet Antarctic Expedition* 2: 9–93.

Jacobs, S.S., Amos, A.F., Bruchhausen, P.M. (1970) Ross Sea oceanography and Antarctic Bottom Water formation. *Journal of Geophysical Research* 17: 935–962.

Jacobs, S.S., Georgi, D.T. (1977) Observation on the southwest Indian Antarctic ocean In a voyage of discovery. M.V. Angel (Ed.) *Deep-Sea Research* (supplement) 24: 43–84.

Jacobs, S.S. (1991) On the nature and significance of the Antarctic Slope Front. *Marine Chemistry* 35(1–4): 9–24.

Jacobs, S.S., Weiss, R.F. (Eds.) *Ocean, Ice, and Atmosphere: Interactions at the Antarctic Continental Margin* 1–27. Washington, DC: American Geophysical Union.

Jackett, D.R., McDougall, T.J. (1997) A neutral density variable for the world's oceans. *Journal of Physical Oceanography* 27(2): 237–263.

Kupetskii, V.N. (1959): O prichinakh anomalii gidrologicheskikh uslovii zaliva Olaf Priuds (On the causes of anomalous hydrological conditions in Prydz Bay). *Izvestiya Vsesoyuznogo Geographicheskogo Obshchestvo (VGO)* 91: 356–357.

Li, W.R., Wang, W., Xiao, S., Ge, Z., Chen, W., Krijgsman. (2017) Productivity-climate coupling recorded in Pleistocene sediments off Prydz Bay (East Antarctica) *Palaeogeography, Palaeoclimatology, Palaeoecology* 485: 260–270.

Lin, L., Hongxi, C., Na, L. (2016) The characteristics of warm water inflowing and its temporal and spatial variation on the Prydz Bay continental shelf. *Antarctic Acta Oceanologica Sinica* 35(9): 51–57.

Lusquinos, A.J. (1963) Extreme temperatures in the Weddell Sea. *Arbok for Universitet: Bergen, Mathemetisk-naturvitenskapelig serie* 23: 19.

Morgan, V.L. (1972) Oxygen isotope evidence for bottom freezing on the Amery Ice Shelf. *Nature* 238: 393–394.

Mosbv, H. (1934) The waters of the Atlantic Antarctic Ocean. *Scientific Results of the Norwegian Antarctic Expeditions, 1927–1928* 1: 131.

Mosby, H. (1934) The waters of the Atlantic Antarctic Ocean. *Scientific Results of the Norwegian Antarctic Expeditions, 1927–1928, Oslo* 1: 131.

Middleton, J.H., Humphries, S.E. (1989) Thermohaline structure and mixing in the region of Prydz Bay Antarctica. *Deep-Sea Research* 36: 1255–1266.

Muench, R.D., Padman, L., Gordon, A.L., Orsi, A.H. (2009) Mixing of a dense water outflow from the Ross Sea, Antarctica: the contribution of tides. *Journal of Marine Systems* 76(1–2): 1–250.

Nunes Vaz, R.A., Lennon, G.W. (1996) Physical oceanography of the Prydz Bay region of Antarctic waters. *Deep-Sea Research Part I: Oceanographic Research Papers* 43(5): 603–641.

Orsi, A.H., Johnson, G.C., Bullister, J.L. (1999) Circulation, mixing, and production of Antarctic Bottom Water. *Progress in Oceanography* 43(1): 55–109.

Orsi, A.H., Wiederwohl, C.L. (2009) A recount of Ross Sea waters. *DeepSea Research Part II: Topical Studies in Oceanography* 56(13–14): 778–795.

Pattyn, F., et al. (2013) Grounding-line migration in plan-view marine ice-sheet models: results of the ice sea MISMIP 3d intercomparison. *Journal of Glaciology* 59: 410–422.

Padman, L., Howard, S.L., Orsi, A.H., Muench (2008) Tides of the northwestern Ross Sea and their impact on dense outflows of Antarctic Bottom Water. In: Gordon, A., Padman, L., Bergamasco, A. (Eds.) *Deep-Sea Research II* [doi:10.1016/jdsr2.2008.10.026]

Padman, L., Siegfried, M.R., Fricker, H.A. (2018) Ocean tide influences on the Antarctic and Greenland Ice Sheets. *Reviews in Geophysics* 364: 821–844.

Roquet, F., Williams G., Hindell M.A., Harcourt R., McMahon C.R., Guinet C., Charrassin J.B., Reverdin G., Boehme L., Lovell P., Fedak M.A. (2014) A Southern Indian Ocean database of hydrographic profiles obtained with instrumented elephant seals. *Nature Scientific Data* 1; 140028.

Ruibo, L. (2010) Annual cycle of land-fast sea ice in Prydz Bay East Antarctica. *Journal of Geophysical Research* 115(C2).

Shaojun, Z., Jiuxin, S., Yutian, J., Renfeng, G.E. (2011) Spatial distribution of ice shelf water in front of the Amery Ice Shelf, Antarctica in summer. *Chinese Journal of Oceanology and Limnology* 29(6): 1325–1338.

Shuzhen, P., Renfeng, G., Zhaoqian, D. (2010) Variability of marine hydrological features at the northern margin of Amery Ice Shelf. *Chinese Journal of Polar Research* (in Chinese) 22(3): 244–253.

Smith, N., Dong, Z., Wright, S. (1984) Water masses and circulation in the region of Prydz Bay. *Antarctica Deep-Sea Research Part A* 31: 1121–1147.

Smith N.R., Zhaoqian, D., Kerry, K.R., Wright, S. (1984) Water masses and circulation in the region of Prydz Bay Antarctica. *Deep-Sea Research* 31: 1121–1147.

Smith, N.R., Treguer, P. (1994) *Physical and Chemical Oceanography in the Vicinity of Prydz Bay, Antarctica.* Cambridge University Press: Cambridge.

Sverdrup, H.U. (1940) *Hydrology, Section 2, Discussion.* Reports of the B.A.N.Z. Antarctic Research Expedition 1921–1931, Series A, 3, Oceanography, Part 2, Section 2, pp. 88–126.

Takashi, Y., Suga, T., Hanawa, K., Matsuoka, K., Kiwada, H., Watanabe, T. (2006) Possible source of the Antarctic Bottom Water in the Prydz Bay region. *Journal of Oceanography* 62: 649–655.

Taylor, F., Mcminn, A. (2002) Late Quaternary Diatom Assemblages from Prydz Bay, Eastern Antarctica. *Quaternary Research* 57: 151–161.

Tolstikov, E.E. (1966) Atlas Antarktiki, Vol. I, G.U.C.K., Moscow (English translation. *Soviet Geography: Reviews and Translations* 8(5–6): 225 pp, American Geographical Society, New York, 1967.

Vaz, R.A.N., Lennon, G.W. (1996) Physical oceanography of Prydz Bay region of Antarctic waters. *Deep-Sea Research (I)* 43(5): 603–641.

Williams, G.D., Herraiz-Borreguero, L., Roquet, F., Tamura, T., Ohshima, K.I., Fukamachi, Y., Fraser, A.D., Gao, L., Chen, H., McMahon, C.R., Harcourt, R., Hindell, M. (2016) The suppression of Antarctic bottom water formation by melting ice shelves in Prydz Bay. *Nature Communications* 7: 1–9.

Whitworth, T., Orsi, A.H., Kim, S.-J., Nowlin Jr., W.D., Locarnini, R.A. (1998) Water masses and mixing near the Antarctic slope front. In: Jacobs, S.S., Weiss, R.F. (Eds.) *Ocean, Ice, and Atmosphere: Interactions at the Antarctic Continental Margin*, pp.1–27. American Geophysical Union: Washington, DC.

Whitworth, T., Orsi, A.H. (2006) Antarctic bottom water production and export by tides in the Ross Sea. *Geophysical Research Letters* 33(12): 1–4.

Wong, A.P.S., Bindoff, N.L., Forbes, A. (1998) Ocean-ice shelf interaction and possible bottom water formation in Prydz Bay, Antarctica. In: S.S. Jacobs, R. F. Weiss (Eds.) *Ocean, Ice, and Atmosphere: Interactions at the Antarctic Continental Margin*, pp. 173–187. Washington, DC: American Geophysical Union.

Worby, A.P., Massom, R.A., Allison I., Lytle, V.I., Heil, P. (1998) East Antarctic sea ice: a review of its structure, properties and drift. *Antarctic Sea Ice Properties, Processes and Variability* 74: 41–67.

Wong, A.P.S., Bindoff, N.L., Forbes, A. (1998) Ocean-ice shelf interaction and possible bottom water formation in Prydz Bay, Antarctica. *Ocean, Ice, and Atmosphere: Interactions at the Antarctic Continental Margin, Antarctic Research Series* 75: 173–187.

Wyrtiki, K. (1971) Oceano*graphic Atlas of the International Indian Ocean Expedition.* National Science Foundation: Washington, D.C.

Yabuki, T., Suga, T., Hanawa, K., Matsuoka, K., Kiwada, H., Watanabe, T. (2006) Possible source of the Antarctic Bottom Water in Prydz Bay region. *Journal of Oceanography* 62(5): 649–655.

Zverev, A.A. (1959) Anomalous seawater temperatures in Olaf Prydz Bay. *Soviet Antarctic Expedition* I: 269–270.

Zverev, A.A. (1963) Currents in the Indian sector of the Antarctic. *Trudy Sovetskoi Antarkticheskoi Ekspeditsii* 17I: 144–155.

11 Understanding the Predictability of East Arctic Sea Ice in a Warming Environment

Suneet Dwivedi

11.1 INTRODUCTION

Sea ice is an important component of the Earth's climate system. It influences the Earth's heat budget and dynamically interacts with the atmosphere, ocean, and land. The global biogeochemical cycles are also affected by changes in sea ice (Vancoppenolle et al., 2013). Global warming has drastically impacted the Arctic sea ice, as a result of which the world's climate and socio-economic activities in the region are being significantly affected (Francis and Vavrus, 2012; Smith and Stephenson, 2013; Wei et al., 2020). Arctic sea ice is rapidly declining with an increase in greenhouse gas (GHG) concentration (Stroeve and Notz, 2018; Serreze and Meier, 2019). Research has argued that the increased GHG concentration will eventually lead to a seasonally ice-free Arctic Ocean by the middle of the 21st century or earlier (Melia et al., 2015; Onarheim et al., 2018; Screen and Deser, 2019; Senftleben et al., 2020).

The Coupled Model Intercomparison Project (CMIP) is an important modelling resource which not only helps us in evaluating the past changes in the Arctic sea ice but also allows us to investigate the future sea ice variability in a warming environment represented by varying levels of GHG concentrations (Massonnet et al., 2012; Melia et al., 2015; Olonscheck and Notz, 2017; Stroeve et al., 2014). Stroeve et al. (2012) found that the CMIP Phase 5 (CMIP5) models perform better than the CMIP Phase 3 (CMIP3) models in simulating the Arctic sea ice extent. Shu et al. (2020) assessed the performance of the state-of-the-art CMIP Phase 6 (CMIP6) models in simulating sea ice variability. The seminal work of Notz et al. (2016) related to the Sea-Ice Model Intercomparison Project (SIMIP) was extended by the Notz & SIMIP Community (2020) for evaluating the fidelity of CMIP6 models. These studies have carried out a detailed process-based evaluation of sea ice simulations in the CMIP6 models and compared the results with CMIP5 models as well as with observations.

India's permanent Arctic research station "Himadri" is located in the eastern part of the Arctic at 11°56′E; 78°55′N. Even though there have been several modelling

DOI: 10.1201/9781003364115-11

studies in which the Arctic sea ice simulation and variability have been studied for the historical and future projection scenarios of the CMIP3, CMIP5, and CMIP6 models, to the best of our knowledge, examples of such studies in the eastern Arctic region covering the Indian station Himadri are not available. This study attempts to evaluate the fidelity of state-of-the-art CMIP6 models in correctly capturing the sea ice area (SIA) of the eastern Arctic region by comparing the model simulations with the corresponding satellite observations. The changes in the SIA of the east Arctic region shall be investigated by utilizing the CMIP6 models for historical and SSP5-8.5 future projection scenarios.

The nonlinear dynamical system techniques have been extensively applied to the meteorological and geophysical time series data for a wide range of uses (Mittal et al., 2003; Rangarajan and Sant, 2004; Dwivedi et al., 2007; Dwivedi, 2012; Baranowski et al., 2015; Hou et al., 2018; Kumar et al., 2018; Agbazo et al., 2019; Chandrasekaran et al., 2019; Kimothi et al., 2019; Mishra et al., 2021; Pandey et al., 2021). These techniques have also been used to estimate the predictability and fractality of the Arctic and Antarctic sea ice data (Chmel et al., 2005; Agarwal et al., 2012; Moon et al., 2019; Rampal et al., 2019). It will be worthwhile to quantify the predictability of the SIA of the east Arctic region (which covers the Indian Arctic station Himadri) in a warming world. One of the objectives of this chapter is to use the historical data of CMIP6 models for quantifying the predictability of the SIA of the east Arctic region in terms of generalized Hurst exponent (Di Matteo et al., 2003) and Climate Predictability Index (Rangarajan and Sant, 2004). The effect of climate change on the predictability of the SIA shall be investigated by analysing the SSP5-8.5 future projection data of CMIP6 models. The predictability of the SIA at the Himadri Indian Arctic station shall also be evaluated for historical and SSP5-8.5 data. We organize the chapter as follows. A brief description of the CMIP6 data and methodology is given in Section 11.2. Section 11.3 describes the simulation of the SIA and its predictability in the eastern Arctic region. The conclusions are presented in Section 11.4.

11.2   DATA AND METHOD

We obtained the historical data of sea ice concentration (SIC) for a total of 31 CMIP6 models (Table 11.1) which were available. Following the suggestion of Notz (2014) and Comiso et al. (2017), we analyse the SIA in this study instead of the SIC so that a meaningful comparison with the observed satellite data may be carried out. The SIA is calculated by multiplying the SIC with the individual grid-cell area and then summing it over the east Arctic region around (0°E–35°E; 65°N–85°N) (Notz & SIMIP Community, 2020). We interpolate the CMIP6 dataset to a 0.5°×0.5° regular grid. We compare the SIA of individual CMIP6 models against the corresponding satellite observations of NOAA/NSIDC Climate Data Record of Passive Microwave Sea Ice Concentration Version 3 (Peng et al., 2013) for the period 1979–2014. The SIA of SSP5-8.5 future projection scenario of CMIP6 models is taken for evaluating the effect of global warming on the variability and predictability of SIA of the region.

We employ fractal dimension analysis to quantify the predictability of SIA. The generalized Hurst exponent technique of Di Matteo et al. (2003) is used to obtain the fractal dimension of the SIA. The Hurst exponent (H) (Di Matteo et al., 2003;

TABLE 11.1
**CMIP6 Models Were Used in the Study. The Models are Classified as Good/
Poor According to their Performance in Realistically Simulating the SIA. The
Third Column Represents those Good Models for which SSP5-8.5 Future
Projection Data Were Available**

S. no.	CMIP6 model	Good/Poor	Good models with SSP5-8.5 data
1	ACCESS-CM2	Poor	–
2	ACCESS-ESM1-5	Good	Available
3	AWI-ESM-1-1-LR	Good	Available
4	BCC-CSM2-MR	Poor	–
5	BCC-ESM1	Poor	–
6	CAMS-CSM1-0	Poor	–
7	CAS-ESM2-0	Poor	–
8	CESM2-FV2	Poor	–
9	CESM2-WACCM-FV2	Poor	–
10	CESM2-WACCM	Poor	–
11	CESM2	Good	Available
12	CIESM	Good	Available
13	CMCC-CM2-SR5	Poor	–
14	E3SM-1-0	Good	Available
15	E3SM-1-1	Good	Available
16	E3SM-1-1-ECA	Good	Not Available
17	EC-Earth3	Poor	–
18	FGOALS-f3-L	Poor	–
19	FIO-ESM-2-0	Good	Available
20	GFDL-CM4	Poor	–
21	GFDL-ESM4	Good	Available
22	GISS-E2-1-H	Poor	–
23	INM-CM5-0	Good	Available
24	IPSL-CM6A-LR	Good	Available
25	MIROC6	Good	Available
26	MPI-ESM1-2-HR	Good	Available
27	MRI-ESM2-0	Good	Available
28	NESM3	Poor	–
29	NorESM2-LM	Good	Available
30	SAM0-UNICON	Poor	–
31	TaiESM1	Good	Available

Peitgen et al., 2004; Rangarajan and Sant, 2004) measures the persistence, i.e. long-term memory, in a data set. The H can be used to obtain the fractal dimension (D) of a time series as $D = 2 - H$. Any time series with $H = 0.5$ (i.e., $D = 1.5$) represents the Brownian motion in which the present state does not correlate with the past or future state. Such processes will have no trend and will be unpredictable. The time series with $0.5 < H \leq 1$ (i.e., $1 \leq D < 1.5$) exhibits "persistence" and an increase in predictability. There will be a known trend in such time series data with future values likely to vary following this trend. The time series with $0 \leq H < 0.5$ (i.e., $1.5 < D \leq 2$) exhibits "anti-persistence". The predictability of such data will also increase since a decrease in the amplitude of the process is more likely to lead to an increase in the future (Rangarajan and Sant, 2004). Rangarajan and Sant (2004) defined the Predictability Index (PI) in terms of fractal dimension (Hurst exponent) as: $PI = 2 |D - 1.5| = 2 |0.5 - H|$, where $|\ |$ denotes the absolute value. The definition of PI suggests that the predictability will increase in both cases: when H is greater than 0.5 (persistence) and when it is less than 0.5 (anti-persistence).

11.3 RESULTS AND DISCUSSION

We begin by evaluating the performance of the 31 CMIP6 models in simulating the mean SIA of the east Arctic region around $0°E–35°E$; $65°N–85°N$. This region covers the Indian Arctic station Himadri. The area-averaged mean SIA of the region is shown in Figure 11.1 for all the CMIP6 models. The corresponding satellite-derived observed SIA value is also shown in the figure. The dashed lower (upper) horizontal line in the figure represents 50% lower (higher) than the observed SIA value. We find that the CMCC-CM2-SR5 model fails to correctly capture the SIA of the region of interest. This model underestimates the mean SIA by showing less than half of the observed value. In contrast to this, the mean SIA in 14 models, namely, ACCESS-CM2, BCC-CSM2-MR, BCC-ESM1, CAMS-CSM1-0, CAS-ESM2-0, CESM2-FV2, CESM2-WACCM-FV2, CESM2-WACCM, EC-Earth3, FGOALS-f3-L, GFDL-CM4, GISS-E2-1-H, NESM3, and SAM0-UNICON, is overestimated by more than 50% of the observed value. We use the remaining 16 models as good models for this study for which the mean SIA remains within 50% of the upper and lower bound of observed SIA. The Multi-Model Mean (MMM) SIA of these models over the region of interest in the east Arctic is shown in Figure 11.2. The corresponding observed SIA is also shown. We see that in the southern part of the domain, the MMM SIA is

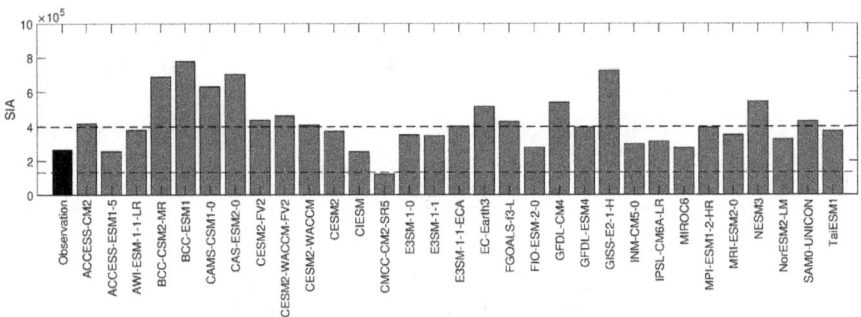

FIGURE 11.1 Mean SIA of the east Arctic region area-averaged over [0°E–35°E; 65°N–85°N] for 31 CMIP6 models (shown as bars) during 1979–2014. The mean SIA of satellite observation of the same period is shown as black bar. The lower (upper) dashed horizontal lines represent SIA values which are 50% less (more) than the mean observed SIA.

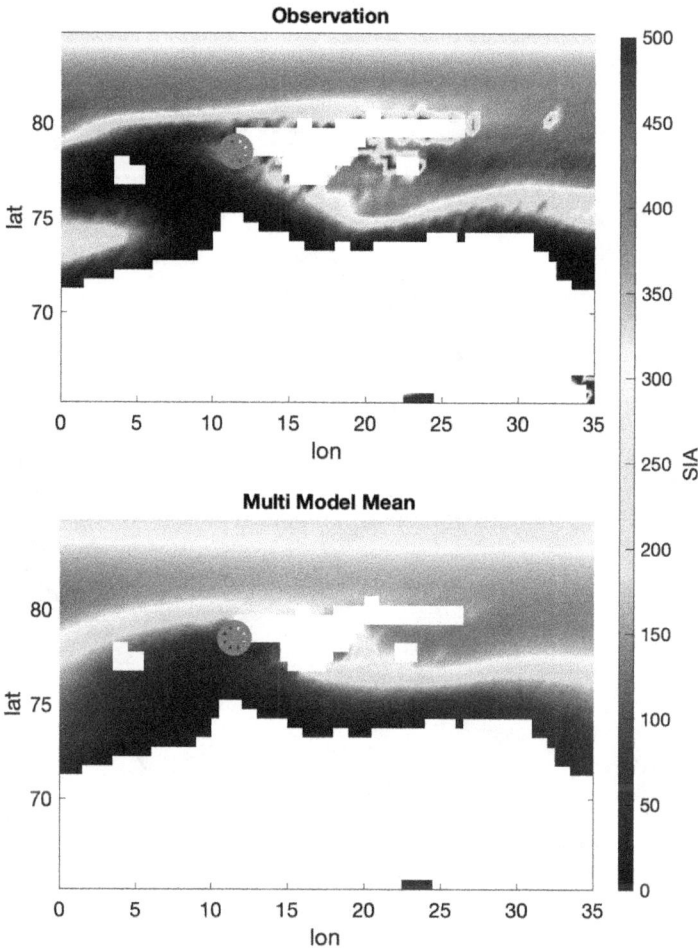

FIGURE 11.2 The mean spatial variability of the SIA for the period 1979–2014 for (a) observed satellite data, and (b) Multi Model Mean (MMM) data of the Historical run of CMIP6 models. The marker denotes the location of the Indian Arctic station Himadri at [11°56′E; 78°55′N].

overestimated as compared to the observation, whereas in the northern and middle parts of the domain, it is largely underestimated. Even though the models fail to produce the correct magnitude of the SIA, their spatial variability matches well with the observation over the region of interest. The correlation coefficient between the observed and MMM SIA is 0.97 (significant at 99.9% level). The annual cycle of SIA of 16 good CMIP6 models as well as their MMM annual cycle is shown in Figure 11.3. This figure also shows the annual cycle of observed data. We can see from the figure that there exists a large inter-model spread in the annual cycle of the models. The maximum (minimum) difference amongst models is seen during Jan–May (Sept). The MMM annual cycle of the SIA matches well with the observed

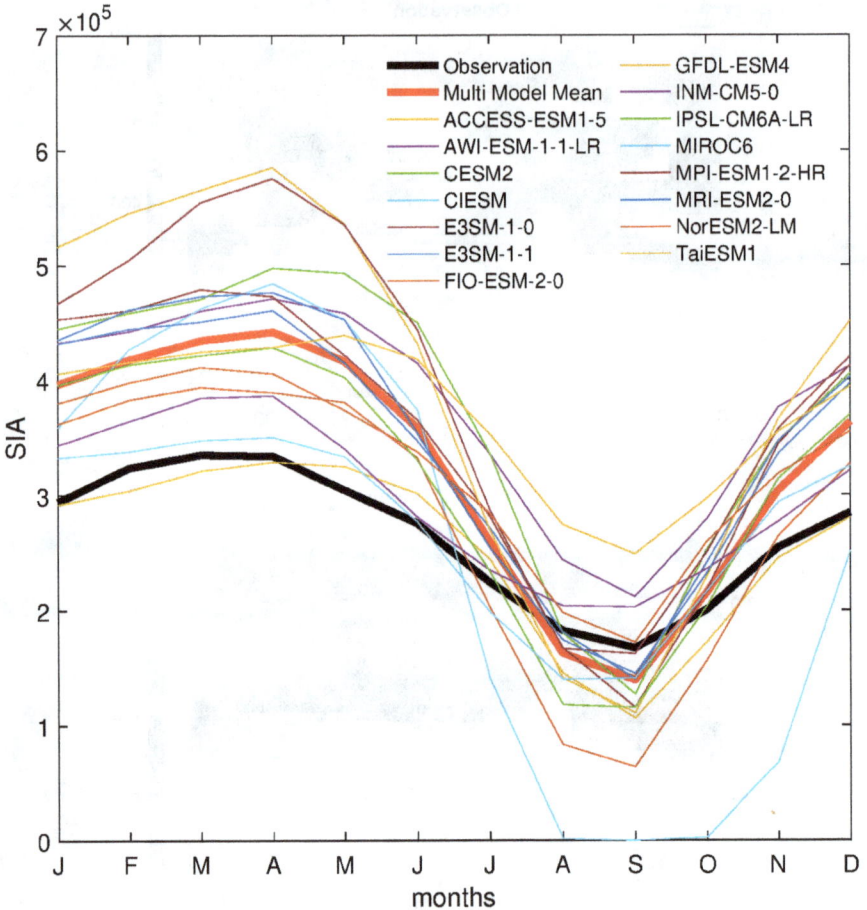

FIGURE 11.3 Annual cycle of area-averaged SIA corresponding to good CMIP6 models. The annual cycle of observed SIA data (thick black curve) and MMM SIA data (thick red curve) are also shown for comparison.

annual cycle, however. The SIA of the east Arctic region remains highest in Mar–Apr (freezing season), whereas, it remains lowest in Sept (melting season).

To understand the effect of global warming on the SIA of the region of interest, we compare the MMM SIA spatial maps of the Historical data with the corresponding maps of SSP5-8.5 data (Figure 11.4). The SIA of the northern part of the domain remains much higher as compared to the southern part. The map of SSP5-8.5 data indicates that the SIA of the region is decreasing as a result of global warming. The SIA difference map in Figure 11.4 suggests that the SIA in the entire northern part of the domain (north of 75°N) will show a high decrease as a result of global warming with the maximum decrease occurring in the east-central (26°E–35°E;75°N–80°N) part of the domain. We generate a long time series of the monthly SIA area averaged over the east Arctic region (0°E–35°E; 65°N–85°N) from 1900 to 2014 (115 years)

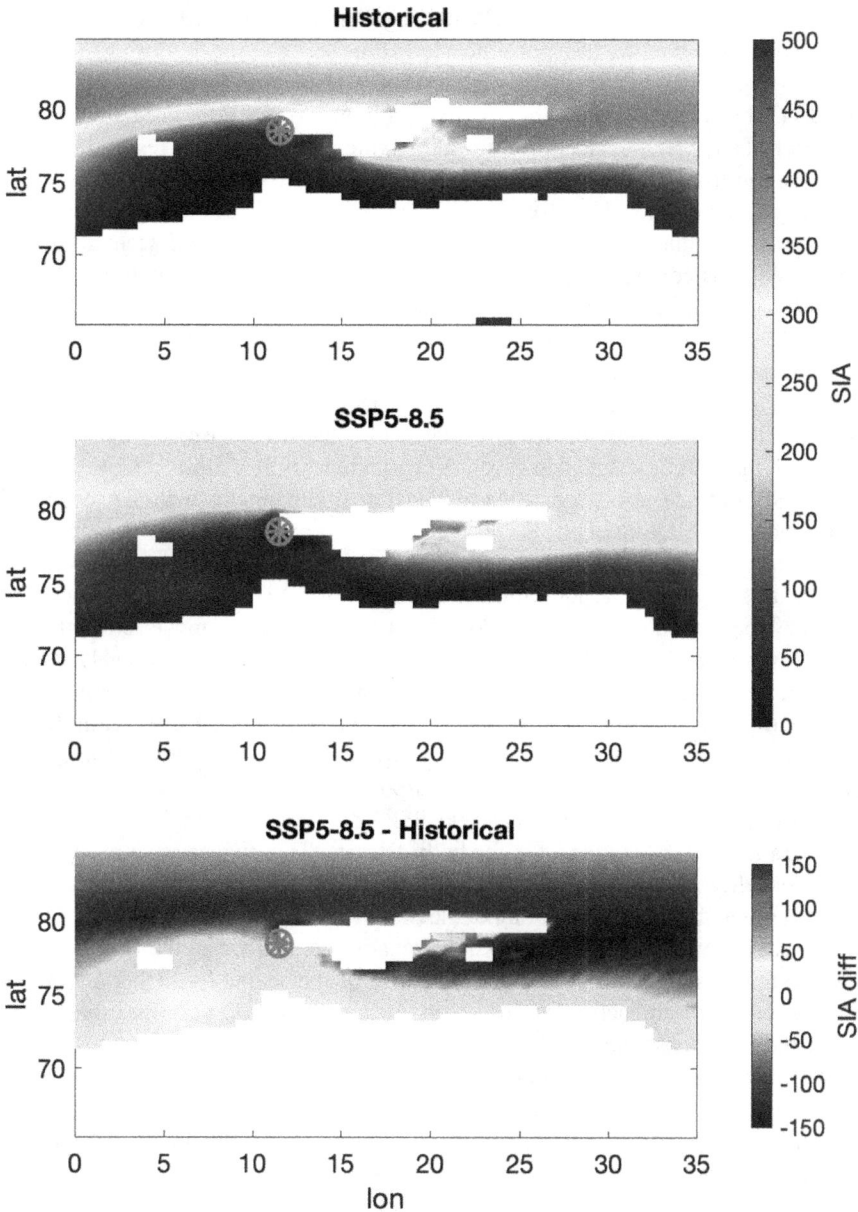

FIGURE 11.4 The mean spatial variability of the (upper panel) Historical data of SIA for the period 1900–2014; (middle panel) SSP5-8.5 data of SIA for the period 2015–2100; (lower panel) difference of SIA between SSP5-8.5 data and Historical data. The marker denotes the location of the Indian Arctic station Himadri.

using the historical data of good CMIP6 models. To gauge the effect of climate change on this region, we also generate a long time series of the SIA using the SSP5-8.5 (very high GHG concentration) future scenario data of the available good CMIP6 models (Table 11.1) for the years 2015–2100 (86 years). The MMM SIA time series of historical and SSP5-8.5 data are shown in Figures 11.5(a) and (b), respectively. It is worth noting from the figure that the SIA shows a decreasing trend in both historical and SSP5-8.5 scenarios. However, it will decrease at a very fast rate in the SSP5-8.5 scenario. For example, the rate of decrease of SIA in the SSP5-8.5 scenario is 4.75 times greater as compared to the historical period. The monthly SIA time series of the SSP5-8.5 projection scenario also suggests that the east Arctic region will become seasonally ice-free (e.g. in Sept) by 2060. Moreover, by the end of the 21st century, much less overall SIA will remain present in this region. The annual cycle of the historical and SSP5-8.5 SIA data is shown in Figure 11.5(c). We find that the annual cycle of SIA derived from the SSP5-8.5 data shows the same seasonal variability as is seen with the historical data. However, the magnitude of SIA shows a large decrease in the SSP5-8.5 scenario (from 2015 to 2100) in all the months with SIA becoming almost negligible during Sept.

To quantify the effect of global warming on the predictability of the SIA, we carry out fractal analysis on the long time series of historical and SSP5-8.5 data derived from the good CMIP6 models. The long-term linear trend is removed from the historical as well as SSP5-8.5 data of the SIA before carrying out the analysis. The detrended SIA time series is used for computing the generalized Hurst exponent in the east Arctic ocean. We followed the method described in Di Matteo et al. (2003) to obtain the generalized Hurst exponent $H(q)$ of the SIA time series. The q-order moments are taken from 1 to 4 for this purpose. The $H(q)$ is plotted as a function of q in Figure 11.6(a) for historical and SSP5-8.5 scenario data. We find that the $H(q)$ is a nonlinear function of q for both datasets. This suggests the existence of multi-fractality in the SIA time series for historical as well as SSP5-8.5 data. The nonlinearity and multi-fractal nature of the SIA in the east Arctic ocean become more prominent under the high GHG emission scenario. We notice that the $H(q)$ of SSP5-8.5 data remains higher as compared to the historical data, but for higher q, both the time series approach each other. The Hurst exponent $H(q=1)$ describes the scaling behaviour of the absolute value of the increments and is commonly used to assign a single fractal dimension to a variable. The $H(q=2)$ is associated with the scaling of the autocorrelation function and is related to the power spectrum (Di Matteo et al., 2003). We use $H(1)$ to determine the fractal dimension and predictability of the SIA. We find that $H(1)$ of historical and SSP5-8.5 data is 0.284 and 0.323, respectively. This gives the corresponding fractal dimension $D = 1.72$ and 1.68, respectively (rounded to two decimal places). The H values are less than 0.5 suggesting the "anti-persistence" and predictable nature of these time series datasets. We argue that the SIA variability of the eastern Arctic region is different from a random unpredictable Brownian motion process.

The $H(q)$ is used to compute the Predictability Index (PI) (Section 11.2) corresponding to each q for a historical run as well as the SSP5-8.5 prediction scenario. The results are shown in Figure 11.6(b). Rangarajan and Sant (2004) suggested that the PI values are very useful in determining how predictable a process is. These

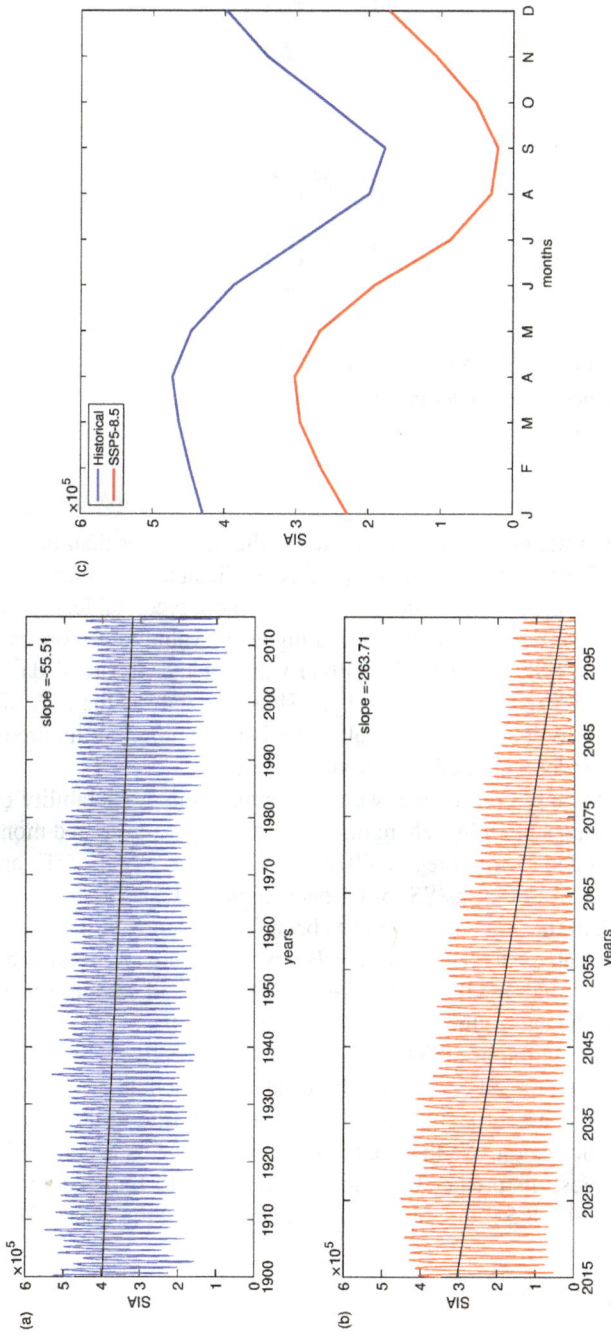

FIGURE 11.5 MMM SIA time series of (a) historical data from 1900–2014 (115 years), and (b) SSP5-8.5 data from 2015-2100 (86 years). The solid black line in (a) and (b) represents linear trend; (c) Annual cycle of the historical (blue) and SSP5-8.5 (red) SIA data.

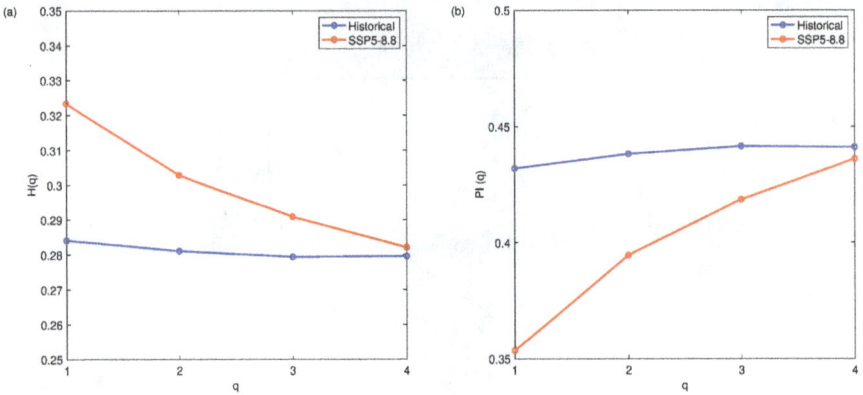

FIGURE 11.6 (a) Generalized Hurst exponent H(q) of the SIA time series over the east Arctic region as a function of q-order moments for historical (blue) and SSP5-8.5 (red) data; (b) Predictability Index (PI) of the SIA time series as a function of q for historical (blue) and SSP5-8.5 (red) data.

values are related to the interannual variability of the SIA rather than its magnitude. It ranges between 0 (unpredictable) and 1 (highly predictable). In other words, as the value of PI increases, the process becomes more and more predictable. Figure 11.6(b) clearly shows that the historical data (with a higher PI value) are more predictable as compared to the SSP5-8.5 data. The PI(1) value of the historical data is 0.43, whereas, for SSP5-8.5 data, its value is 0.35. Thus, not only is the SIA of the east Arctic region likely to decrease as a result of global warming, but its predictability will also decrease under a high GHG emission scenario.

To put the results into perspective, we also compute the predictability of SIA of Himadri Indian Arctic station in a changing climate. The area-averaged monthly SIA time series is generated over the region (10.5°E–12.5°E; 77.5°N–79.5°E) around the Himadri for historical as well as SSP5-8.5 data (Figure 11.7). We can see from the figure that the SIA of the Himadri region has been fast depleting with the presence of seasonally ice-free values in recent decades. But what is more alarming to note is the fact that the Himadri region will become free of sea ice (in all seasons) as a result of climate change by the end of the 21st century. The H(1), D, and PI(1) values of SIA around the Himadri station are 0.33, 1.67, and 0.33, respectively, for the historical data (Figure 11.8a,b). For SSP5-8.5 data, these values are obtained as 0.30, 1.70, and 0.40, respectively (Figure 11.8a,b). These results are opposite to those obtained for the entire east Arctic region. We find that in contrast to the SIA predictability of the total region of interest, its predictability around the Himadri Indian station is likely to increase in the SSP5-8.5 future projection scenario during the years 2015–2100.

11.4 CONCLUSIONS

The SIC data of the eastern Arctic region (0°E–35°E; 65°N–85°N) are converted to the corresponding SIA data for the historical and SSP5-8.5 projection scenario of 31

CMIP6 models. The performance of these models in simulating the spatio-temporal variability of SIA is evaluated against the corresponding satellite data. We find that even though there is a large inter-model spread in the annual cycle of SIA over the east Arctic region covering the Indian station Himadri, the multi-model mean annual cycle of the SIA matches well with the corresponding observed satellite data. We find that there will be a huge impact of global warming on the SIA of the region. The SIA will decrease nearly five times faster in the SSP5-8.5 scenario during 2015–2100 as compared to its rate of decrease during the years 1900–2014. Moreover, the region may become free from sea ice in some months by the year 2060. The effect of climate change on the predictability of east Arctic SIA is investigated in terms of the generalized Hurst exponent and associated Predictability Index (PI). The Hurst exponent values for the historical and SSP5-8.5 data suggest that the SIA of the east Arctic region is predictable. The PI analysis suggests that the historical SIA is more predictable as compared to the corresponding SSP5-8.5 SIA data. The predictability of SIA of the east Arctic ocean will decrease with an increase in global warming (high GHG concentration). We find that there will be no sea ice (in any season) around the Himadri Indian Arctic station by the end of this century if global warming occurs according to the SSP5-8.5 projection scenario. The predictability of SIA around the Himadri station is likely to increase in the SSP5-8.5 projection scenario under the concomitant effect of climate change.

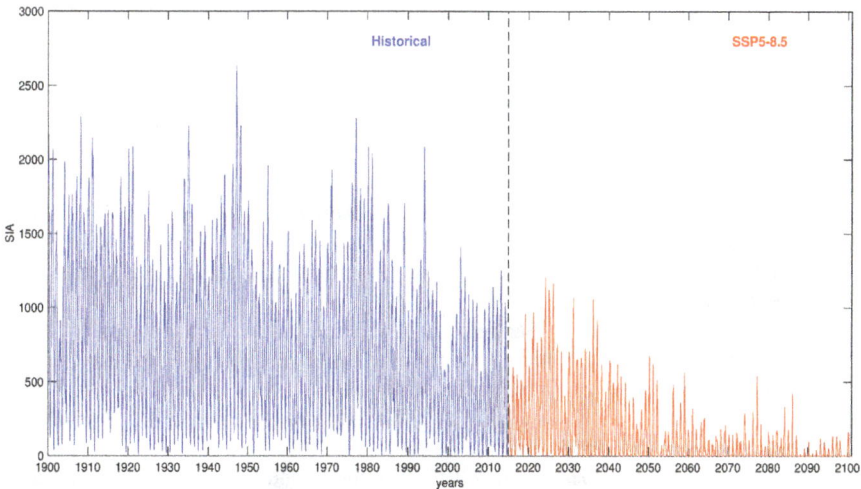

FIGURE 11.7 Area-averaged monthly SIA time series over the region [10.5°E–12.5°E; 77.5°N–79.5°E] around the Himadri Indian Arctic station. The historical (SSP5-8.5) time series for the period 1900–2014 (2015–2100) is shown in blue (red) colour.

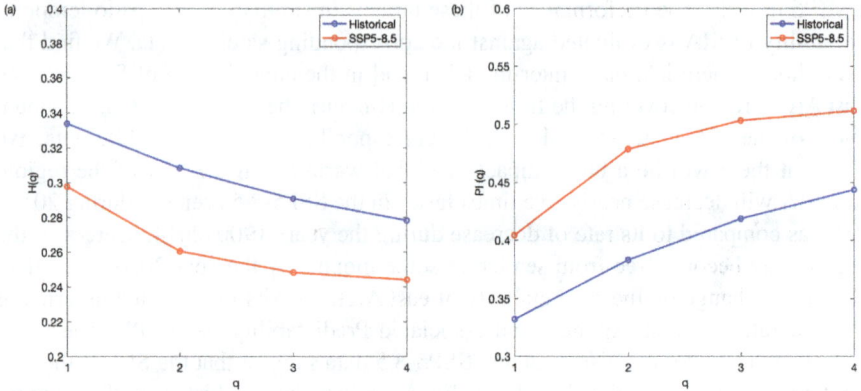

FIGURE 11.8 (a) Generalized Hurst exponent H(q) of the SIA time series around the Himadri Indian Arctic station as a function of q-order moments for historical (blue) and SSP5-8.5 (red) data; (b) Predictability Index (PI) of the SIA time series around the Himadri station as a function of q for historical (blue) and SSP5-8.5 (red) data.

ACKNOWLEDGEMENTS

SD is thankful for the DST-PURSE grant (SR/PURSE Phase 2/43) for infrastructural support to the University of Allahabad. The CMIP6 data of the SIC were downloaded from https://esgf-node.llnl.gov/search/cmip6/.

REFERENCES

Agarwal, S., Moon, W., Wettlaufer, J.S. (2012) Trends, noise and re-entrant long-term persistence in Arctic sea ice. *Proceedings of the Royal Society A* 468: 2416–2432.

Agbazo, M., Koto N'gobi, G., Alamou, E., Kounouhewa, B., Afouda, A., Kounkonnou, N. (2019) Multifractal behaviors of daily temperature time series observed over Benin synoptic stations (West Africa). *Earth Sciences Research Journal* 23(4): 365–370.

Baranowski, P., Krzyszczak, J., Slawinski, C., Hoffmann, H., Kozyra, J., Nieróbca, A., Siwek, K., Gluza, A. (2015) Multifractal analysis of meteorological time series to assess climate impacts. *Climate Research* 65: 39–52.

Chandrasekaran, S., Poomalai, S., Saminathan, B., Suthanthiravel, S., Sundaram, K., Abdul Hakkim, F.F. (2019) An investigation of the relationship between the Hurst exponent and the predictability of a rainfall time series. *Meteorological Applications* 26: 511–519.

Chmel, A., Smirnov, V.N., Astakhov, A.P. (2005) The Arctic sea-ice cover: fractal space-time domain. *Physica A* 357; 556–564.

Comiso, J.C., Meier, W.N., Gersten, R. (2017) Variability and trends in the Arctic sea ice cover: Results from different techniques. *Journal of Geophysical Research: Oceans* 122: 6883–6900.

Di Matteo, T., Aste, T., Dacorogna, M.M. (2003) Scaling behaviours in differently developed markets. *Physica A: Statistical Mechanics and its Applications* 324: 183–188.

Dwivedi, S., Mittal, A.K., Pandey, A.C. (2007) Effect of averaging timescale on a forced Lorenz model. *Atmosphere-Ocean* 45(2): 71–79.

Dwivedi, S. (2012) Quantifying predictability of Indian summer monsoon intraseasonal oscillations using nonlinear time series analysis. *Meteorologische Zeitschrift* 21(4); 413–419.

Francis, J.A., Vavrus, S.J. (2012) Evidence linking Arctic amplification to extreme weather in mid-latitudes. *Geophysical Research Letters* 39: L06801.

Hou, W., Feng, G., Yan, P., Li, S. (2018) Multifractal analysis of the drought area in seven large regions of China from 1961 to 2012. *Meteorology and Atmospheric Physics* 130: 459–471.

Kimothi, S., Kumar, A., Thapliyal, A., Ojha, N., Soni, V.K., Singh, N. (2019) Climate predictability in the Himalayan foothills using fractals. *MAUSAM* 70: 357–362.

Kumar, A., Dwivedi, S., Pandey, A.C. (2018) Quantifying predictability of sea ice around the Indian Antarctic stations using coupled ocean sea ice model with shelf ice. *Political Science* 18: 83–93.

Massonnet, F., Fichefet, T., Goosse, H., Bitz, C. M., Philippon-Berthier, G., Holland, M.M., Barriat, P.Y. (2012) Constraining projections of summer Arctic sea ice. *The Cryosphere* 6(6): 1383–1394.

Melia, N., Haines, K., Hawkins, E. (2015) Improved Arctic sea ice thickness projections using bias-corrected CMIP5 simulations. *The Cryosphere* 9(6): 2237–2251.

Mishra, A.K., Dwivedi, S., Di Sante, F. (2021) Performance of the RegCM-MITgcm coupled regional model in simulating the Indian Summer Monsoon rainfall. *Pure and Applied Geophysics* 178: 603–617.

Mittal, A.K., Dwivedi, S., Pandey, A.C. (2003) A study of the forced Lorenz model of relevance to monsoon predictability. *Indian Journal of Radio and Space Physics* 32(4): 209–216.

Moon, W., Nandan, V., Scharien, R.K., Wilkinson, J., Yackel, J.J., Barrett, A., Lawrence, I., Segal, R.A., Stroeve, J., Mahmud, M., Duke, P.J. (2019) Physical length scales of wind-blown snow redistribution and accumulation on relatively smooth Arctic first-year sea ice. *Environmental Research Letters* 14: 104003.

Notz, D. (2014) Sea-ice extent and its trend provide limited metrics of model performance. *The Cryosphere* 8(1): 229–243.

Notz, D., Jahn, A., Holland, M., Hunke, E., Massonnet, F., Stroeve, J., Vancoppenolle, M. (2016) Sea Ice Model Intercomparison Project (SIMIP): Understanding sea ice through climate-model simulations. *Geoscientific Model Development Discussions* 9: 3427–3446.

Notz, D., SIMIP Community (2020) Arctic sea ice in CMIP6. *Geophysical Research Letters* 47: e2019GL086749. https://doi.org/10.1029/2019GL086749

Olonscheck, D., Notz, D. (2017) Consistently estimating internal climate variability from climate model simulations. *Journal of Climate* 30(23): 9555–9573.

Onarheim, I.H., Eldevik, T., Smedsrud, L.H., Stroeve, J.C. (2018) Seasonal and regional manifestation of Arctic sea ice loss. *Journal of Climate* 31(12): 4917–4932.

Pandey, L.K., Dwivedi, S., Martin, M. (2021) Short-term predictability of the Bay of Bengal region using a high-resolution Indian Ocean model. *Marine Geodesy*, DOI: 10.1080/01490419.2021.1894273

Peitgen, H.-O., Jurgens, H., Saupe, D. (2004) *Chaos and Fractals: New Frontiers of Science*, pp 1–864. Springer-Verlag: New York.

Peng, G., Meier, W.N., Scott, D., Savoie, M. (2013) A long-term and reproducible passive microwave sea ice concentration data record for climate studies and monitoring. *Earth System Science Data* 5: 311–318.

Rampal, P., Dansereau, V., Olason, E., Bouillon, S., Williams, T., Korosov, A., Samaké, A. (2019) On the multi-fractal scaling properties of sea ice deformation. *The Cryosphere* 13: 2457–2474.

Rangarajan, G., Sant, D.A. (2004) Fractal dimensional analysis of Indian climatic dynamics. *Chaos, Solitons and Fractals* 19: 285–291.

Screen, J.A., Deser, C. (2019) Pacific Ocean variability influences the time of emergence of a seasonally ice-free Arctic Ocean. *Geophysical Research Letters* 46: 2222–2231.

Senftleben, D., Lauer, A., Karpechko, A. (2020) Constraining uncertainties in CMIP5 projections of September Arctic sea ice extent with observations. *Journal of Climate* 33(4): 1487–1503.

Serreze, M.C., Meier, W.N. (2019) The Arctic's sea ice cover: trends, variability, predictability, and comparisons to the Antarctic. *Annals of the New York Academy of Sciences* 1436: 36–53.

Shu, Q., Wang, Q., Song, Z., Qiao, F., Zhao, J., Chu, M., Li, X. (2020) Assessment of sea ice extent in CMIP6 with comparison to observations and CMIP5. *Geophysical Research Letters* 47: e2020GL087965.

Smith, L.C., Stephenson, S.R. (2013) New Trans-Arctic shipping routes navigable by midcentury. *Proceedings of the National Academy of Sciences USA* 110(13): E1191–1195.

Stroeve, J., Kattsov, V., Barrett, A., Serreze, M., Pavlova, T., Holland, M., Meier W.N. (2012) Trends in Arctic sea ice extent from CMIP5, CMIP3 and observations. *Geophysical Research Letters* 39: L16502.

Stroeve, J., Barrett, A., Serreze, M., Schweiger, A. (2014) Using records from submarines, aircraft and satellites to evaluate climate model simulations of Arctic sea ice thickness. *The Cryosphere* 8(5): 1839–1854.

Stroeve, J., Notz, D. (2018) Changing state of Arctic sea ice across all seasons. *Environmental Research Letters* 13: 103001.

Vancoppenolle, M., Meiners, K.M., Michel, C., Bopp, L., Brabant, F., et al. (2013) Role of sea ice in global biogeochemical cycles: emerging views and challenges. *Quaternary Science Reviews* 79: 207–230.

Wei, T., Yan, Q., Qi, W., Ding, M., Wang, C. (2020) Projections of Arctic sea ice conditions and shipping routes in the twenty-first century using CMIP6 forcing scenarios. *Environmental Research Letters* 15: 104079.

12 Spatio-Temporal Variation in Surface Mass Balance Patterns of Vestre Brøggerbreen Glacier, Ny-Ålesund, Svalbard Since 2011

Surendra Jat, Pradeep Kumar, Abhishek Verma, Deepak Y. Gajbhiye, and Vikash Chandra

12.1 INTRODUCTION

Climate change in the Arctic region over the past few hundred years was dominated by the generally cool Little Ice Age and subsequent warming up to several degrees marking the termination of this cold period (Fagan, 2008). Fluctuations in the recession rate of glaciers during recent years have initiated widespread discussions, especially in the context of global warming and its effects (Dyurgerov and Meier, 1997). Glaciers provide multi-proxy information on regional and global climate, as changes in glacier extent and mass can reflect changes in temperature and precipitation (Oerlemans and Fortuin, 1992). The climate of the Arctic region is characterized by cold winters and cool summers. The Arctic glaciers quickly respond to the climate, and therefore serve as the best and early indicators of climatic variations.

Glacier mass balance studies are concerned with fluctuations in annual cycles for glacier mass (Cogley, 2011). Each year, a glacier gains ice from snowfall but also loses ice by melting and other processes including sublimation and basal melting. If the gains and losses are not equal, the size of the glacier, its dimensions, and the mass of ice it contains will change over time. Mass balance studies assess such changes and seek an understanding of the processes involved. The magnitude of ablation is controlled by many factors, such as meteorological conditions, glacier exposition, and height above mean sea level, among others (Eriksson, 2014; Hock and Holmgren, 2005). The phase transition of snow and ice and latent heat contributes to the increase in the quantity of meltwater in a glacier system and thus may favour the development

DOI: 10.1201/9781003364115-12

of areas with different thermal conditions and considerable variability in the intensity of melting, referred to as "glacial zones" (Alley, 2011).

The mean air temperature at Ny-Ålesund was from 2011 to 2020 varied from – 1.1°C (2016) to –4.1°C (2013). The maximum air temperature was between 16°C (2019) and 10.5°C (2012), and the annual mean precipitation varied from 1003.17 mm (2016) to 331.38 mm (2019) (The Norwegian Meteorological Institute, Norway).

These investigations involved traditional stake measurements and snow cover measurements, and frequent glacierized area mapping by deploying GNSS techniques. Most of the long-term mass balance investigations in Svalbard have been carried out on small cirque glaciers at relatively low altitude. These investigations show a negative net balance and mainly thinning in the period of 1950–88 (Hagen et al., 1999).

The main objectives of this research are: (1) to estimate the net and specific surface mass balance of Vestre Brøggerbreen glaciers based on in situ observations and (2) to assess the changing climatic impact over glaciers by revealing the relationship between climatic parameters in the Arctic region.

12.2 STUDY AREA

The Vestre Brøggerbreen (78°53'–78°56'N; 11°36'–11°48'E) is a small high Arctic NE-flowing valley glacier that is 4 km west of Ny-Ålesund (Figure 12.1). A medial moraine divides the glacier into Vestre Brøggerbreen I (VB-I) and Vestre Brøggerbreen II (VB-II) with distinct accumulation catchments (Figure 12.2). The melt water streams from Vestre Brøggerbreen and nearby Austre Brøggerbreen glaciers join a short distance from the present snout to form the Bayelva stream which finally drains into the Kongs Fjord (Kongsfjorden). The altitudinal variation ranges from 31 m to 46 8m and 114 m to 366 m in VB-I and VB-II, respectively. VB-I is a larger glacier that has a prominent ice wall with a length of 3.8 km and area of 2.68 km². VB-II is a curvilinear cirque glacier which is 2 km long and comprises of an area of 0.85 km².

12.3 METHOD AND MATERIALS

The surface mass balance refers to mass balance measured according to the glaciological method and is obtained from repeated field visits; it comprises the studies of density profiling by digging pits and manual volume estimation by cylinder and repeated height readings of an array of stakes. Snow accumulation and ablation measurements in VB-I and VB-II glaciers are done by stake measurement during the winter and summer periods. The sum of gains and losses of mass over an entire glacier, from the end of one summer melt season to the end of the next, is known as the net surface mass balance (Brugman and Ostrem, 1991). This method quantifies the surface mass balance, i.e., the mass changes at the surface of the glacier, exposed to the atmosphere, and within near-surface layers, but does not include internal mass changes below the last summer surface, frontal ablation due to calving, and submarine melting at the front of tidewater glaciers (Cogley, 2011).

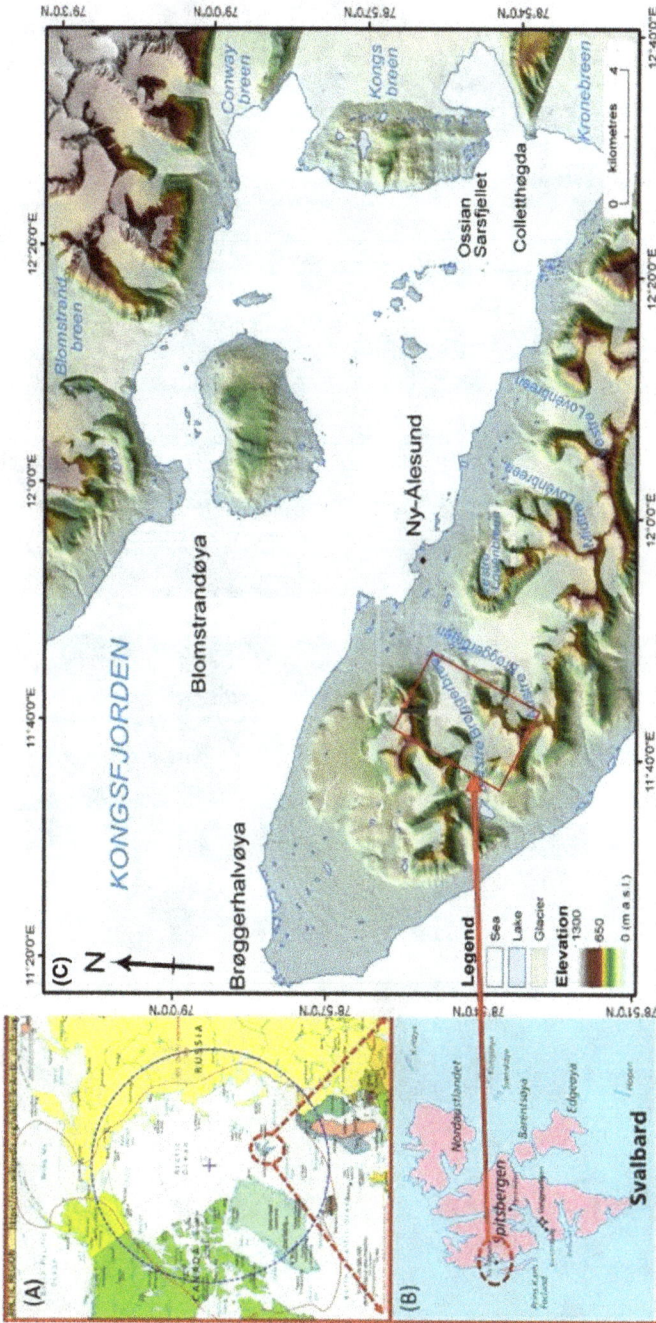

FIGURE 12.1 (A) Regional map of the Arctic circle. (B) Major island of Svalbard. (C) Study area in the Vestre Brøggerbreen glacier (in red box) base data from Norwegian Polar Institute (http://geodata.npolar.no/).

FIGURE 12.2 Monitoring of VB glaciers with DGPS and overview of snout.

Average annual air temperature and precipitation have been measured by the Ny-Ålesund observatory (station no. SN99910) and data are taken from https://sekl ima.met.no/observations/. The data used for the current study have average annual meteorological reflections from 2011 to 2019.

To obtain the net surface mass balance, traditional stratigraphic methods (Brugman and Ostrem, 1991) are used, where surface stakes, snow probing, snow pit methods, snow coring, and survey locations data are summarized for data analysis since 2011. The accumulation and ablation measurements are plotted on a geomorphological map of Vestre Brøggerbreen I and II.

In the field, mapping of the glacier terminus/snout was carried out with the help of a differential global positioning system (DGPS) by taking a series of observation points along the periphery of the terminus since 2011. The snout of VB-I glacier is spread like a fan. The frontal portion of the glacier shows thinned ice which often breaks due to high surface melting in the snout region, and several small rivulets are generated which flow into the pro-glacial region resulting in the marshy nature of the ground moraines.

The surface mass balance was calculated from 2011 to 2019 with a similar stratigraphic glacial zone method. The elevation of glacial topography may change with time. There is no remarkable trend in the winter balance, and the year to year variability is very low. Consequently, the net mass balance variability is determined almost entirely by the variability of the summer climate. Winter balance plays almost no part in the changing values of net balance from year to year (Koerner 2005).

This study will be helpful for the validation of the surface mass balance models to estimation the behaviour of the glacier. The following steps are exercises for calculation of net and specific surface mass balance of Vestre Brøggerbreen I and II using Arc-GIS 10.2 and MS Excel.

i.   Summarized the measured available accumulation and ablation data since 2011 from previous unpublished Geological Survey of India reports.

ii. In VB I stakes were installed in the ablation zone up to 270 m elevation, while in the accumulation zone stakes were not installed due to it being an inaccessible area. Mainly summer data have been used for mass balance calculation except the accumulation zone of VB-I where snow pit data have been analysed. In VB-II, stakes are installed throughout the glacier. For initial monitoring of VB in 2011, a total of 23 stakes are installed of which 12 stakes are on VB-I and nine stakes are on VB-II glacier, with data measured each year in the summer season. Annual ablation/accumulation is calculated to measure the exposed height of the stakes during the summer expedition and is estimated by comparing with the previous year's observations. Snow cover on the surface is measured to estimate the annual snow thickness in the winter season.

iii. The values of surface ablation were converted into water equivalent (weq). An ice density of 910 kgm^{-3} was used to convert ablation thickness (cm) to water equivalent for stakes data (Cogley, 2011). Snow pits were dug to measure the snow density at different levels at 5 cm intervals and converted into water equivalent (weq) by multiplying with the measured snow density.

The manual mode method was used to measure the density of the snow using a fixed-volume cylinder filled with snow and weighed to estimate the density of the snow at the particular horizon by using the empirical formula [ρ=m/v, or density (ρ) is equal to mass (m) divided by volume (v)]. Sometimes, a snow fork also was used to estimate the snow density.

iv. The equilibrium line of altitude (ELA) was marked during field work at 270 m (±30) by Kumer and Gajbhiye (2018). Between 270–300 m there is a sub-vertical scarp with a steep slope. During the field work on the above sub-vertical scarp at 300 m elevation snow accumulation was observed.

v. To estimate the surface mass balance of VB I and II the entire glacier is divided into altitudinal zones at contour intervals of 50 m from 2011 to 2016 (www.toposvalbard.npolar.no.). After 2017 the altitudinal contour interval for VB I and II was estimated at 30 m on the basis of DGPS mapping.

vi. For VB-I glacier annual measurements of stakes and pits differences were calculated from 2011 to 2019. Snow and ice ablation/accumulation data (weq) are plotted using Arc-GIS-10.2.

Annual net mass balance (Ba) = ΔSummer (weq) (Ablation zone in VB-I) + ΔWinter (weq) (Accumulation zone in VB-I)

where ΔWinter = yearly snow pits differences, ΔSummer = yearly stakes differences.

vii. The ablation/accumulation data (weq) for the individual stakes was extrapolated to cover the glacier area using the inverse distance weighting (IDW) method in Arc-GIS 10.2. The resultant contours are generated throughout the glacier where stakes are installed. Ablation/accumulation contour intervals are kept depending on the variation of ablation.

viii. The ablation/accumulation contours and elevation contours are bounded by the glacier's boundary and build the glacial zones using Arc-GIS 10.2 (Figure 12.4).

ix. For mass balance calculation, the glacial zone area is multiplied by the average values of the elevation contour. Subsequently, the sum of the area of the ablation/accumulation zone throughout the glacier is used to calculate net surface and specific mass balance.

The surface mass balance is calculated as

$$b = \int_{t0}^{t1} \begin{cases} \left(c_{wto} - c_{wt1} \right) \times \rho_{snow} \times Area\ of\ Accumulation\ contour \Big\} + \\ \left(a_{sto} - a_{st1} \right) \times \rho_{ice} \times Area\ of\ Ablation\ contour \Big\} \end{cases}$$

where c_{wto} and c_{wt1} are the accumulation (cm) value of the accumulation zone and a_{sto} and a_{st1} are the stake data in the ablation zone, t_0 and t_1 are one year of time, ρ_{snow} is the density of snow, and ρ_{ice} is the density of ice.

The curves are drawn smooth for clarity, but in reality the mass increases in steps corresponding to snowfall events and decreases at an irregular rate as changing weather conditions modulate the ablation. The mass added to the glacier surface generally increases over the winter season, attaining a maximum in late spring or early summer ("End of winter"). Summer melting then removes mass, giving a minimum in late summer (Anonymous, 1969). In the diagram, T1s and T2s denote the times of two successive time intervals of the summer season and T1w and T2w time intervals of the winter season. Accordingly, Δablation and Δaccumulation are calculated in difference values between two successive minima in summer and maxima in the winter season (Figure 12.3).

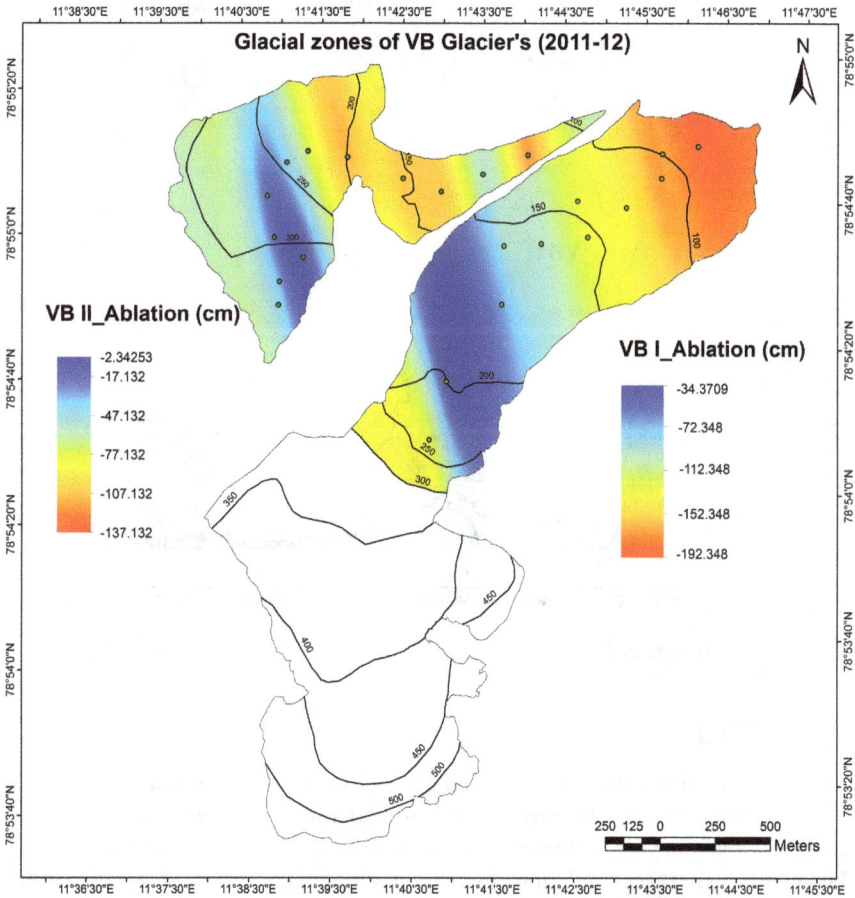

FIGURE 12.3 An idealized seasonal cycle of winter and summer mass balance of glaciers (modified after Anonymous, 1969).

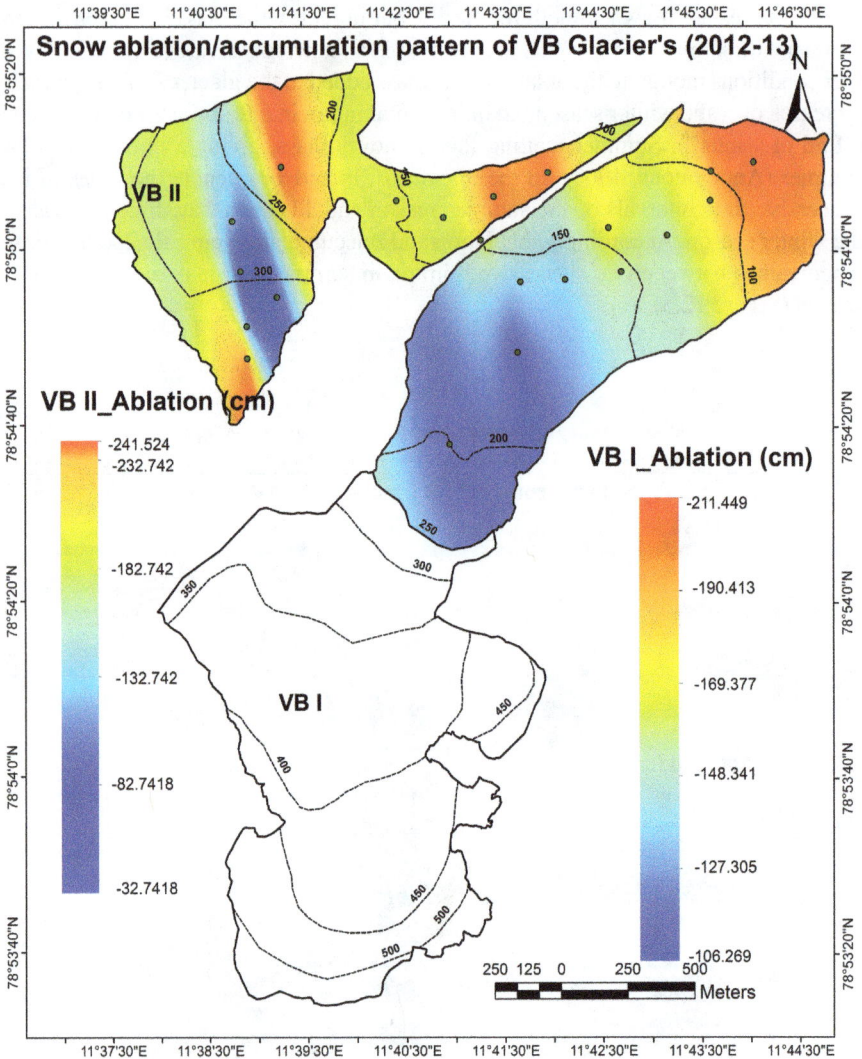

FIGURE 12.3   (Continued)

12.4   RESULT

The highest ablation values are observed as –377 cm in 2017–18, while in 2014–15, 11 cm accumulation are observed in VB-I around 150 msl nearer to snout. In VB-II the highest ablation values are observed as –359cm in 2015–17 and +99 cm in 2012–13 being the highest accumulation observed (Table 12.1). In 2011–12, ablation ranges of VB-I glacier vary from –41.5 cm to –206cm with a specific mass balance of –1.12 mwe. In VB-II ablation values vary from –3.69 cm to –150.61 cm with a specific mass balance (SMB) of –0.77 mweq (Table 12.2).

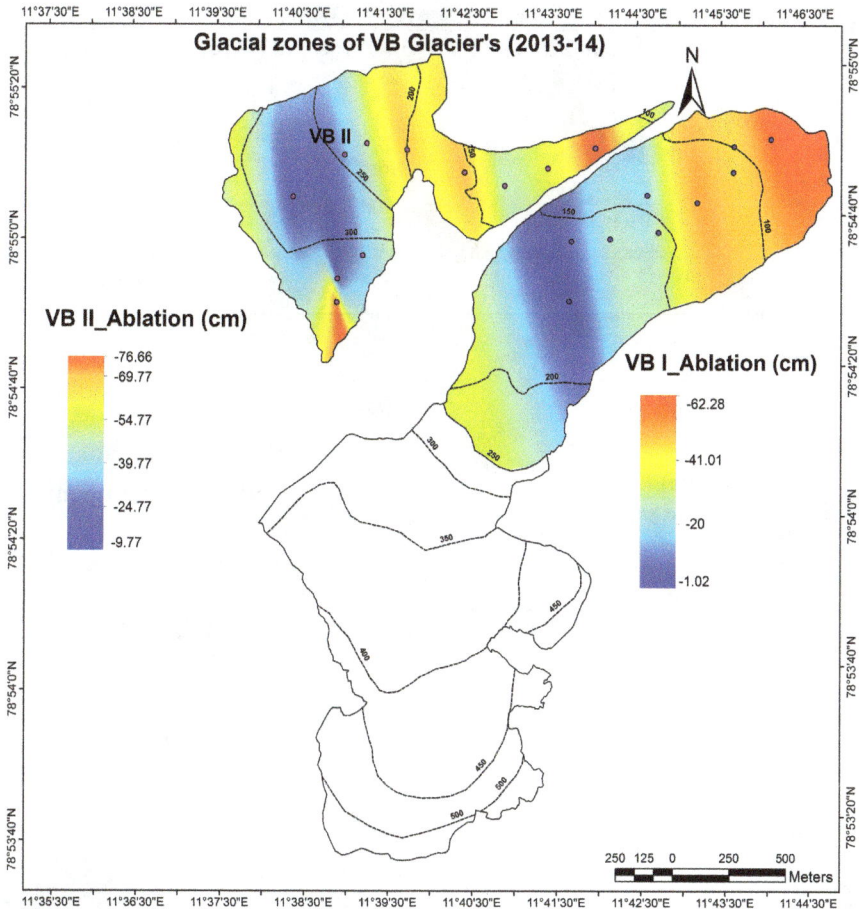

FIGURE 12.3   (Continued)

The ablation values in 2012–13 are somewhat more than the previous year, possibly due to higher temperature and precipitation in both glaciers. In 2012–13, in total 12 stakes' data from VB-I were measured and calculated (Kumar, 2011). The stakes data in ablation zones of VB-I glacier vary from –137 cm to –237 cm for different stakes. The total net mass balance is –2.17 mweq and the specific mass balance is –0.74 mweq. The VB-II stakes data show quite high variation from +99 cm to –263.5 cm and the net and specific mass balances are –1.45 mwe and –1.53 mwe, respectively. In 2012 and 2013, the ablation values are high in all stakes but gradually decrease from the snout towards higher elevations.

The least ablation was observed in 2013–14 for both glaciers. The maximum ablation values are observed in VB-I from 2015 to 2017 near 240 msl and may be due to scooping out of snow by natural events. No measurement was made in the summer of 2016 so that a 2-year balance from the end of summer 2015 to summer 2017 was

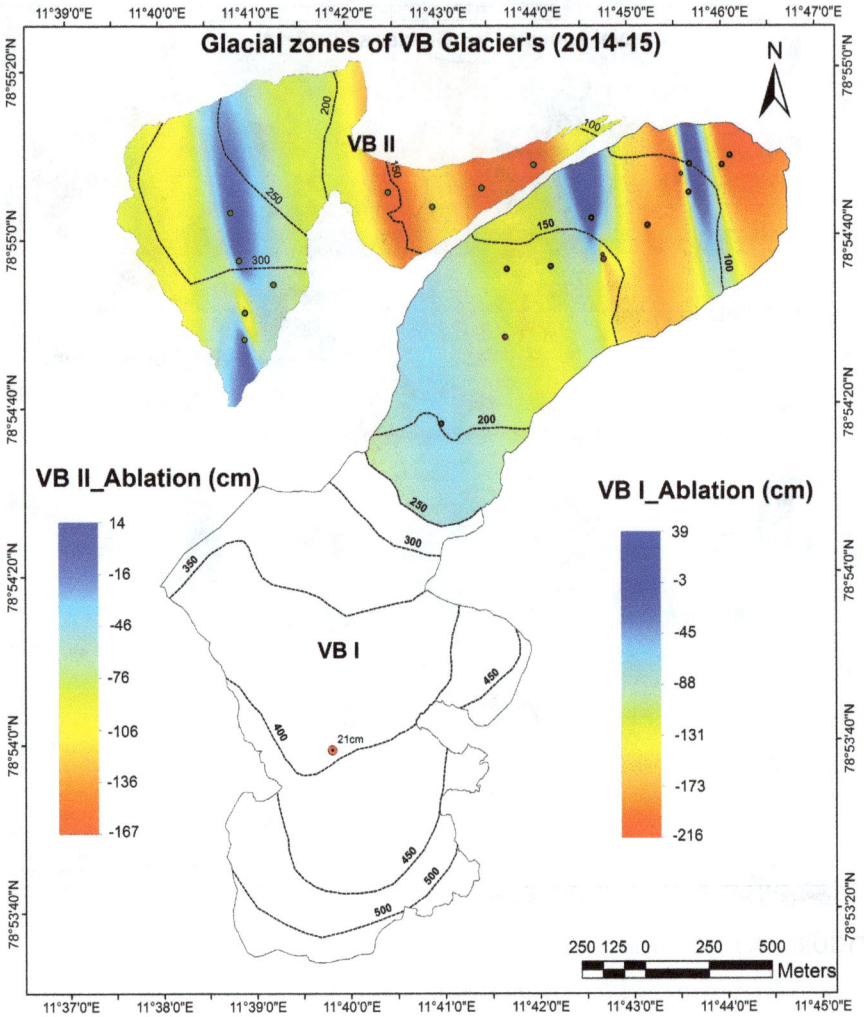

FIGURE 12.3 (Continued)

calculated for a mass balance calculation for 2015–17. In 2015–17, intensive ablation was observed in both glaciers. In VB-I the highest ablation was near 210 m elevation and in VB-II it was around 240 msl, and the intensity of ablation decreased towards lower and higher altitudes from the intense ablation area.

In 2018–19, stakes data varied from 67.9 cm to 189 cm for VB-I. The net mass balance was calculated as –1.15 mweq and specific mass balance was –0.43 mweq. Higher ablation was shown near 180 msl in both glaciers and the intensity of ablation decreased towards higher altitudes. In VB-I, the specific mass balance fluctuated slightly from 2017–18 to 2018–19, but in VB-II it shows quite significant variation (Table 12.2). The average temperature and precipitation are increased accordingly when slightly less ablation was observed.

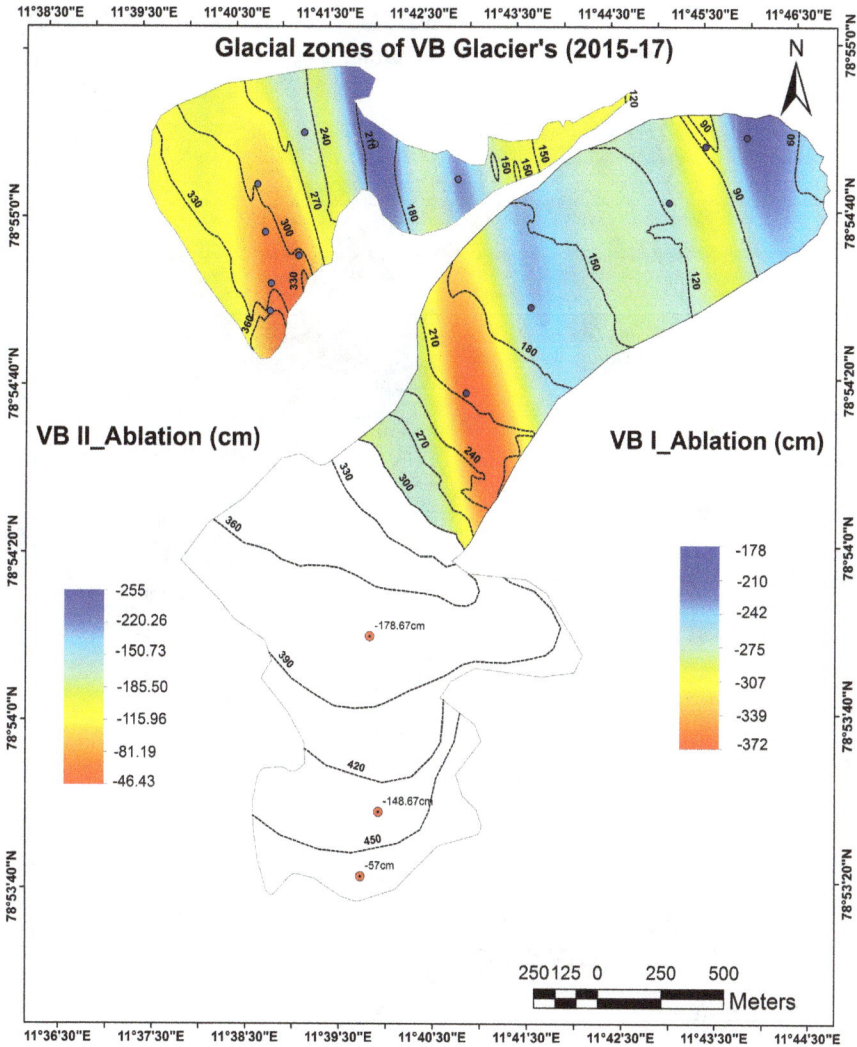

FIGURE 12.3   (Continued)

In the higher altitude (accumulation zone) it was flatter with wet snow, insulating the ice and reducing the ablation rate. At the highest altitude it decreases again in the accumulation basin (Wójcik and Sobota, 2020). This indicates that the local conditions can affect the distribution of ablation.

12.5   DISCUSSION

In Svalbard, due to the summer temperature being above 0°C, especially in lower parts of the glacier, the ablation gradient reached quite a high value (Sobota, 2013). The ablation zones were diversely spatially distributed from the snout to higher

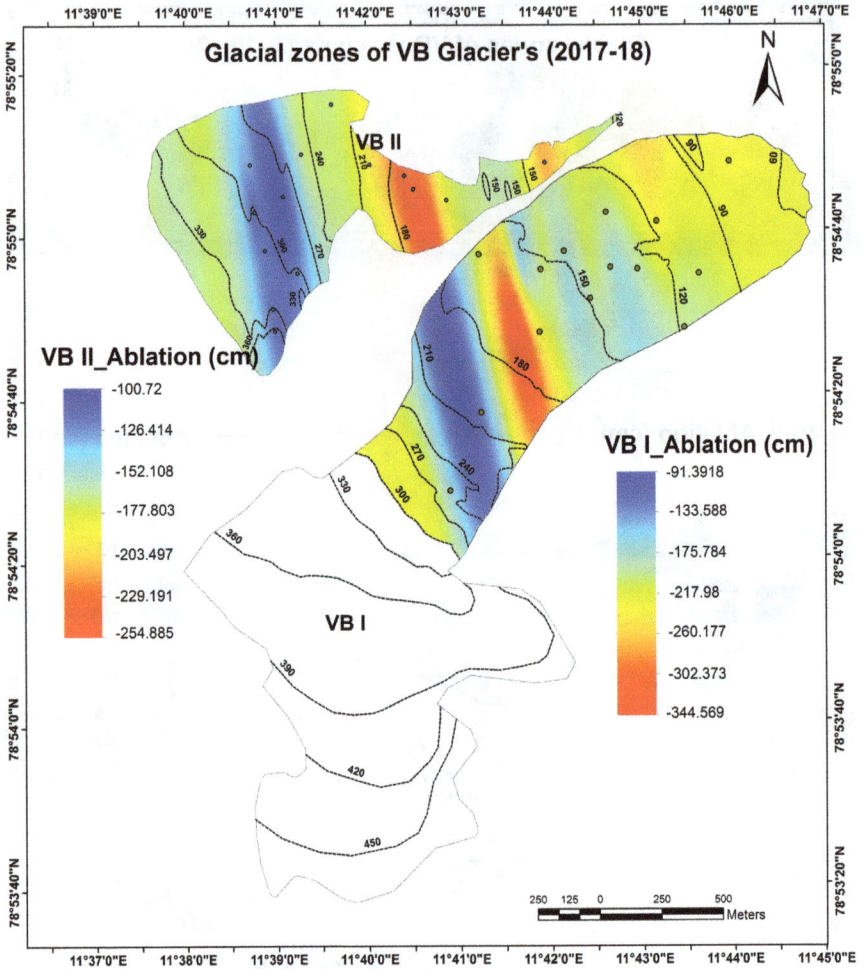

FIGURE 12.3   (Continued)

elevations. The presence of a dense network of transverse crevasses on the ablation area at a lower elevation of the glacier reflected its steeper gradient and higher
movement of glacier ice towards the terminus. In order to accommodate higher
supply of ice from the large and wide accumulation area, the ablation of the area has
fractured into a number of open crevasses (Mir and Majeed, 2018). On the snout of
the glacier melting occurred more intensively and those areas responded more clearly
to changes in temperature. As the elevation increases, the responses were delayed and
the ablation rate decreased. However, the elevation factor can be modified by local
topography.

This study was based on the observations for the period 2011–2019, and previous
studies also reported similar results from the decadal climate data analysis (Hanssen-
bauer, 2000; López-Moreno et al., 2016). If the atmospheric temperature is also

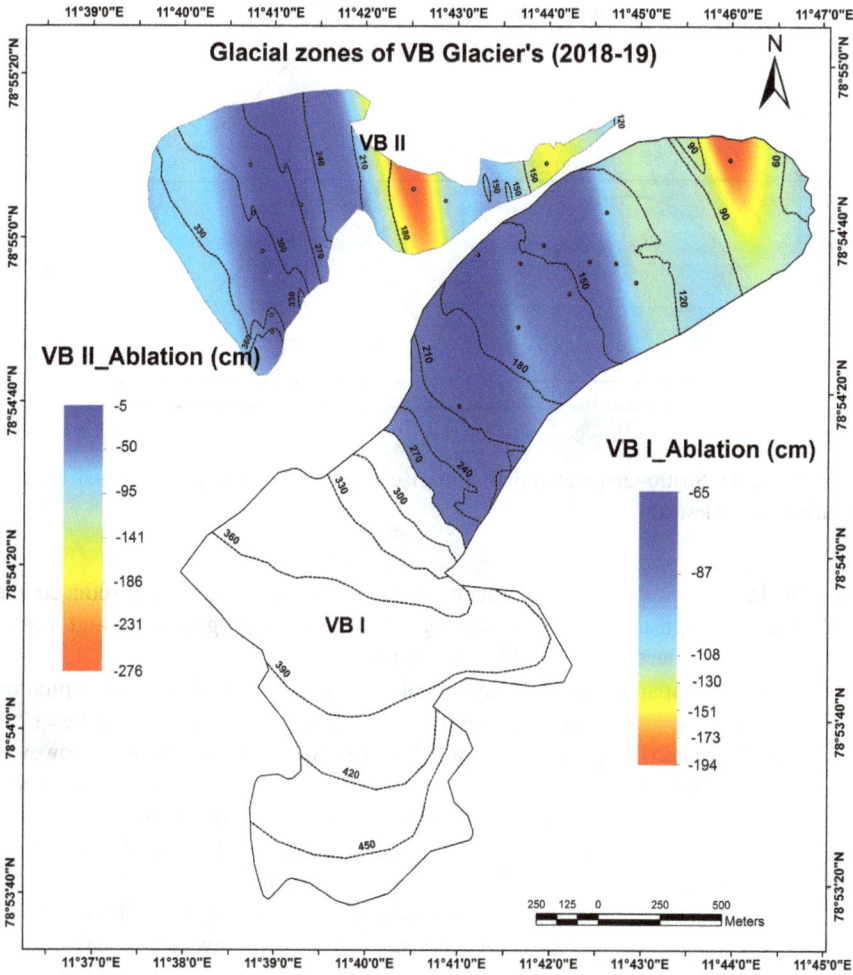

FIGURE 12.3   (Continued)

high during precipitation, there could be a chance of increased precipitation. In Ny-Ålesund, temperature and precipitation have been increasingly controlled by southerly winds (Maturilli, 2013), particularly in winter. As the southerly winds are warm and moist, enhanced precipitation may be recorded. Available data reflect that this glacier has experienced a negative annual net balance since 2011 and lost an SMB of –0.53 mwe of glacier ice over the past 9 years. The study revealed that SMB is one of the most significant factors controlling the mass budget of the studied glaciers in Ny-Ålesund (Figure 12.6).

   Although the surface fresh snow over the glacier protects it from incoming radiation, the presence of the refreezing layers over the glacier leads to the incoming radiation being trapped because of the reduced albedo of the refreezing mass, which reinforces the ablation processes over the glacier surface. The precipitation increased

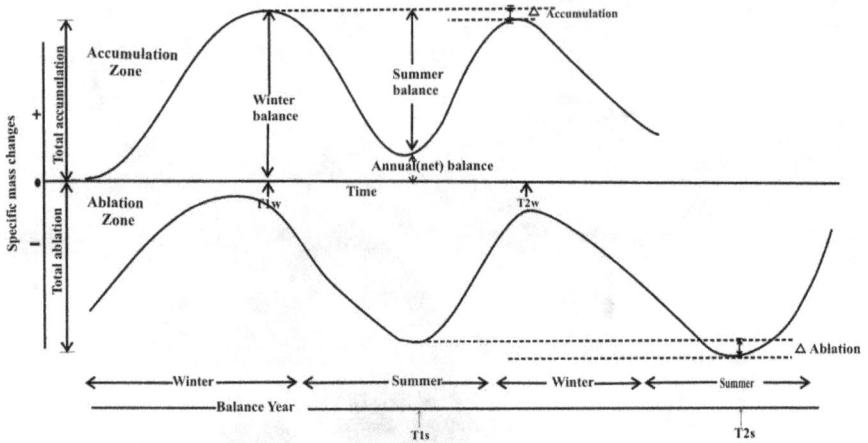

FIGURE 12.4 Spatio-temporal variation in glacial zones in Vestre Brøggerbreen I and II glaciers in Ny-Ålesund.

over the last 9 years, and a significant increase in precipitation has reduced the SMB and increased the total mass wastage of the Vestre Brøggerbreen and Austre Brøggerbreen glaciers in the Ny-Ålesund region.

This study compared the SMB pattern of VB-I and VB-II with the precipitation and temperature data and obtained a more-or-less positive correlation (Figures 12.6, 12.7). For the year 2013, precipitation was low and the SMB was also low. However, for the years 2014–2016, precipitation was progressively increased and vice versa SMB negatively declined from 2014. In 2014, the SMB and precipitation correlation was negative, which may be due to wet precipitation. The results showed a significant correlation with the period of 2017–2019 with precipitation data. However, the correlation was negative, indicating that high precipitation leads to lower SMB.

Available long-term glaciological mass balance records reveal a negative mass balance correlation in Svalbard glaciers (Figure 12.9). This observation reveals that smaller glaciers (<10 km²) are rapidly losing mass (Asutre Brøggerbreen, Middle Lovénbreen, Waldemarbreen, and Irenebreen) compared with larger glaciers (>50 km²) like Kongsvegen and Kronebreen-Holtedahlfonna (Rye et al., 2010; Schuler et al., 2020; Sobota, 2017).

In VBI, the specific mass balance is lower than the net mass balance because of the higher elevated accumulation zone. In VBII the specific mass balance is moreover equal to the surface net balance. The Vestre Brøggerbreen glacier has shown a negative mass balance since 2011 except in the 2014 when a slight positive trend was observed in VB-I. A similarly strong positive correlation of mass balance is also observed over the neighbouring Austre Brøggerbreen glacier, monitored by the Norwegian Polar Institute (NPI) (Figure 12.5). In other words, there is a regular decline in the health of the glaciers despite some fluctuations. The trend of the ablation rate has been quite variable from 2011 to 2015 and the high fluctuation observed from 2015 to 2017 followed a similar trend in tempreature.

TABLE 12.1
Surface Mass Balance of Vestre Brøggerbreen (VB) II Glacier

Surface mass balance of Vestre Brøggerbreen II glacier, Ny-Ålesund

Year	Area (km²)	Stake data minimum (cm)	Stake data maximum (cm)	Total ablation (MB)	Net balance (mweq)	Specific balance (mweq)
2011–12	1.01	–3.69	–150.61	–0.782	–0.782	–0.769
2012–13	0.944	99	–263.5	–1.453	–1.453	–1.539
2013–14	0.942	20.25	–68	–0.269	–0.269	–0.286
2014–15	0.942	–55.2	–184	–0.839	–0.839	–0.890
2015–17	0.826	–209	–359	–2.187	–2.187	–2.646
2017–18	0.824	–111	–280	–1.388	–1.388	–1.684
2018–19	0.857	–9	–173.5	–0.470	–0.470	–0.548

TABLE 12.2
Surface Mass Balance of Vestre Brøggerbreen (VB) I Glacier

					Surface mass balance of Vestre Brøggerbreen I glaciers, Ny-Ålesund		
Year	Area (km²)	Stake data minimum (cm)	Stake data maximum (cm)	Total ablation (MB)	Total accumulation	Net balance (mweq)	Specific balance (mweq)
2011–12	2.922	−41.5	−206	−1.53	–	–	−1.125
2012–13	2.920	−137	−237	−1.913	−0.258	−2.171	−0.743
2013–14	2.804	10	−82	−0.459	1.221	0.762	0.272
2014–15	2.878	11	−123	−0.633	1.157	−0.633	−0.220
2015–17	2.74	−59	−278	−1.505	−1.471	−2.976	−1.110
2017–18	2.712	−101	−377	−1.070	0.075	−0.995	−0.371
2018–19	2.680	−67.9	−189	−1.269	0.114	−1.155	−0.431

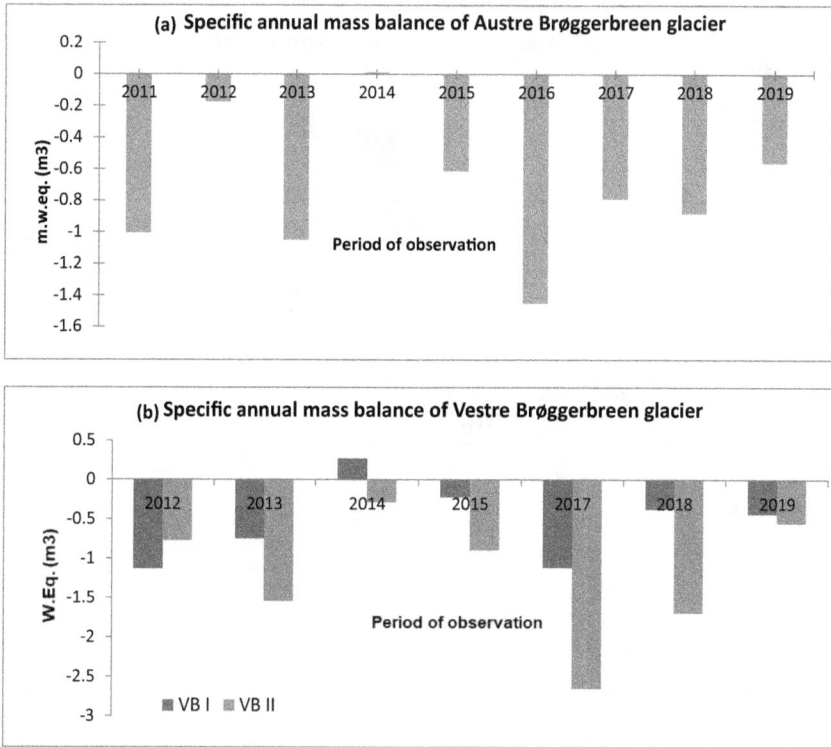

FIGURE 12.5 (a) Specific annual mass balance of Austre Brøggerbreen glacier and (b) specific annual mass balance of Vestre Brøggerbreen glacier since 2011.

The spatial distribution of mass changes in Vetsre Brøggerbreen responded to weather conditions. The annual mass balance is governed by many climatic factors such as temperature, precipitation, radiation, wind conditions, etc., and the relationship between specific mass balance and major climatic parameters like temperature and precipitation has been established. The relationship between area, slope, and SMB indicates that smaller glaciers lose mass at a much higher rate than larger glaciers due to their hypsometric distributions of the area with elevation (Schuler et al., 2020). Most Arctic glaciers have experienced predominantly negative mass balance over the last few decades, although within a context of higher interannual variability. More than 80% of Arctic glaciers have shown a negative mean net balance (Dowdeswell et al., 1997). The relationship between specific mass balance and major climatic parameters like temperature and precipitation has been established. When the temperature rises it increases the mass balance negatively (Figure 12.8a,c) the effect of the temperature is stronger ($R^2=0.9$) in the VB-II glacier than the VB-I glacier. Similarly, when precipitation in the form of snow fall increases and after some time it starts melting, it releases the specific heat from the snow crystals and during this process the temperature of the snow increases and enhances the melting compaction in the snow pack. The combined effect of the snow fall and simultaneous increase of

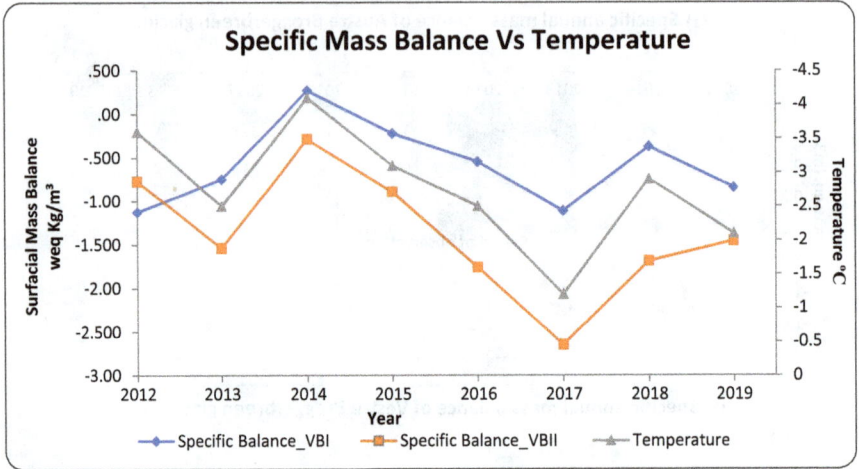

FIGURE 12.6 Temporal variability of specific mass balance and temperature in Vestre Brøggerbreen glaciers.

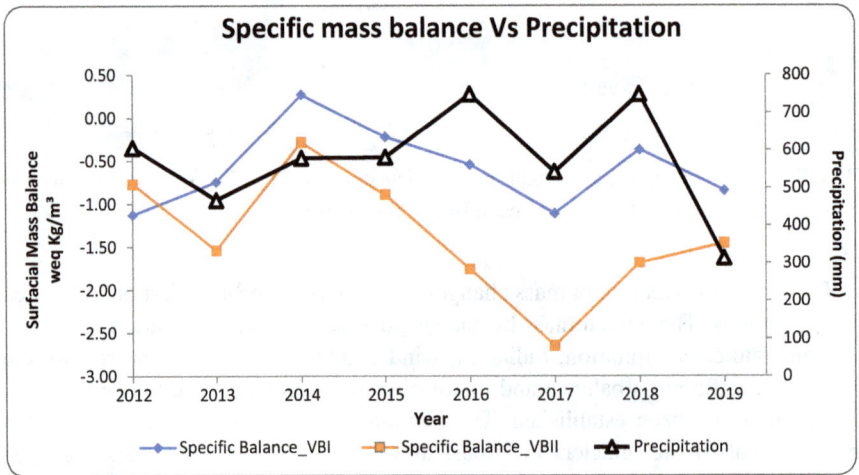

FIGURE 12.7 Temporal variability of specific mass balance and precipitation in Vestre Brøggerbreen glacier.

temperature in the snow pack does not have a remarkable effect on the accumulation/ablation pattern of the glaciers in annual monitoring.

12.6 CONCLUSION

This study provides an overview over the different estimations of Svalbard-wide SMB and net mass balance. In the research work modified mass balance calculation

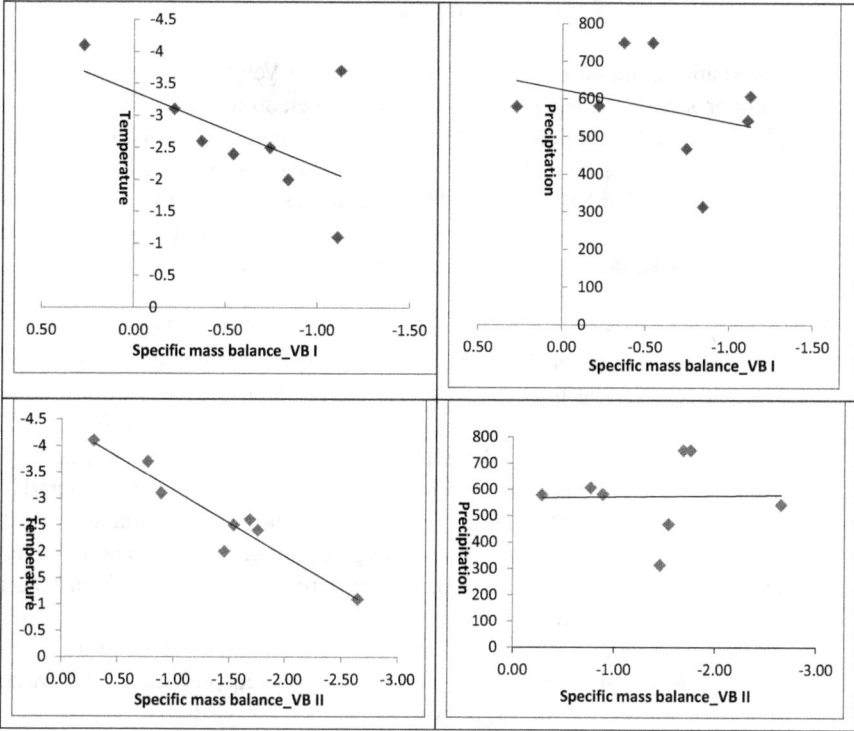

FIGURE 12.8 Linear correlation of specific mass balance with temperature and precipitation.

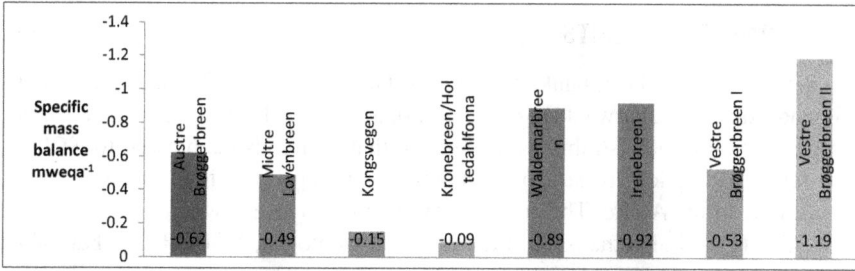

FIGURE 12.9 Comparative study of specific mass balance of Svalbard glaciers with Vestre Brøggerbreen.

techniques are described that's suited for Vestre Brøggerbreen glacier and dependent on the stakes density, area and geomorphological condition (Jon Ove Hagen et al. 2003).

The present study concludes the following:

a) The spatio-temporal distribution of ablation on Vetsre Brøggerbreen in the summer season is dependent primarily on meteorological parameters with added spatial distribution caused by local topographic conditions. The ablation intensity increases from the snout to higher elevations in Vetsre Brøggerbreen as in other Svalbard glaciers, but this correlation showed a certain weakness at about 270–300 msl due to subvertical scarp. The higher altitude of the glaciers revealed dynamic changes and a lower rate of ablation, and responded to changes in weather conditions.

b) Comparative ablation studies in both glaciers reveal that the smaller VB-II has been melting faster than VB-I since 2011 but their ablation trends are the same. It reveals that both glaciers have similar metrological parameters but their topographical factors cannot be ruled out. VB-I has a large frontal area directly facing seaward winds, whereas VB-II is less affected by these high winds. VB-I has a large open area in contrast with VB-II which is sheltered by the surrounding mountain (from north to west) and medial moraine (towards the south). The highest elevation and large accumulation area and relatively short, narrow, and very gentle slope may have prevented the VB-I from experiencing rapid recession.

c) The net balance of VB-I shows less negative SMB than VB-II. From 2011–2019 the mean specific mass balance is –0.53 mweq/year and –1.19 mweq/year for VB-I and VB-II, respectively.

d) The spatio-temporal variiation in specific mass balance was similar to other glacier trends obtained from long-term research. Vetsre Brøggerbreen like other Svalbard glaciers is significantly melting.

ACKNOWLEDGEMENTS

The authors extend their thanks to Director General, GSI, ADG, and Head M-IV, GSI and Amit A. Dharwadkar, Director Polar Studies Division for their support and permission to publish this work. Sincere thanks are also expressed to NCPOR, Goa, for providing logistic support and permission to participate in Indian Scientific Expeditions to the Arctic. The authors are thankful to senior workers for providing the field data, and also thank Sh. Raghuram Sr. Geologist, NCEGR GSI Faridabad, for comments on the manuscript.

REFERENCES

Anonymous (1969) Mass-balance terms. *Journal of Glaciology* 8(52): 3–7.
Alley, R. (2011) *The Physics of Glaciers.* Fourth Edition. Amsterdam, Academic Press.
Brugman, M., Ostrem, G. (1991) *Mass-Balance Measurements: A Manual for Field and Office Work.* Saskatoon, Saskatchewan: Environment Canada.
Cogley, J.G., Hock, R., Rasmussen, L.A., Arendt, A., Bauder, A., Braithwaite, R., Jansson, P., Kaser, G., Möller, M., Nicholson, L., Zemp, M. (2011) Glossary of glacier mass balance and related terms. *IHP-VII Technical Documents in Hydrology* 86: 124.

Dowdeswell, J.A., et al. (1997) The mass balance of circum-Arctic glaciers and recent climate change. *Quaternary Research* 48(1): 1–14.

Dyurgerov, M.B., Meier, M.F. (1997) Year-to-year fluctuations of global mass balance of small glaciers and their contribution to sea-level changes. *Arctic and Alpine Research* 29(4): 392–402.

Eriksson, P. (2014) Meteorological differences between Rabots Glaciär and Storglaciären and its impact on ablation. *Physical Geography and Quaternary Geology.*

Fagan, B. (2008) Grove, J.M. 1988: The Little Ice Age. London: Routledge, Xxii + 498 Pp. *Progress in Physical Geography* 32(1): 103–106.

Hagen, J.O., et al. (1999) Mass balance methods on Kongsvegen, Svalbard. *Geografiska Annaler, Series A: Physical Geography* 81(4): 593–601.

Hagen, J.O., Melvold, K., Pinglot, F., Dowdeswell, J.A. (2003) On the net mass balance of the glaciers and ice caps in Svalbard, Norwegian Arctic. *Arctic, Antarctic, and Alpine Research* 35(2): 264–270.

Hanssen-bauer, I. (2000) Increased Precipitation in the Norwegian Arctic: True or False? *Climatic Change* 46(4): 485–509.

Hock, R., Holmgren, B. (2005) A distributed surface energy-balance model for complex topography and its application to Storglaciären, Sweden. *Journal of Glaciology* 51(172): 25–36.

Koerner, R.M. (2005) Mass balance of glaciers in the Queen Elizabeth Islands, Nunavut, Canada. *Annals of Glaciology* 42: 417–423.

López-Moreno, J. I., Boike, J., Sanchez-Lorenzo, A., Pomeroy, J.W. (2016) Impact of climate warming on snow processes in Ny-Ålesund, a polar maritime site at Svalbard." *Global and Planetary Change* 146: 10–21.

Mir, R.A., Majeed, Z. (2018) Frontal recession of Parkachik Glacier between 1971–2015, Zanskar Himalaya using remote sensing and field data. *Geocarto International* 33(2): 163–177.

Oerlemans, J., Fortuin, J.P.F. (1992) Sensitivity of glaciers and small ice caps to greenhouse warming. *Science* 258(5079): 115–117.

Rye, C.J., Arnold, N.S., Willis, I.C., Kohler, J. (2010) Modeling the surface mass balance of a high Arctic glacier using the ERA-40 reanalysis. *Journal of Geophysical Research: Earth Surface* 115(F2): 1–18.

Schuler, T.V., et al. (2020) Reconciling Svalbard glacier mass balance. *Frontiers in Earth Science* 8(May): 1–16.

Sobota, I. (2017) Selected problems of snow accumulation on glaciers during longterm studies in North-Western Spitsbergen, Svalbard. *Geografiska Annaler, Series A: Physical Geography* 99(2): 177–192.

Wójcik, K.A., Sobota, I. (2020) Spatial and temporal changes in ablation, distribution and evolution of glacial zones on Irenebreen, a small glacier of the High Arctic, Svalbard. *Polar Science* 23.

13 Arctic Amplification and the Monsoon Variability

*Rajib Chattopadhyay, Mahesh Kalshetti,
R. Phani, Pulak Guhathakurta, and A.K. Sahai*

13.1 INTRODUCTION

The Arctic sea ice shows significant decreasing trends in the century scale and explosive decline in the last few decades with strong signatures of near-surface warming (Perovich, 2011; Cohen et al., 2014; Cai et al., 2021; Meredith et al., 2019). A monthly plot of the trend underlining the decline is shown in Figure 13.1. A recent NASA estimate put the decline at 13.1% per decade, with a strong decrease in September when the sea-ice is at its climatological minimum state (https://climate.nasa.gov/vital-signs/arctic-sea-ice/). Similarly, decreases in other months are also documented, especially a drastic reduction in northern hemispheric snow cover extent during May–June is well documented (Derksen and Brown, 2012). The Intergovernmental Panel on Climate Change (IPCC), in a recent report (https://www.ipcc.ch/srocc/chapter/chapter-3-2/), summarizes that (Meredith et al., 2019): "During the winters (January to March) of 2016 and 2018, surface temperatures in the central Arctic were 6°C above the 1981–2010 average, contributing to unprecedented regional sea ice absence." Since the change in the surface temperature and the global warming signatures are amplified in recent decades and especially the last decade, there has been speculation of its significant impact on several weather and climatic features and their predictability as well as for high-latitude eco-systems (Vihma, 2014; Cohen et al., 2014; Overland et al., 2014; Balmaseda et al., 2010; Hoskins and Woollings, 2015; Coumou et al., 2018; Schneider et al., 2015). One of the main reasons for the modulation of weather and climate is hypothesized to be due to a change in the air–sea interaction with a decline in sea ice cover (e.g., thickness, latitudinal extent) change in the albedo, and weakening of the air–sea coupling at the polar region (Mallett et al., 2021).

The accelerated polar loss of sea ice, also termed *polar amplification*, which in the context of the North Pole is the *Arctic amplification*, refers to the change in surface temperature associated with the change in temperature at the poles due to climate forcing related to global anthropogenic warming. Studies suggest the modulation of midlatitude weather and climate in two different ways. Several studies link the amplification and midlatitude weather in terms of modulation of low-frequency variability forcing manifested as changes in low-frequency variability such as Rossby waves and associated phenomena (like transport process and blocking) due to modulation in the basic state (i.e., change in baroclinicity or temperature gradients or radiative forcing). The other approach studies the modulation by assuming *Arctic amplification*

DOI: 10.1201/9781003364115-13

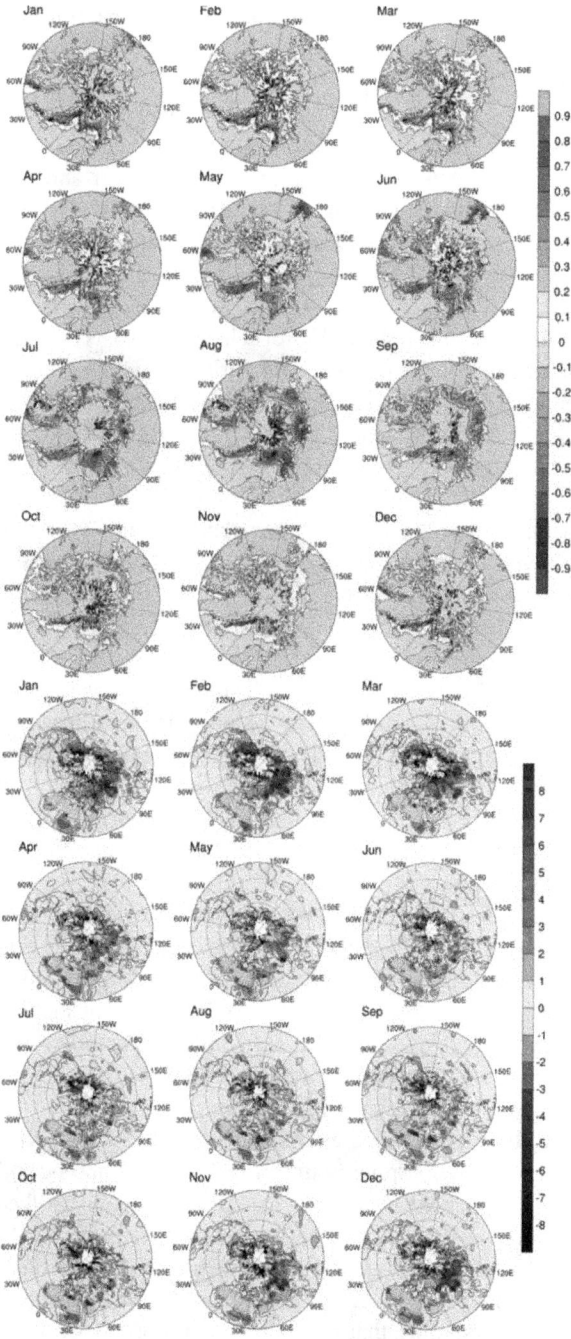

FIGURE 13.1 The monthly trend in sea-ice concentration and skin (surface) temperature for the period 1871–2004. The plot is based on HADCRU data (Morice et al., 2012; Rayner et al., 2003). Months are mentioned at the top of each panel.

induced modulation on *Arctic oscillation* and *Antarctic multidecadal oscillations,* which modulate mid-latitude weather through similar methods (i.e., changing storm tracks, Rossby waves, and blocking).

Several modeling studies are available in the former category which points out that the impact of such a loss of sea-ice and induced warming on global and regional scales is far-reaching in controlling the midlatitude weather and extremes (Collins et al., 2012; Cohen et al., 2014; Smith et al., 2019; Levine et al., 2021; Schneider et al., 2015). Such warming destabilizes the meridional heat transport process, and the equatorial heat source and the polar heat sink gradient enter an unbalanced phase (Mallett et al., 2021). Such a change in balance could lead to a change in the general atmospheric circulation, especially the Ferrel cell, which exists due to the heat transfer by the eddies arising due to the existence of meridional temperature gradients (Qian et al., 2016; Levine et al., 2021). Also, this points to a change in the baroclinic setup in the atmosphere, leading to a change in the eddy forcing baroclinic instability generation and propagation of baroclinic and barotropic Rossby waves (Gong et al., 2020; Budikova, 2009; Deser et al., 2010; Coumou et al., 2018; Hoskins and Woollings, 2015; Riboldi et al., 2020). The alternate approach is to study the *Arctic amplification* as an external forcing bringing change in the *Arctic oscillation* or *northern annular mode* and its impact on midlatitude extreme events and blocking events (Chylek et al., 2009; Trenberth and Shea, 2006; Hassanzadeh and Kuang, 2015).

Despite the studies mentioned above, it is to be noted that a recent IPCC report (Meredith et al.. 2019) mentions clearly that "There is only low to medium confidence in the current nature of Arctic/mid-latitude weather linkages because conclusions of recent analyses are inconsistent ...". The report emphasizes that the atmosphere–ocean–cryosphere coupled system represents a host of physical processes (both resolved and unresolved), and the atmosphere interacts with the ocean and cryosphere through direct and indirect radiations with varying degrees of reflection (i.e., albedo), transport and transfer of heat and moisture, wind circulations and precipitations, and through convergence and divergence brought about by the planetary waves. A complete understanding of such complex interconnected physical processes is not easily derivable. Many studies suggest that the Arctic forcing on the atmosphere from the loss of sea ice and terrestrial snow is increasing. Still, the capability of such forcing to impact the mid-latitude weather is not apparent in studies, for example, indicating that the Arctic/mid-latitude weather linkages vary for different jet stream patterns (Grotjahn et al., 2016; Overland and Wang, 2018a,b). Meredith et al. (2019) speculated that this causal connectivity is not apparent because of the influence of chaotic internal natural variability and other explicitly unresolved tropical and oceanic forcing. Part of the scientific disagreement is due to irregular connections in the Arctic to mid-latitude linkage pathways within and between years (Overland and Wang, 2018a). Similarly, the linkage of Arctic amplification with mid-latitude blocking is unclear (Hassanzadeh and Kuang, 2015). Nevertheless, a wide range of modeling studies do indicate one thing in common: that prescribing appropriate levels of Arctic sea ice provides improved mean climate state over the midlatitudes (Dethloff et al., 2019) and hence sea-ice concentration and its latitudinal extent as a driver of mid-latitude circulation are well explained.

The impacts of Arctic amplification on non-polar low latitudes are also studied in some studies, indicating an impact on winter severity over East Asia and Eurasia (Kim et al., 2014). There are, however, much less clear indications of the impacts of polar amplification on the monsoonal flow over the Indian region. Although monsoonal circulation lies in the deep tropics dominated by equatorial wave dynamics, several studies point out the role of extratropical wave trains reaching over the Indian region, leading to breaks, droughts, and extreme rainfall events, especially over the Himalayas (Hunt et al., 2018; Kalshetti et al., 2020; Fadnavis and Chattopadhyay, 2017; Hoskins and Woollings, 2015). Although several studies relate monsoon to global warming, the Arctic amplification's impact, especially extratropical teleconnection on monsoonal variability, still needs to be fully understood in a warming scenario. Recent studies indicate extratropical teleconnections as decisive drivers in monsoon variability. Extratropical wave trains intrude over the Indian region and can cause synoptic disturbances to invigorate or impact the monsoon low-pressure systems (Borah et al., 2020). Also, North Atlantic or Pacific decadal oscillations correlate with both summer and winter monsoonal variabilities, especially extreme events (Chatterjee et al., 2021; Midhuna and Dimri, 2019). These extratropical systems are linked to the phases of Arctic oscillations, which are strongly impacted by polar amplification. Near-future simulations suggest a clear and distinct linkage between Arctic amplification and low-frequency variability (Serreze and Francis, 2006).

The current work will show whether or how Arctic amplification can be linked to monsoonal variability in a warming scenario in a climate-scale (i.e., 30–40 years) simulation runs. The analysis would be based on a hybrid approach in which we study the monsoon modulation occurring as a change in the teleconnection with the extratropical oscillations (Atlantic oscillations or Arctic oscillations) and the change in the extratropical wave-induced modulations modulating the wave-flux transport over the Indian region. These approaches are complimentary as both can be assumed as external forces driving the monsoon dynamics in addition to the canonical equatorial or tropical dynamics. Unlike many similar analyses, the current research would not only study the monsoon in terms of warming. We would equally study what would happen in an opposite (i.e., cooling) scenario based on sea-ice concentration. This analysis, as we shall see below, will put monsoons in a unique scenario, indicating that the existence of monsoons relies on an equator to pole gradient or balance of heat transfer rather than its existence depending only on equatorial dynamics itself.

13.2 DATA AND METHODS

13.2.1 OBSERVATION DATA

Several long-term century-scale reanalyses are available now. We use the Hadley Centre-CRU sea-ice concentration data (Rayner et al., 2003) and skin temperature data (Morice et al., 2012), 20th-century reanalysis data for air temperature at *sigma995* (~10 m above surface) level (Compo et al., 2011) for the current analysis. Long-term averages and trends are computed between 1871 and 2004, during which uniform sample comparison is possible.

13.2.2 MODEL DATA

The study is conducted closely following the objectives of the *Polar Amplification Model Intercomparison Project* (PAMIP) protocol and recent studies showing that the prescribed extent of Arctic sea ice improves the description of midlatitude circulations (Smith et al., 2019; Dethloff et al., 2019). We have conducted two sensitivity experiments, which we will refer to as **perpetual *WARM*** and **perpetual *COLD*** experiments. The idea is that it is assumed that climate will evolve with average sea ice reduced (increased) to the lowest (highest) extent in ***WARM (COLD)*** runs. The sensitivity experiments are conducted with the Global Forecast System (GFS) model, which has a realistic monsoon and good extratropical climatology (plots are shown in relevant sections). The GFS is a global numerical weather prediction system containing a global computer model and variational analysis run by the US National Weather Service (NWS). More details on GFS are available at www.emc.ncep.noaa.gov/emc/pages/numerical_forecast_systems/gfs.php. The model description is provided in Table 13.1. The ***WARM*** experiment selects a month from the 1981–2016 period when the sea ice concentration was at a minimum latitudinal extent.

Similarly, the ***COLD*** experiment selects a month during which the sea ice concentration is maximum. The sea surface temperature is also chosen for the corresponding months for both experiments. These runs are referred to as *sensitivity runs* in the text. We also have created a control (***CTRL***) run during which the model is forced with SST and sea ice climatological values. The sensitivity runs are conducted with the hypothesis that a decrease in the sea-ice cover is related to increased anthropogenic forcing.

In contrast, an increase is associated with the cooling anthropogenic radiative forcing. This control run would serve as a baseline for comparison. The model is run for ~50 years, and we have omitted the first 10 years and the last 5 years from our analysis. The composite plots and relevant research are shown for these intermediate 30 years. The details of the experiments are given in Table 13.1. The sea ice concentration plot is shown in Figure 13.2.

WARM and ***COLD*** experiments are reported to understand how the monsoon flow, as a response to meridional heat distribution, changes under extreme scenarios, and how they complement each other.

13.3 CHANGE IN MODEL SIMULATED TEMPERATURE AND CIRCULATION MEAN PATTERNS

13.3.1 SOME GENERAL CHANGES IN THE CIRCULATION AND TEMPERATURE PATTERN

Figure 13.3a shows the 31-year (1984–2014) surface temperature climatology at level sigma=0.995. Figure 13.3b shows the same plot for the **CTRL** runs. Figures 13.3c and 13.3e show the same for the ***COLD*** and the ***WARM*** runs. As expected, it can be seen that the cooler temperature isolines extend to more latitudes for the COLD runs, whereas, for the ***WARM*** runs, they are much more concentrated at the poles. The figure shows that the model is sensitive to the change in sea ice concentration. Also,

TABLE 13.1
The Experimental Setup for the Model Runs

		SST (NOAA daily OISST v2)	Sea ice concentration (NOAA daily)	Total run (IC19811020)
Control experiment		Climatology (1981–2015)	Climatology (1981–2015)	51 year (intermediate 31 year considered)
Arctic	**Maximum extent**	Maximum extension day identified on 17 Feb 2010, used for forcing	Maximum extension day identified on 17 Feb 2010, used for forcing	56 year (intermediate 31 year considered)
	Minimum extent	Minimum extension day identified on 11 August 2012, used for forcing	Minimum extension day identified on 11 August 2012, used for forcing	56 year (intermediate 31 year considered)

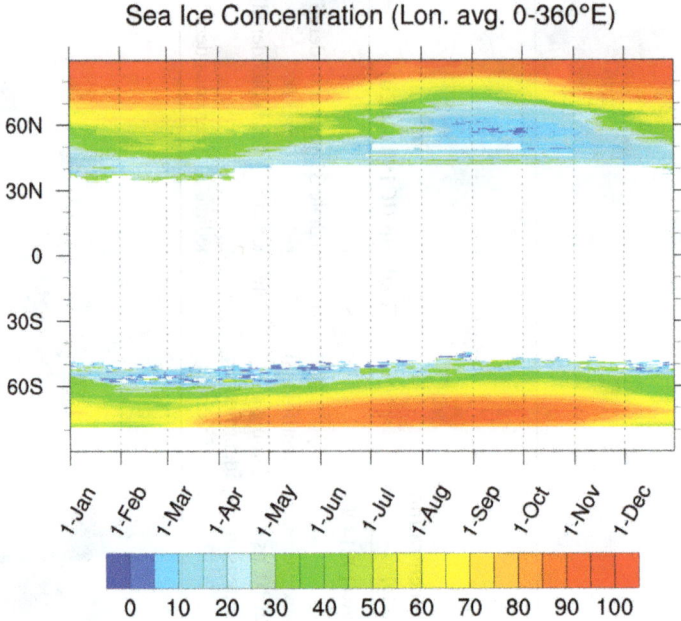

FIGURE 13.2 Plot showing global longitude (0–360°E) averaged daily sea ice concentration for the period 1981–2015. The seasonal sea ice maxima and minima attained during respective hemispheric winter and summer seasons. Over the Arctic region, 17 February 2010 and 11 August 2012 show maximum sea ice concentration.

the warming experiment (Figure 13.3f) shows that the effect of warming is asymmetrical or not circumpolar (i.e., not zonal), with continental (meridional) extent over the United States and Canada being more as compared to other parts. Such extension and asymmetries are essential when the regional impact of climate change is to be studied.

Next, we plot the global mean zonal circulation shown in Figure 13.4. The plot shows zonal wind (contours) and, vertical velocity (-ω, shaded), and vector wind (v, -ω). The difference plot shows subsidence between 15–45°N and weakening of the midlatitude jet stream. Both the above features have far-reaching consequences. The jet stream's weakening implies the weakening of the north–south temperature gradient and meridional flux transport, which are the critical features of mid-latitude circulations. The weakening of the jet stream would also indicate that it would impact the mid-latitude Rossby wave propagation in the zonal and the meridional direction. The phase speed would become modulated, making the wave phase move more slowly.

Similarly, the large subsidence over the region around 15°N indicates that the equatorial convection would probably weaken. The most important aspect of this experiment is that in both the **COLD** and the **WARM** runs, the subsidence increases, and the subsidence increases more in the cooling runs. This indicates that the equatorial convection will be severely affected in both increasing and decreasing scenarios. This

Annual Mean Temperature (°C)

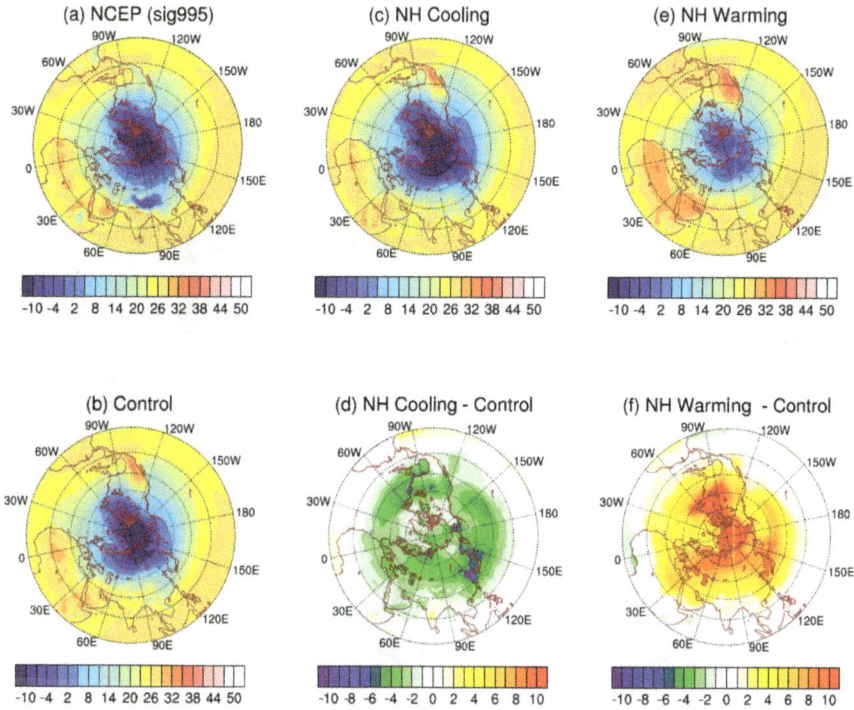

FIGURE 13.3   (a) Plot showing the annual average air temperature (°C) at sigma995 level from the 20Cv2 data. (b) Same but for "CTRL" run, (c) COLD run, and (d) bias from CTRL. (e) WARM run and (f) bias from CTRL.

analysis suggests that the features of general circulation are not only affected over the polar region as a result of Arctic amplification but the equatorial region also gets affected. Although this is not unexpected from the viewpoint of general circulation theory, which depends on meridional heat transport, it is not explored further. We focus on the change in the monsoonal circulation. Figure 13.5 shows the model **CTRL** basic state for the June–September averaged monsoon season. The plots show that the GFS control run effectively captures the general features of the monsoonal region. The wind at 850 hPa and sea level pressure show the monsoonal flow and monsoonal troughs over India and adjoining areas. The upper-level wind and the geopotential height show the broad Tibetan anticyclone extending over the northern Indian region, and the right panel shows the rainfall distribution pattern. The rainfall distribution is comparable to observation with the coastal part of western India, the Head Bay of Bengal, and adjoining elevated regions of Myanmar and Bangladesh. The model also shows higher rainfall over the foothills of the Himalayas. These plots indicate that the GFS **CTRL** faithfully represents the monsoon features over the Indian region. The sensitivity to sea ice forcing is discussed next and compared to the **CTRL** runs.

Annual Mean u-wind (contour m/s), omega (shading $\times -10^{-2} Pa/s$)

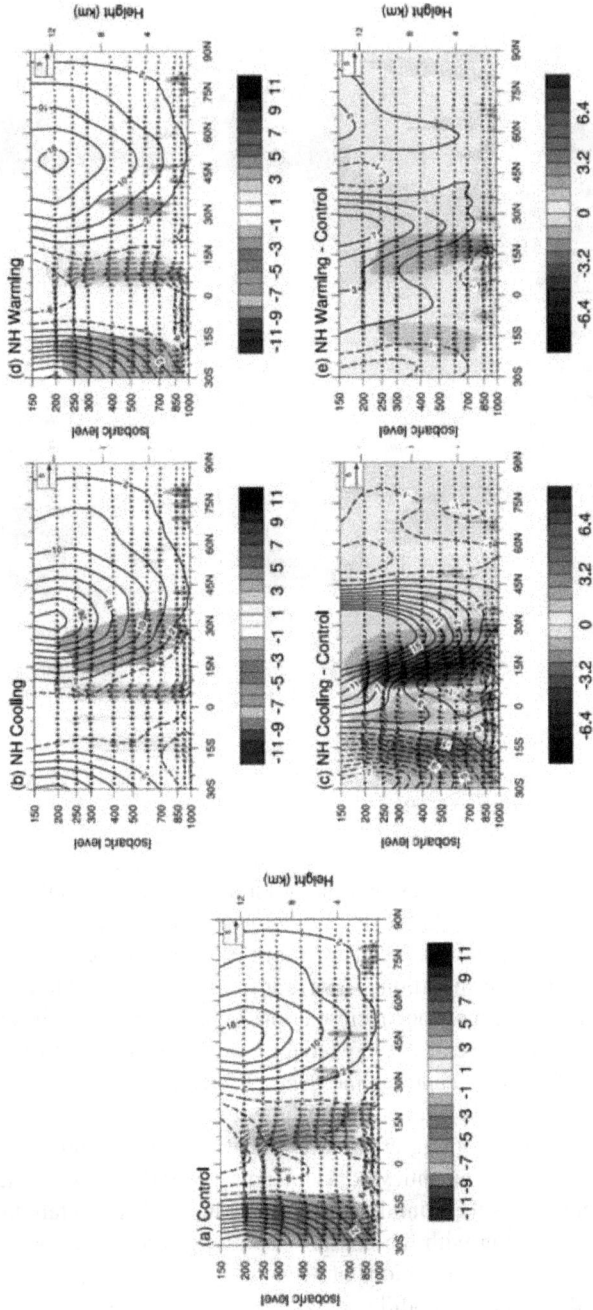

FIGURE 13.4 (a) Plot showing global longitude (0–360°E) averaged height-latitude profile of vertical velocity (shaded) and zonal wind (contour) from the CTRL run, (b) represents the same profile plot for the COLD run and (c) bias from CTRL, (d) represents the same profile plot for the WARM runs and (e) bias from CTRL. (Units: zonal wind: m/s.)

Control, JJAS mean

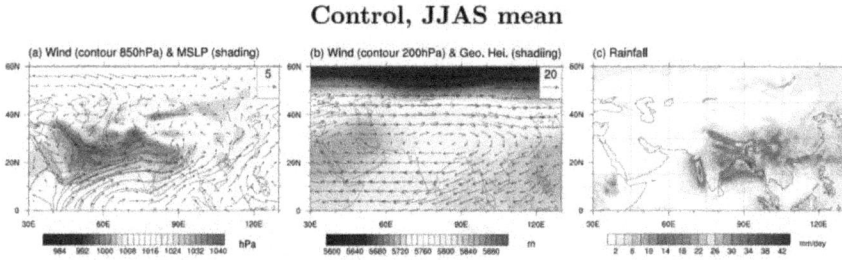

FIGURE 13.5 Plot showing the mean sea level pressure, wind at 850 hPa in (a), 200 hPa wind and geopotential (b) and rainfall (c) for the CTRL run. These plots show that CTRL creates a realistic mean state climatology. (Units: wind: m/s, mean sea level pressure: hPa, and rainfall: mm/day.)

13.3.2 CHANGES IN THE MONSOONAL CIRCULATION

Figure 13.6 shows the changes in the rainfall pattern as a result of the sea ice modulations. The plot clearly shows a decrease in rainfall in both the *COLD* and *WARM* runs. The **COLD** runs provide a greater decrease in rainfall. The regional response is the same in both the warming and cooling runs. The foothills of the Himalayas are most affected, and the Head Bay of Bengal is mainly affected. The west coast of India and the adjoining Arabian Sea also show a considerable difference. The regional Hadley Cell (variables averaged between 60°–90°E) is plotted in Figure 13.7. The plot shows increased subsidence over the Indian region, with the same global pattern (Figures 13.7d, 13.7e, and 13.4). The most essential and exciting difference is perhaps the weakening of the jet stream, which is much weakened compared to the *CTRL* run in both sensitivity runs. The decline is much more substantial in the *WARM* run, indicating the Rossby wave's slower phase speed. This weakening also shows a change in the waveguide nature of the jet stream and may involve a change in the critical latitude for absorption of the extratropical Rossby wave. However, these are not discussed here.

The wind response at 850 hPa and 200 hPa along with sea level pressure and geopotential height, is shown in Figures 13.8 and 13.9, respectively. The plots clearly show the weakening of the low-pressure region, the appearance of low-level north-easterlies instead of monsoon southwesterlies, and the weakening of the Tibetan anticyclone in the sensitivity runs (Figures 13.8c, e and 13.9c, e). The plots from Figures 13.6–13.9 demonstrate that (a) there is a substantial change in the midlatitude jet stream with a remarkable weakening of the westerlies in the upper level, (b) weakening of rainfall, i.e., development of a dry bias, and (c) weakening of the monsoon flow at the low level and weakening of upper-level anticyclone. In the next section, we describe the change in the teleconnection pattern and a possible linkage of rainfall dry bias developing in the sensitivity run because of the weakening of the extratropical teleconnection.

Rainfall, JJAS

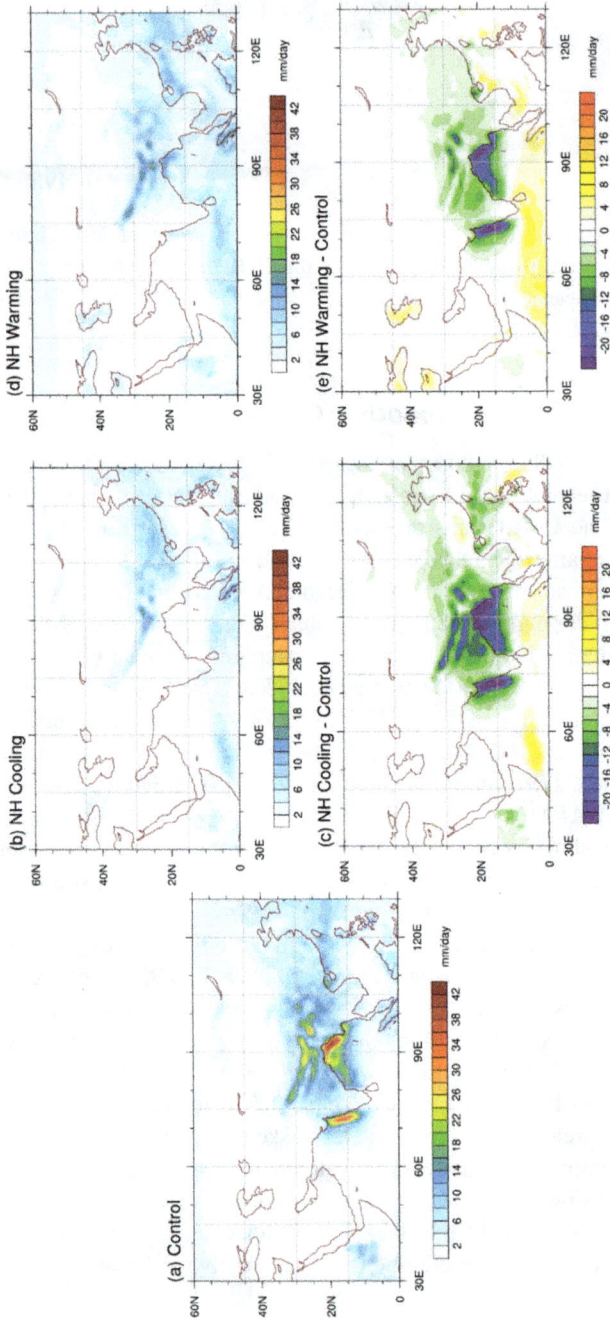

FIGURE 13.6 (a) The CTRL rainfall and (b) the COLD scenario rainfall and (c) bias from CTRL. (d) WARM scenario rainfall and (e) bias from CTRL. (Units: rainfall: mm/day.)

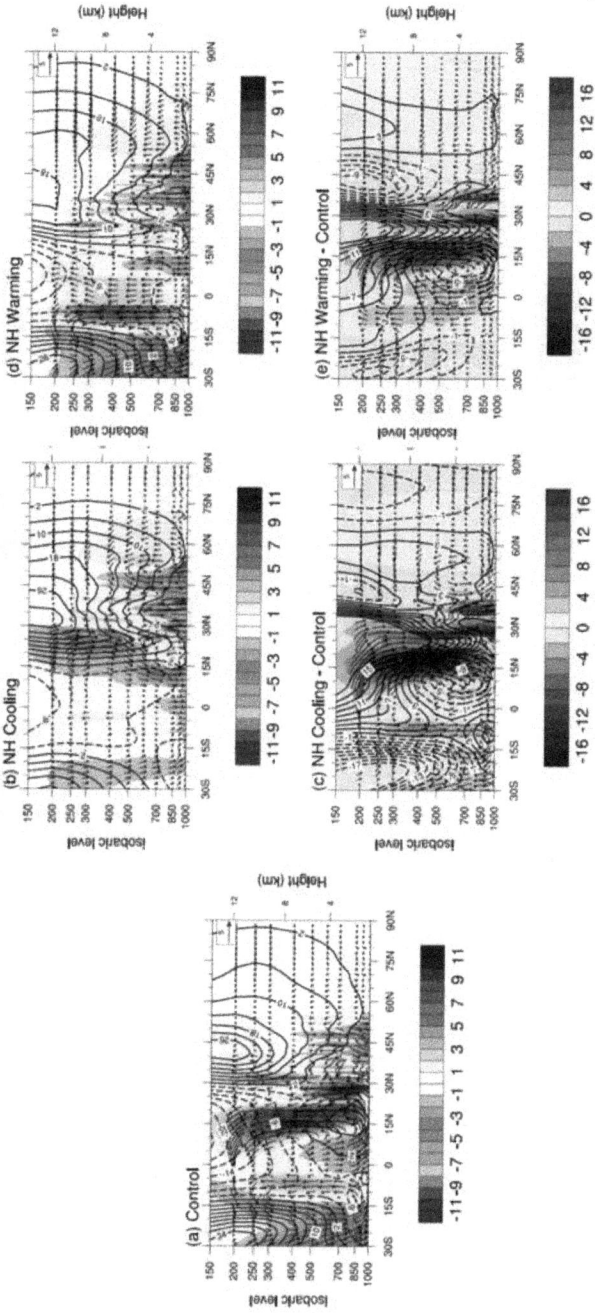

FIGURE 13.7 Same as Figure 13.4 but showing the 60–90°E longitude averaged height–latitude profile.

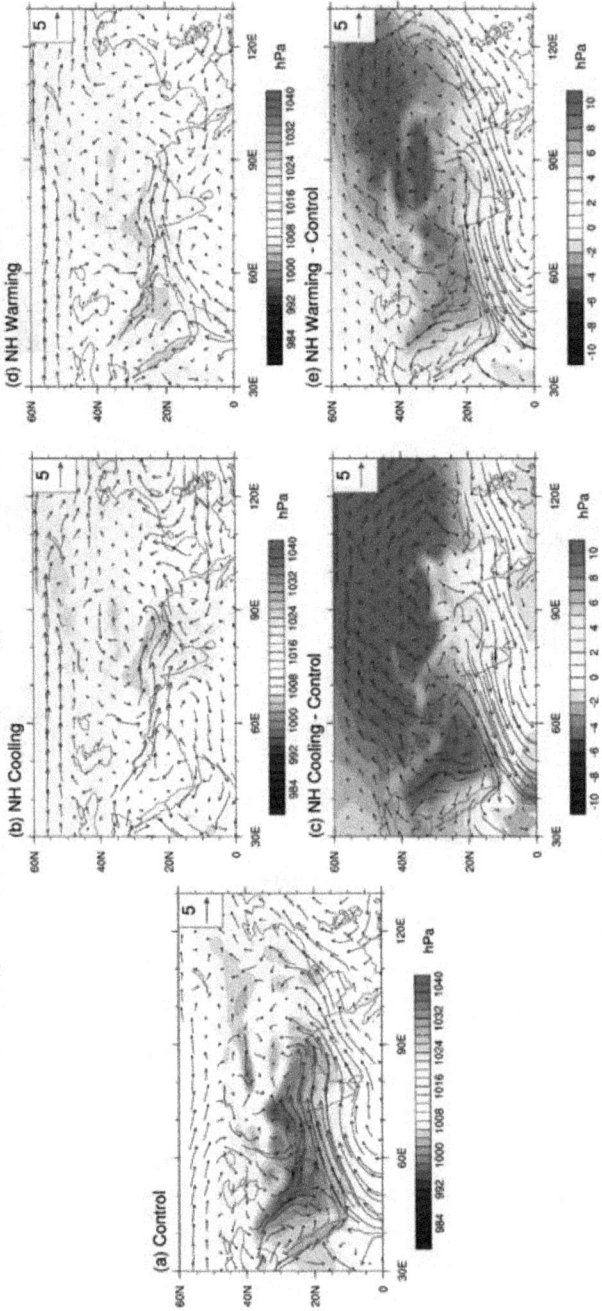

FIGURE 13.8 (a) CTRL mean sea level pressure and 850 hPa winds, (b) shows the same but for the COLD run and (c) bias from CTRL. (d) The same but for WARM run and (e) basis from CTRL.

Wind (contour 200hPa) & Geopotential Height (shading 500hPa)

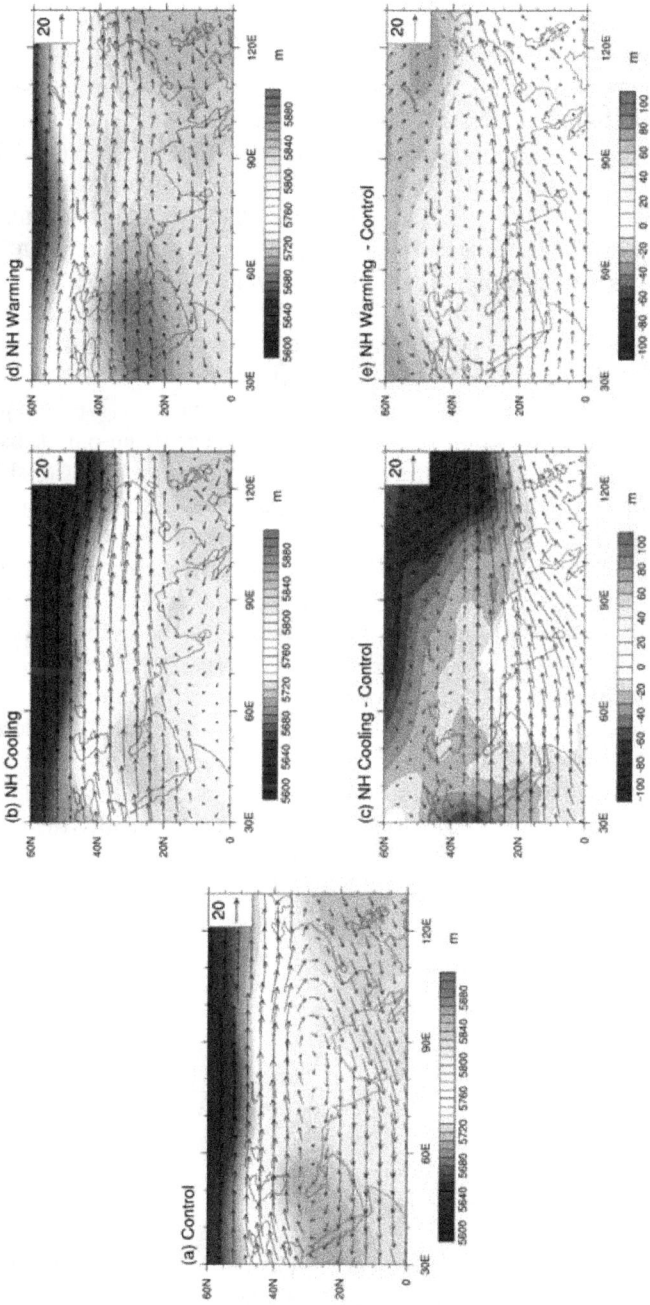

FIGURE 13.9 Same as Figure 13.8 but showing the wind vectors at 200 hPa and geopotential height (shaded) at 500 hPa.

13.4 CHANGES IN MODEL SIMULATED TELECONNECTION AND EDDY TRANSPORT

As depicted in the last section, the change in the rainfall and circulation pattern implies a difference in the Arctic circulation patterns. The dominant mode of the Arctic region is the Arctic oscillation, which is represented in several ways. One of the simple ways proposed by Hurrell and Deser (2009) is described in detail in the *UCAR* climate data guide (https://climatedataguide.ucar.edu/climate-data/hurrell-north-atlantic-oscillat ion-nao-index-pc-based) and this is the first empirical orthogonal function (EOF1) mode computed using the mean sea level pressure for the latitude ranging from 20°–90°N for the global longitudes (0°–360°E). The plot is shown in Figure 13.10a. The figure indicates the opposing nature of pole to midlatitude variability. The strongest mid-latitude zone of variability lies along with the Atlantic sector and another over the Pacific sector. The EOF1 for the control is shown in Figure 13.10b, and the EOF1 calculated from the sensitivity runs is shown in Figures 13.10c and 13.10d. The change in the large-scale pattern is apparent in these patterns, with the plot showing that the north–south mode of variability has disappeared and only substantial variability in the mid-latitude persists. In the sensitivity runs, opposite shading (indicating the opposing mode of polarity appears) appears in the east–west direction.

The power spectra plot of the first principal component (PC1) variance is shown in Figures 13.11a, b, and c for the **CTRL** and the two sensitivity runs, respectively. The dotted lines denote the 95% (top), mean (middle), and 5% (bottom) level of significance. The plot interestingly shows a clear shift in the variance periodicity. The **CTRL** run shows low-frequency decadal (~100 months) variability. In contrast, the **COLD** experiment shows that the variability is almost insignificant, while **WARM** experiments show that higher frequency variability (~10 months) is more significant in the warming scenario. The plot suggests that the reduction in sea ice concentration increases the oscillation frequency, as is evident from the PC1 plot. Thus, large-scale structures such as EOF1 would have high-frequency variability (less than a year), indicating a rapid change in the teleconnection phase, leading to an increase in climate system variability. The probability distribution plot for the PC1 for all the runs is shown in Figure 13.11d. The **CTRL** distribution, while showing clear unimodality with peak along with zero, the **WARM** and **COLD** experiments show clear bimodality, indicating a periodic phase shifting with sign reversal of the dominant EOF1 patterns (Figure 13.10).

$$E = \left(\overline{v'^2 - u'^2}, \ \overline{-u'v'}, \frac{f}{\partial\theta/\partial p} \overline{\theta'v'} \sim \overline{T'v'} \right) \tag{13.1}$$

where $\overline{v'^2 - u'^2}$ = eddy anisotropy or asymmetry, $\overline{u'v'}$ = eddy momentum flux transport (EMF), and $\overline{T'v'}$ = eddy heat flux transport (EHF)

To understand how the teleconnection relation with the rainfall over the Indian region changes, we plot the regressed rainfall pattern over the Indian region for the **CTRL** and the sensitivity runs in Figure 13.12. The plot clearly shows that the sensitivity runs show an increased positive relationship over the Himalayan region,

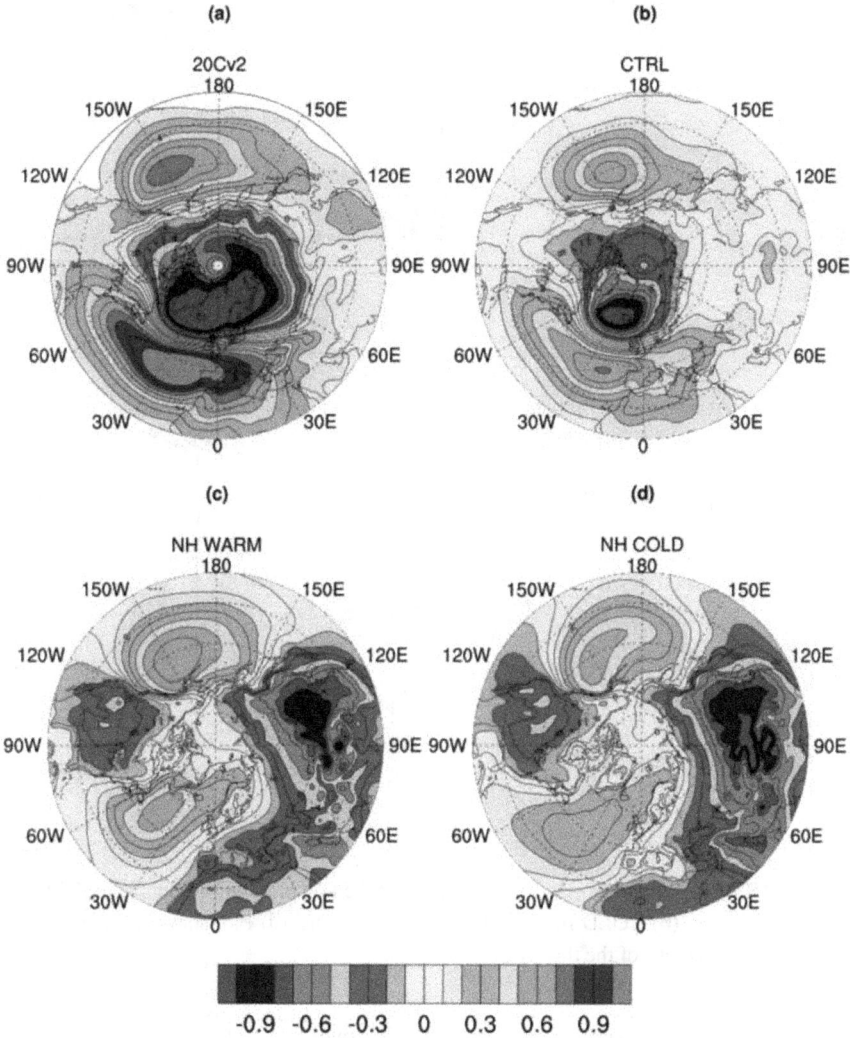

FIGURE 13.10 The first empirical orthogonal function (EOF1) using sea level pressure data from (a) 20Cv2 reanalysis, (b) CTRL run, (C) WARM experiments, and (d) COLD experiments.

indicating increased regulation of rainfall by the extratropical variability over this region. It is clearly seen that while the "**OBS**" or "**CTRL**" do not show any obvious climatological persistence, i.e. there is no regime-like behavior, the sensitivity runs show them, implying a climatological (or multidecadal) persistence or regime-like behavior in the sensitivity runs. Although both runs show increased relationships over the Indian region, the ***COLD*** experiments show increased positive relationships. It is speculated here that the relationship indicates that more extratropical systems would intrude over the Indian region in both runs. To clarify this, we plot the probability density function rainfall over a region north of 20°N in Figure 13.13. The plot clearly

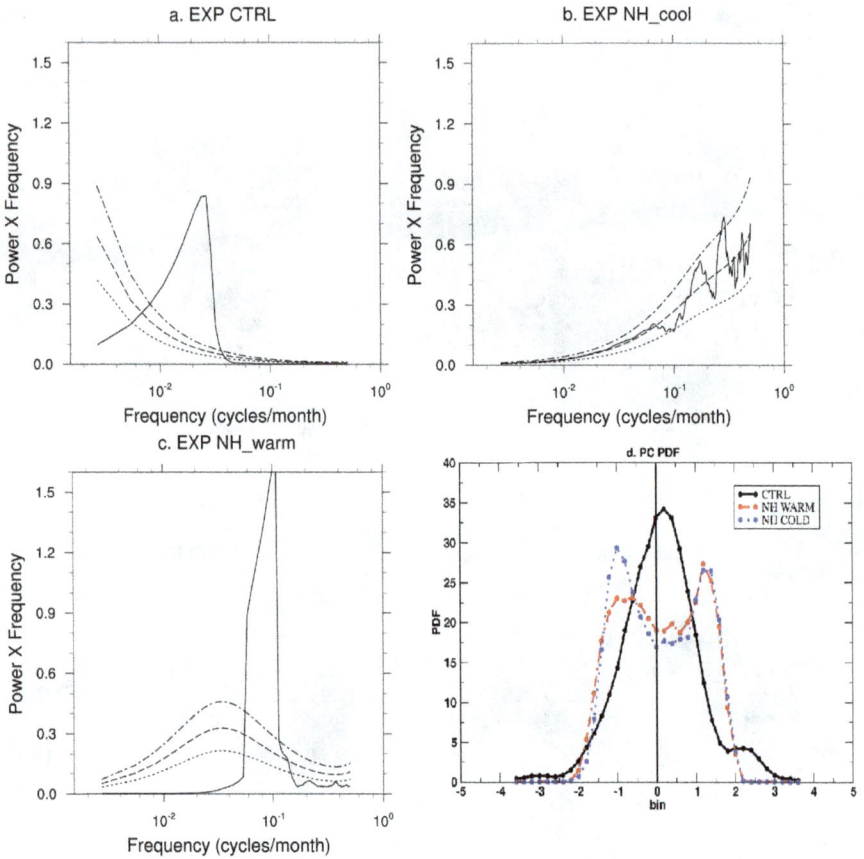

FIGURE 13.11 Plot showing the power spectra of the first principal components (PC1) from the (a) CTRL run, (b) COLD run, and the (c) WARM runs. (d) Plot showing the probability density function (PDF) of the three PCs.

shows the increased expectation value of rains over the positive bins. Thus, in the warming or cooling scenario, the rainfall over the northernmost part of India is likely to increase.

Another way to understand the change in the teleconnection pattern and the model produced sensitivity to the change in sea ice concentration is to diagnose this change using eddy transport analysis and plot the components of eddy transports of a generalized vector field called the **E-vector**. The components of the E-vectors are defined in Eq. 13.1.

The components of *E-vectors* give a measure of eddy asymmetry and momentum and heat transports, respectively. The positive sign of E-vector components (EMF and EHF) indicate northward transport, and a negative sign indicates southward transport. The probability distribution functions of the EMF and EHF components averaged between 20°–40°N; 65°–100°E are shown in Figure 13.14. The EMF plots show

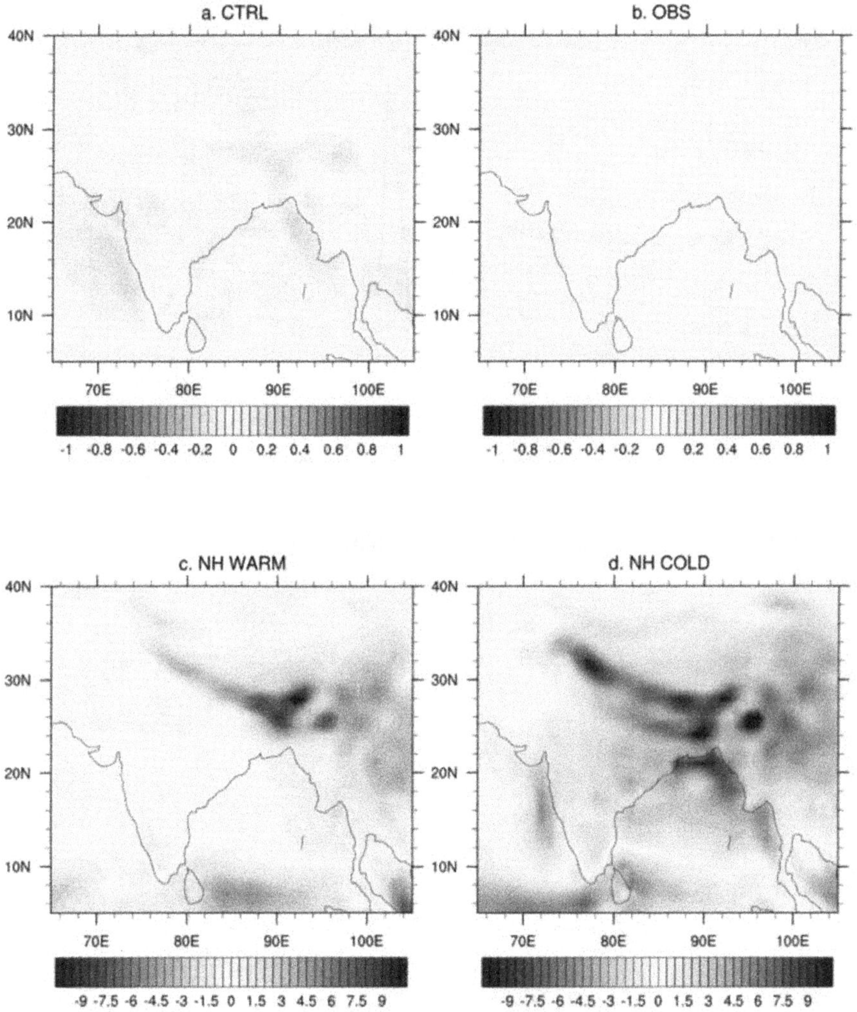

FIGURE 13.12 Plot showing the rainfall (shaded) spatial pattern regressed with respect to the PC1 of (a) CTRL run, (b) WARM run, and (C) COLD runs as reference series.

peaking on the positive side for the **WARM** runs and peaking on the negative side in **COLD** runs. The control shows a neutral situation indicating (i.e., no favoring of northward or southward transport). The shifts in the EMF peaks suggest that there will be more northward eddy transport in the warming scenario. Northward eddy transport indicates southward phase propagation of Rossby waves, which would imply more extratropical intrusions over the tropical region. The EHF shows no such preferential peak, although warming shows more southward eddy heat transport (increased bulge in the negative side of the x-axis of the PDF). The increased bulge indicates that there will be southward transfer with more frequencies as compared to the **CTRL** scenario.

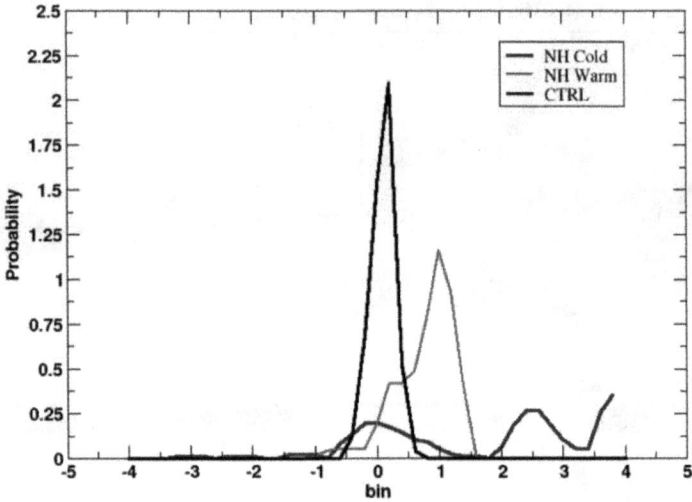

FIGURE 13.13 Probability distribution function (pdf) of the north India averaged regressed rainfall (as shown in Figure 13.12) for the CTRL, WARM, and COLD runs.

JJAS (avg. 20-40°N, 65-100°E)

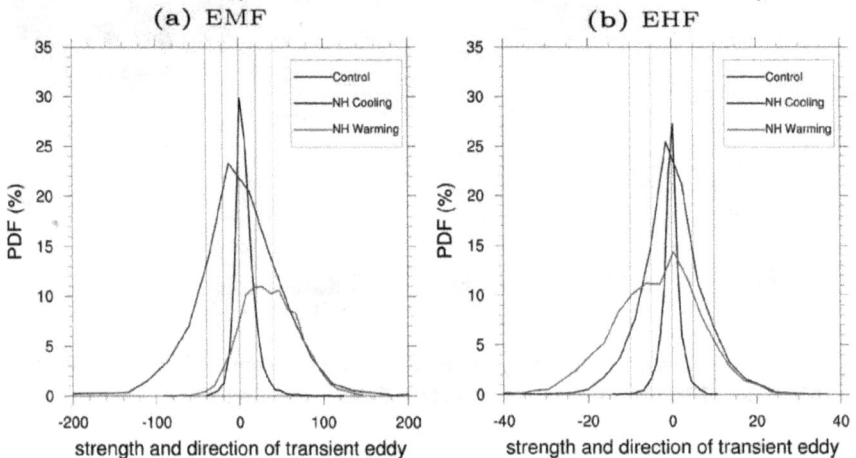

FIGURE 13.14 Probability distribution function (pdf) of eddy momentum flux (a: EMF, left panel) and eddy heat flux (b: EHF, right panel) for the CTRL, WARM, and COLD runs. Positive and negative values of the bins represent northward and southward transport, respectively (see text).

In order to understand the change in the transfer profile, we plot the interannual variability of the transfer by separating them in two transfer bands: F_L (30–60-day filtered data or the low-frequency subseasonal range) and F_H (2–30-day filtered data or the high-frequency range) and plot an area-averaged (20°–40°N; 65°–100°E) vertical profile for the simulation years in Figure 13.15. The plot is separately shown for EMF and EHF variables and different frequency bands: the F_L or F_H (mentioned at the top of each panel). The ***CTRL*** is shown for the actual range, and the ***CTRL***'s bias is shown in other panels. For the ***WARM*** case (middle panels), the difference plot for the EMF F_L band shows strong northward (i.e., positive) transfer and less intense transfer (i.e., a weak difference from climatology) in the 2–30-day (F_H) band. On the other hand, EHF transfer shows dominant southward transfer in the F_L band at all levels for most of the simulation years. EMF shows decadal variability for the COLD case, and EHF in the F_L band shows substantial southward flux transfer like the ***WARM*** counterpart (Figure 13.15h).

Thus, from these plots, we conclude that (a) there is a change in the extratropical circulation pattern, and (b) this change in the extratropical circulation pattern changes the teleconnection pattern through a modification of eddy transport. Modification of eddy transport indicates a change in the eddy forcing of the mean flow over the monsoonal region in the warming (or equivalently cooling) scenario.

13.5 DISCUSSION AND CONCLUSIONS

The reduction in Arctic sea ice has been very drastic in recent decades. The increased melting of ice due to an increase in climate (greenhouse) forcing has been studied in several recent studies in the context of modulation of mid-latitude circulation and general circulation. The current study describes a pathway for modulation of monsoonal circulation due to a reduction in sea ice components in the Arctic (northern hemispheres). We use the GFST126 model to simulate the response of the general circulation as well as monsoonal flow. Unlike typical warming studies, the current study describes two contrasting scenarios: (a) a decrease in sea latitudinal extent of ice concentration, which is referred to as a ***WARM*** scenario, and (b) another experiment with an increase in the latitudinal extent of sea ice concentration which is referred as a ***COLD*** scenario. The two experiments reflect how the monsoon circulation, which implies poleward transport of heat and momentum results due to multiple forcing, gets changed under contrasting climate change scenarios. The most important result from these experiments is that there is a significant response to the warming or cooling in rainfall amplitude and variability, especially over the Himalayan region. Such changes are due to changes in the midlatitude circulation or Arctic amplification and changes in the eddy forcing patterns that modulated the tropical–extratropical teleconnection. In particular, the effect of heat and momentum transport in the warming and cooling scenarios can be used to explain the increased rainfall activity over northern India, especially over the Himalayas and the foothills of the Himalayas. The difference in transport can be understood through a change in the Arctic oscillation pattern represented as the first empirical orthogonal function (EOF1) of sea level pressure. The EOF1 pattern shows a drastic change in the polar sea level pressure variance, and the pattern becomes more zonal (i.e., east–west) oriented. Such a change in the pressure distribution reflects changes in north–south transport. The analysis shown here clearly indicates changes in the Hadley cell and eddy transports. Monsoon flow represents a complicated balance in mass and energy transports during the boreal

JJAS (avg. 20–40°N, 65–100°E)

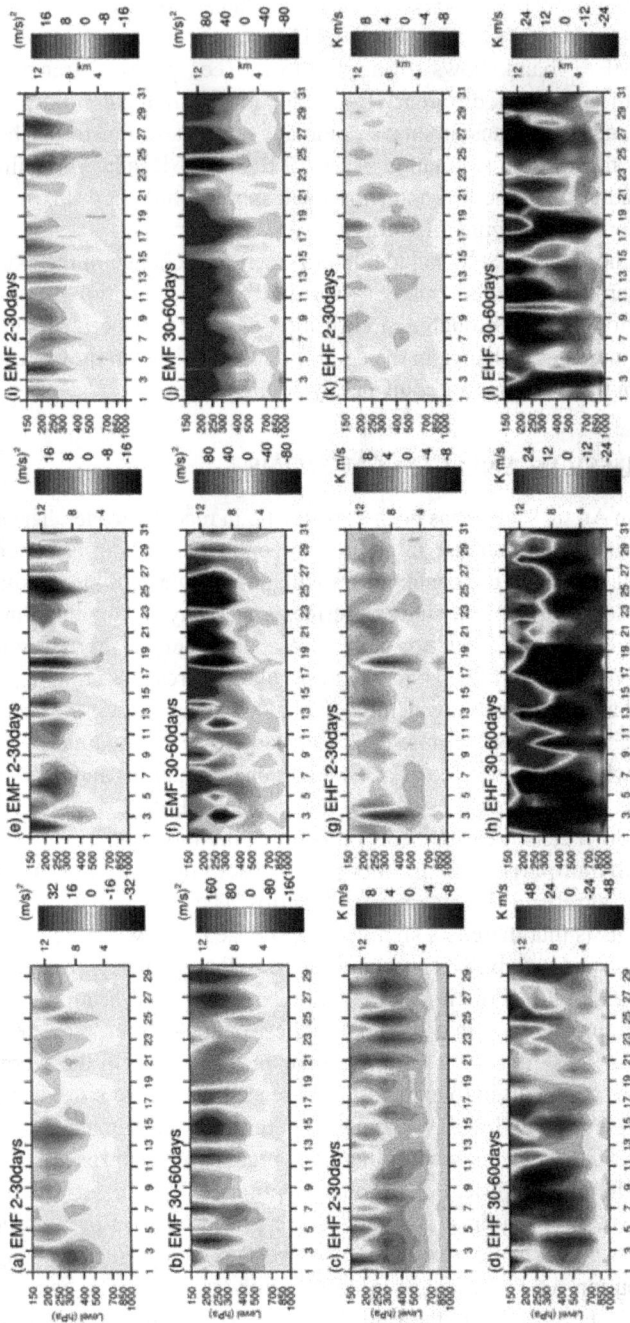

FIGURE 13.15 Interannual variability of height profiles of JJAS averaged and box averaged EMF and EHF in the high- (2–30-day) and low-frequency (30–60-day) bands. Panels (a)–(d) are for the CTRL run, (e)–(h) are for the WARM runs, and (i)–(l) are for the COLD runs. Frequency bands are mentioned at the top of each panel.

summer. The analysis suggests that monsoon transport would require the polar sink and equatorial source, i.e., north–south transport gradients, in addition to equatorial flow dynamics. Any untoward change in this balanced condition would lead to the weakening of the monsoon. In the warming scenario (the most likely scenario), the monsoon is expected to respond strongly to the change in equator to pole gradients. The results here indicate an increase in the rainfall in the Himalayan region and the north-eastern part of India. The full impact of such changes must be assessed using impact-based modeling studies.

ACKNOWLEDGEMENTS

IITM is an autonomous institute funded by the Ministry of Earth Sciences (MoES), Govt. of India. Reanalysis data used in this analysis are downloaded from several freely available reanalysis data sources. We acknowledge the availability of the data via www. Model runs are conducted using the GFS model run on IITM high-performance computers. All the plots are prepared using freely available NCAR Command Language (NCL) and XMGRACE software.

REFERENCES

Balmaseda, M. A., Ferranti, L., Molteni, F., Palmer, T.N. (2010) Impact of 2007 and 2008 Arctic ice anomalies on the atmospheric circulation: Implications for long-range predictions. *Quarterly Journal of the Royal Meteorological Society* 136: 1655–1664.

Barnes, E.A., Screen, J.A. (2015) The impact of Arctic warming on the midlatitude jet-stream: Can it? Has it? Will it? *WIREs Climate Change* 6: 277–286.

Borah, P., Venugopal, V., Sukhatme, J., Muddebihal, P., Goswami, B.N. (2020) Indian monsoon derailed by a North Atlantic wavetrain. *Science* 370: 1335–1338.

Budikova, D. (2009) Role of Arctic sea ice in global atmospheric circulation: A review. *Global and Planetary Change* 68: 149–163.

Cai, Q., Wang, J., Beletsky, D., Overland, J., Ikeda, M., Wan, L. (2021) Accelerated decline of summer Arctic sea ice during 1850–2017 and the amplified Arctic warming during the recent decades. *Environmental Research Letters* 16: 034015.

Chatterjee, S., Ravichandran, M., Murukesh, N., Raj, R.P., Johannessen, O.M. (2021) A possible relation between Arctic sea ice and late season Indian Summer Monsoon Rainfall extremes. *npj Climate and Atmospheric Science* 4: 36.

Chylek, P., Folland, C.K., Lesins, G., Dubey, M.K., Wang, M. (2009) Arctic air temperature change amplification and the Atlantic Multidecadal Oscillation. *Geophysical Research Letters* 36.

Cohen, J., et al. (2014) Recent Arctic amplification and extreme mid-latitude weather. *Nature Geoscience* 7: 627–637.

Collins, M., Chandler, R.E., Cox, P.M., Huthnance, J.M., Rougier, J., Stephenson, D.B. (2012) Quantifying future climate change. *Nature Climate Change* 2: 403–409.

Compo, G.P., et al. (2011) The Twentieth Century Reanalysis Project. *Quarterly Journal of the Royal Meteorological Society* 137: 1–28.

Coumou, D., Di Capua, G., Vavrus, S., Wang, L., Wang, S. (2018) The influence of Arctic amplification on mid-latitude summer circulation. *Nature Communications* 9: 2959.

Derksen, C., Brown, R. (2012) Spring snow cover extent reductions in the 2008–2012 period exceeding climate model projections. *Geophysical Research Letters* 39.

Deser, C., Tomas, R., Alexander, M., Lawrence, D. (2010) The seasonal atmospheric response to projected Arctic sea ice loss in the late twenty-first century. *Journal of Climate* 23: 333–351.

Dethloff, K., Handorf, D., Jaiser, R., Rinke, A., Klinghammer, P. (2019) Dynamical mechanisms of Arctic amplification. *Annals of the New York Academy of Sciences* 1436: 184–194.

Fadnavis, S., Chattopadhyay, R. (2017) Linkages of subtropical stratospheric intraseasonal intrusions with Indian summer monsoon deficit rainfall. *Journal of Climate* 30: 5083–5095.

Gong, T., Feldstein, S.B., Lee, S. (2020) Rossby wave propagation from the Arctic into the midlatitudes: Does it arise from in situ latent heating or a trans-Arctic wave train? *Journal of Climate* 33: 3619–3633.

Grotjahn, R., et al. (2016) North American extreme temperature events and related large scale meteorological patterns: a review of statistical methods, dynamics, modeling, and trends. *Climate Dynamics* 46: 1151–1184.

Hassanzadeh, P., Kuang, Z. (2015) Blocking variability: Arctic Amplification versus Arctic Oscillation. *Geophysical Research Letters* 42: 8586–8595.

Hoskins, B., Woollings, T. (2015) Persistent extratropical regimes and climate extremes. *Current Climate Change Reports* 1: 115–124.

Hunt, K.M.R., Turner, A.G., Shaffrey, L.C. (2018) The evolution, seasonality and impacts of western disturbances. *Quarterly Journal of the Royal Meteorological Society* 144: 278–290.

Hurrell, J.W., Deser, C. (2009) North Atlantic climate variability: The role of the North Atlantic Oscillation. *Journal of Marine Systems* 78: 28–41.

Kalshetti, M., Chattopadhyay, R., Phani, R., Joseph, S., Sahai, A.K. (2020) Climatological patterns of subseasonal eddy flux transfer based on the co-spectral analysis over the Indian region and the derivation of an index of eddy transfer for operational tracking. *International Journal of Climatology* 41 (Suppl. 1): E1906–E1925.

Kim, B.-M., Son, S.-W., Min, S.-K., Jeong, J.-H., Kim, S.-J., Zhang, X., Shim, T., Yoon, J.-H. (2014) Weakening of the stratospheric polar vortex by Arctic sea-ice loss. *Nature Communications* 5: 4646.

Levine, X.J., Cvijanovic, I., Ortega, P., Donat, M.G., Tourigny, E. (2021) Atmospheric feedback explains disparate climate response to regional Arctic sea-ice loss. *npj Climate and Atmospheric Science* 4: 28.

Mallett, R.D.C., Stroeve, J.C., Tsamados, M., Landy, J.C., Willatt, R., Nandan, V., Liston, G.E. (2021) Faster decline and higher variability in the sea ice thickness of the marginal Arctic seas when accounting for dynamic snow cover. *The Cryosphere* 15; 2429–2450.

Meredith, M., et al. (2019) IPCC special report on the ocean and cryosphere in a changing climate. *Geneva: Intergovernmental Panel on Climate Change.*

Midhuna, T.M., Dimri, A.P. (2019) Impact of arctic oscillation on Indian winter monsoon. *Meteorology and Atmospheric Physics* 131: 1157–1167.

Morice, C.P., Kennedy, J.J., Rayner, N.A., Jones, P.D. (2012) Quantifying uncertainties in global and regional temperature change using an ensemble of observational estimates: The HadCRUT4 data set. *Journal of Geophysical Research: Atmospheres* 117.

Overland, J.E., Wang, M. (2018a) Resolving future Arctic/midlatitude weather connections. *Earth's Future* 6: 1146–1152.

Overland, J.E., Wang, M. (2018b) Arctic-midlatitude weather linkages in North America. *Polar Science* 16: 1–9.

Overland, J.E., Wang, M., Walsh, J.E., Stroeve, J.C. (2014) Future Arctic climate changes: Adaptation and mitigation time scales. *Earth's Future* 2: 68–74.

Perovich, D. (2011) The changing Arctic sea ice cover. *Oceanography* 24: 162–173.

Qian, W., Wu, K., Leung, J.C.-H., Shi, J. (2016) Long-term trends of the Polar and Arctic cells influencing the Arctic climate since 1989. *Journal of Geophysical Research: Atmospheres* 121: 2679–2690.

Rayner, N.A., Parker, D.E., Horton, E.B., Folland, C.K., Alexander, L.V., Rowell, D.P., Kent, E.C., Kaplan, A. (2003) Global analyses of sea surface temperature, sea ice, and night marine air temperature since the late nineteenth century. *Journal of Geophysical Research: Atmospheres* 108.

Riboldi, J., Lott, F., D'Andrea, F., Rivière, G. (2020) On the linkage between Rossby wave phase speed, atmospheric blocking, and Arctic amplification. *Geophysical Research Letters* 47: e2020GL087796.

Schneider, T., Bischoff, T., Potka, H. (2015) Physics of changes in synoptic midlatitude temperature variability. *Journal of Climate* 28: 2312–2331.

Serreze, M.C., Francis, J.A. (2006) The Arctic amplification debate. *Climatic Change* 76: 241–264.

Smith, D.M., et al. (2019) The Polar Amplification Model Intercomparison Project (PAMIP) contribution to CMIP6: investigating the causes and consequences of polar amplification. *Geoscientific Model Development* 12: 1139–1164.

Trenberth, K.E., Shea, D.J. (2006) Atlantic hurricanes and natural variability in 2005. *Geophysical Research Letters* 33.

Vihma, T. (2014) Effects of Arctic sea ice decline on weather and climate: A review. *Surveys in Geophysics* 35: 1175–1214.

Glossary

Accumulation/Ablation On a glacier, the accumulation zone is the area above the fern line, where snowfall accumulates and exceeds the losses from ablation, (melting, evaporation, and sublimation). The annual equilibrium line separates the accumulation and ablation zone annually.

Albedo Albedo is the measure of the diffuse reflection of solar radiation out of the total solar radiation and measured on a scale from 0, corresponding to a black body that absorbs all incident radiation, to 1, corresponding to a body that reflects all incident radiation.

Antarctic bottom water The Antarctic bottom water (AABW) is a type of water mass in the Southern Ocean surrounding Antarctica with temperatures ranging from −0.8 to 2°C (35°F) and salinities from 34.6 to 34.7 psu. As the densest water mass of the oceans, AABW is found to occupy the depth range below 4000 m of all ocean basins that have a connection to the Southern Ocean at that level.

Aperture A space through which light passes in an optical or photographic instrument, especially the variable opening by which light enters a camera.

Arctic amplification Polar amplification is the phenomenon that any change in the net radiation balance (for example, greenhouse intensification) tends to produce a larger change in temperature near the poles than in the planetary average.

Atlantic meridional overturning circulation The Atlantic meridional overturning circulation (AMOC) is part of a global thermohaline circulation in the oceans and is the zonally integrated component of surface and deep currents in the Atlantic Ocean. It is characterized by a northward flow of warm, salty water in the upper layers of the Atlantic, and a southward flow of colder, deep waters.

Biogas iogas is a mixture of gases, primarily consisting of methane, carbon dioxide, and hydrogen sulphide, produced from raw materials such as agricultural waste, manure, municipal waste, plant material, sewage, green waste, wastewater, and food waste.

Biogeography Biogeography is the study of the distribution of species and ecosystems in geographic space and through geological time.

Buoyancy Buoyancy, or upthrust, is an upward force exerted by a fluid that opposes the weight of a partially or fully immersed object.

Catchment A drainage basin is an area of land where all flowing surface water converges to a single point, such as a river mouth, or flows into another body of water, such as a lake or ocean.

Circumpolar Deep Water Circumpolar deep water (CDW) is a designation given to the water mass in the Pacific and Indian Oceans that is a mixing of other water masses in the region.

Continental shelf A continental shelf is a portion of a continent that is submerged under an area of relatively shallow water, known as a shelf sea. Much of these shelves were exposed by drops in sea level during glacial periods.

Crevasse A crevasse is a deep crack that forms in a glacier or ice sheet that can be a few inches across to over 40 feet.

Cryosphere The cryosphere is the part of the Earth's climate system that includes solid precipitation, snow, sea ice, lake and river ice, icebergs, glaciers and ice caps, ice sheets, ice shelves, permafrost, and seasonally frozen ground.

CTD A CTD or sonde is an oceanography instrument used to measure the electrical conductivity, temperature, and pressure of seawater (the D stands for "depth", which is closely related to pressure).

Dakshin Gangotri Dakshin Gangotri was the first scientific base station of India situated in Antarctica, part of the Indian Antarctic Programme.

Deglaciation Deglaciation is the transition from full glacial conditions during ice ages, to warm interglacial, characterized by global warming and a sea level rise due to a change in continental ice volume.

Digital elevation model A digital elevation model (DEM) or digital surface model (DSM) is a 3D computer graphics representation of elevation data to represent terrain or overlaying objects, commonly of a planet, moon, or asteroid.

Discharge A discharge is a measure of the quantity of any fluid flow over unit time.

El Niño El Niño is the warm phase of the El Niño–Southern Oscillation (ENSO) and is associated with a band of warm ocean water that develops in the central and east-central equatorial Pacific (approximately between the International Date Line and 120°W), including the area off the Pacific coast of South America.

Equilibrium line of altitude The climatic snow line is the boundary between a snow-covered and snow-free surface. The annual equilibrium line separates the accumulation and ablation zone annually.

Expedition An exploration, journey, or voyage undertaken by a group of people especially for discovery and scientific research.

Fjord In physical geography, a fjord or fiord is a long, narrow inlet with steep sides or cliffs, created by a glacier.

Gangotri glacier Gangotri glacier initiates from Uttarkashi district, Uttarakhand, India, in a region bordering Tibet, as one of the primary sources of the Ganges. This glacier is found between 30°43'22"–30°55'49" (lat.) and 79°4'41"–79°16'34" (long.), extending in height from 4120 to 7000 m.a.s.l., and has a volume of over 27 km^3.

Garhwal Himalayas The Garhwal Himalayas are mountain ranges located in the Indian state of Uttarakhand. The cities which are included in these ranges are Pauri, Tehri, Uttarkashi, Rudraprayag, Chamoli, and Chota Char Dham pilgrimage, namely Gangotri, Yamunotri, Badrinath, and Kedarnath.

Geopotential height Geopotential height or geopotential altitude is a vertical coordinate referenced to the Earth's mean sea level, an adjustment to geometric height (altitude above mean sea level) that accounts for the variation of gravity with latitude and altitude.

Geothermal heating Geothermal heating is the direct use of geothermal energy for some heating applications. Geothermal energy originates from the heat retained within the Earth since the original formation of the planet, from radioactive decay of minerals, and from solar energy absorbed at the surface.

Glacial lake expansion A glacial lake outburst flood is a sudden release of water from a lake fed by glacier melt that has formed at the side, in front, within, beneath, or on the surface of a glacier.

Glacial retreat This refers to the process of a glacier shrinking or receding in size over time due to a decrease in ice accumulation or an increase in ice melt.

Glacierets A small glacier, or small mass of ice somewhat resembling a glacier.

Global navigation satellite system A satellite navigation system with global coverage may be termed a global navigation satellite system (GNSS).

Global warming The ongoing increase in global average temperature—and its effects on Earth's climate system.

GLOFs A glacial lake outburst flood (GLOF) is a type of outburst flood caused by the failure of a dam containing a glacial lake.

GPS Survey The Global Positioning System (GPS), originally Navstar GPS, is a satellite-based radionavigation system owned by the United States government and operated by the United States Space Force. It is one of the global navigation satellite systems (GNSS) that provides geolocation and time information to a GPS receiver anywhere on or near the Earth where there is an unobstructed line of sight to four or more GPS satellites.

Gravimetry Gravimetry is the measurement of the strength of a gravitational field. Gravimetry may be used when either the magnitude of a gravitational field or the properties of matter responsible for its creation are of interest.

Greenhouse gas A greenhouse gas (GHG or GhG) is a gas that absorbs and emits radiant energy at thermal infrared wavelengths, causing the greenhouse effect.

Heat flux Heat flux or thermal flux, sometimes also referred to as heat flux density, heat-flow density, or heat flow rate intensity, is a flow of energy per unit area per unit time.

HEC-RAS HEC-RAS is simulation software used in computational fluid dynamics—specifically, to model the hydraulics of water flow through natural rivers and other channels.

Hoarfrost layer Hoar frost, also hoarfrost, radiation frost, or pruina, refers to white ice crystals deposited on the ground or loosely attached to exposed objects, such as wires or leaves. They form on cold, clear nights when conditions are such that heat radiates into outer space faster than it can be replaced from nearby warm objects or brought in by the wind.

Hydrograph A hydrograph illustrates a type of activity of water during a specific time frame. Salinity and acidity are sometimes measured, but the most common types are stage and discharge hydrographs.

Hydrostatic equilibrium Hydrostatic equilibrium (hydrostatic balance, hydrostasy) is the condition of a fluid or plastic solid at rest, which occurs when external forces, such as gravity, are balanced by a pressure-gradient force.

Hysteresis Hysteresis is the dependence of the state of a system on its history. For example, a magnet may have more than one possible magnetic moment in a given magnetic field, depending on how the field changed in the past.

Insolation Solar irradiance is the power per unit area (surface power density) received from the Sun in the form of electromagnetic radiation in the wavelength range of the measuring instrument.

Isopycnal Isopycnals are layers within the ocean that are stratified based on their densities and can be shown as a line connecting points of a specific density or potential density on a graph.

Jhelum basin The Jhelum basin is fed by the river Jhelum, the main drainage channel. The Jhelum basin comprises about 147 glaciers covering an area of about 75 square kilometres. The river Jhelum has 24 tributaries, some of which flow from the Pir Panjal range's slopes and join the river on the left bank, while others flow from the Himalayan range and join the river on the right bank.

La Niña La Niña is an oceanic and atmospheric phenomenon that is the colder counterpart of El Niño, as part of the broader El Niño–Southern Oscillation (ENSO) climate pattern.

Laminar flow In fluid dynamics, laminar flow is characterized by fluid particles following smooth paths in layers, with each layer moving smoothly past the adjacent layers with little or no mixing.

Land use land cover Land cover indicates the physical land type such as forest or open water, whereas land use documents how people are using the land.

Landsat The Landsat program is the longest-running enterprise for acquisition of satellite imagery of Earth, jointly run by NASA and USGS.

Last glacial maximum The Last Glacial Maximum (LGM), also referred to as the Late Glacial Maximum, was the most recent time during the Last Glacial Period that ice sheets were at their greatest extent 26 ka – 20 ka, during an interval of low obliquity.

Mass balance In physics, a mass balance, also called a material balance, is an application of conservation of mass to the analysis of physical systems.

Monsoon A monsoon is traditionally a seasonal reversing wind accompanied by corresponding changes in precipitation but is now used to describe seasonal changes in atmospheric circulation and precipitation associated with annual latitudinal oscillation of the Intertropical Convergence Zone (ITCZ) between its limits to the north and south of the equator.

Multifractal analysis A multifractal system is a generalization of a fractal system in which a single exponent (the fractal dimension) is not enough to describe its dynamics; instead, a continuous spectrum of exponents (the so-called singularity spectrum) is needed.

Paleoclimate Paleoclimatology is the study of past climatic conditions for which direct measurements were not taken. Paleoclimatology uses a variety of proxy methods from Earth and life sciences to obtain data previously preserved within rocks, sediments, boreholes, ice sheets, tree rings, corals, shells, and microfossils.

Patagonia Patagonia is a geographical region that encompasses the southern end of South America, governed by Argentina and Chile. The region comprises the southern section of the Andes Mountains with lakes, fjords, temperate rainforests, and glaciers in the west, and deserts, tablelands, and steppes to the east.

Perennial ice This is snow that persists on the ground year after year.

Permafrost Permafrost is soil or underwater sediment which continuously remains below 0°C (32°F) for two or more years. Land-based permafrost can include the surface layer of the soil, but if the surface is too warm, it may still occur within a few centimetres of the surface down to hundreds of metres.

Perturbation This is any departure introduced into an assumed steady state of a system. In synoptic meteorology, this term is used for any departure from zonal flow within the major zonal currents of the atmosphere. It is especially applied to the wavelike disturbances within the tropical easterlies.

Polar Amplification Model Inter Comparison Project Polar amplification is the phenomenon that external radiative forcing produces a larger change in surface temperature at high latitudes than the global average. The Polar Amplification Model Intercomparison Project (PAMIP) seeks to improve our understanding of this phenomenon through a coordinated set of numerical model experiments as it finds that sea ice decline would weaken the jet stream by 10% and increase the probability of atmospheric blocking.

Polarisation Polarization is a property of transverse waves which specifies the geometrical orientation of the oscillations. In a transverse wave, the direction of the oscillation is perpendicular to the direction of motion of the wave.

Precipitation In meteorology, precipitation is any product of the condensation of atmospheric water vapor that falls from clouds due to gravitational pull. The main forms of precipitation include drizzle, rain, sleet, snow, ice pellets, graupel, and hail.

Predictability index Predictability is the degree to which a correct prediction or forecast of a system's state can be made, either qualitatively or quantitatively.

Regression A marine regression is a geological process occurring when areas of submerged seafloor are exposed above the sea level. The opposite event, marine transgression, occurs when flooding from the sea covers previously exposed land.

Retreating rate The retreat of glaciers since 1850 affects the availability of fresh water for irrigation and domestic use, mountain recreation, animals and plants that depend on glacier-melt, and, in the longer term, the level of the oceans.

Rossby waves Rossby waves, also known as planetary waves, are a type of inertial wave naturally occurring in rotating fluids. They are observed in the atmospheres and oceans of Earth and other planets, owing to the rotation of Earth or of the planet involved.

Salinity Salinity is the saltiness or amount of salt dissolved in a body of water, called saline water (see also soil salinity).

Satellite data Satellite data or satellite imagery is understood as information about Earth and other planets in space, gathered by man-made satellites in their orbits. The most common use for satellite data is Earth Observation (EO): satellites deliver information about the surface and weather changes on the planet Earth.

Sea ice area Sea ice covers about 7% of the Earth's surface and about 12% of the world's oceans. Sea ice area is the integral sum of the product of ice concentration and area of all grid cells with at least 15% ice concentration.

Sea ice concentration Sea ice concentration is a useful variable for climate scientists and nautical navigators. It is defined as the area of sea ice relative to the total at a given point in the ocean.

Sea ice extent Sea ice extent is the integral sum of the areas of all grid cells with at least 15% ice concentration. Arctic sea ice extent hit an all-time low in September 2012, when the ice was determined to cover only 24% of the Arctic Ocean, offsetting the previous low of 29% in 2007.

Shortwave radiation Shortwave radiation (SW) is radiant energy with wavelengths in the visible (VIS), near-ultraviolet (UV), and near-infrared (NIR) spectra.

Snout The lowest end of a glacier; also called the glacier terminus or toe.

Supraglacial lake A supraglacial lake is any pond of liquid water on the top of a glacier. Although these pools are ephemeral, they may reach kilometres in diameter and be several metres deep. They may last for months or even decades at a time, but can empty in the course of hours.

Surface velocity Superficial velocity of multiphase flows in porous media is a hypothetical (artificial) flow velocity calculated as if the given phase or fluid were the only one flowing or present in a given cross-sectional area.

Sustainable Development Goal Sustainable Development Goals or Global Goals are a collection of 17 interlinked objectives designed to serve as a "shared blueprint for peace and prosperity for people and the planet, now and into the future".

Temperature Temperature is a physical quantity that expresses quantitatively the perceptions of hotness and coldness. Temperature is measured with a thermometer.

Thermal conductivity The thermal conductivity of a material is a measure of its ability to conduct heat.

Thermal emissivity The emissivity of the surface of a material is its effectiveness in emitting energy as thermal radiation. Thermal radiation is electromagnetic radiation that most commonly includes both visible radiation (light) and infrared radiation, which is not visible to human eyes.

Thermocline A thermocline (also known as the thermal layer or the metalimnion in lakes) is a distinct layer based on temperature within a large body of fluid (e.g. water, as in an ocean or lake; or air, e.g. an atmosphere) with a high gradient of distinct temperature differences associated with depth.

Thermodynamics Thermodynamics is a branch of physics that deals with heat, work, and temperature, and their relation to energy, entropy, and the physical properties of matter and radiation.

Thermohaline Thermohaline circulation (THC) is a part of the large-scale ocean circulation that is driven by global density gradients created by surface heat and freshwater fluxes.

Topography Topography is the study of the forms and features of land surfaces. The topography of an area may refer to the land forms and features themselves, or a description or depiction in maps.

Water masses An oceanographic water mass is an identifiable body of water with a common formation history which has physical properties distinct from surrounding water.

Younger Dryas The Younger Dryas, which occurred circa 12,900 to 11,700 years before the present, was a return to glacial conditions which temporarily reversed the gradual climatic warming after the Last Glacial Maximum (LGM), which lasted from circa 27,000 to 20,000 years BP.

Index

Note: Page numbers in **bold** refer to tables and those in *italic* refer to figures.

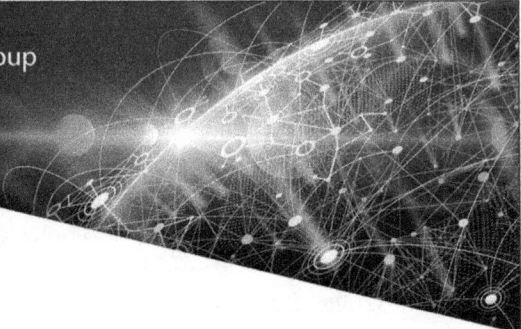

For Product Safety Concerns and Information please contact our EU
representative GPSR@taylorandfrancis.com
Taylor & Francis Verlag GmbH, Kaufingerstraße 24, 80331 München, Germany

9 781032 427478